An Introduction to Linear Programming and Game Theory

An Introduction to Linear Programming and Game Theory

PAUL R. THIE
Department of Mathematics
Boston College

JOHN WILEY & SONS

New York
Chichester
Brisbane
Toronto

Library of Congress Cataloging in Publication Data:

Thie, Paul R 1938–
An introduction to linear programming and game theory.

1. Linear programming. 2. Game theory. I. Title.

T57.74.T44 519.7′2 78-15328
ISBN 0-471-04248-X

Printed in the United States of America

10 9 8 7 6 5 4 3 2 1

To Mary Lou

Preface

PURPOSE

This textbook provides, at an introductory level, a development of some of the theoretical concepts and computational techniques of linear programming and game theory; it also offers some discussion on the applications of these topics in the social, life, and managerial sciences. Closely related to this development, it presents an introduction to the process of mathematical model building, which is discussed in two distinct settings. The chapters on linear programming contain various examples of real-world situations involving a single decision maker faced with some sort of deterministic (except in Section 7.4) optimization problem. In the two chapters on game theory the emphasis is on the development of a different type of model, a model of a conflict situation involving two participants with opposing interests.

LEVEL AND PREREQUISITES

The text is written for students in mathematics, science, economics, and operations research. The presentation is, for the most part, mathematically complete, that is, in terms of definitions, theorems, and proofs. However, examples are used frequently, not only to motivate new ideas, but also to assist in the understanding of the theory and the associated proofs. The goal is to provide a book that the student will find rigorous and challenging, yet readable and helpful.

The prerequisites for reading the text are minimal. The material should be accessible to any student who has successfully completed one or two undergraduate mathematics courses. No use is made of the theoretical concepts from linear algebra such as dimension and basis of a vector space or linear independence of vectors. Matrices and vectors are used only as notational tools, so any student familiar with these tools and their operations of addition and multiplication can read the text. Appendix A contains a brief list of the topics from linear algebra that are used in the book.

LENGTH AND ORGANIZATION

The book probably contains more material than can be taught in a one-semester course. However, Chapters 5–8 are independent of each other. Thus, once the central ideas of Chapters 3 and 4 have been developed, the instructor has considerable latitude in the selection of other topics to be discussed. The following diagram indicates more precisely the relationships among the chapters.

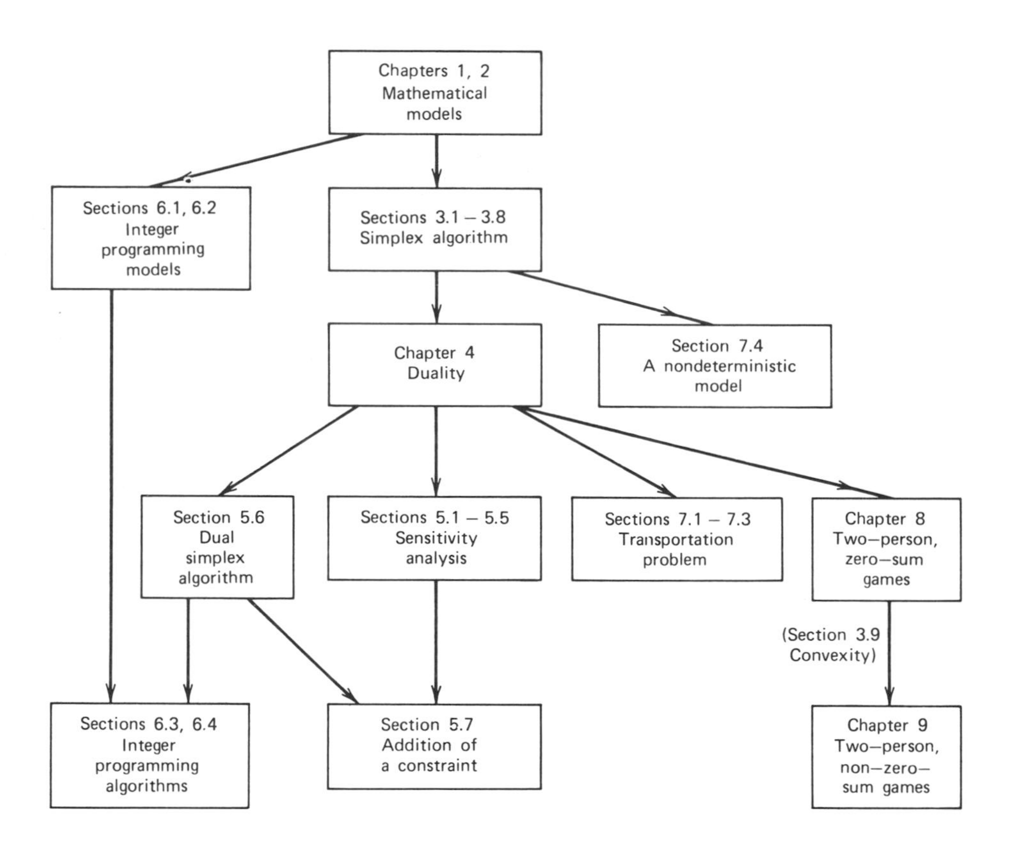

CONTENTS

Linear programming and game theory are introduced in Chapter 1 by means of examples. This chapter also contains some discussion on the application of mathematics and on the roles that linear programming and game theory can play in such applications. To introduce the reader to the broad scope of the theory, Chapter 2 (on model building) presents various real-world situations that lead to mathematical models involving linear optimization problems. Several of these examples are returned to later in the text as the tools are developed to resolve the questions raised here.

Chapters 3 and 4 are the core of the book. The simplex algorithm is presented in Chapter 3, and the concept of duality in Chapter 4. The development of the simplex algorithm is motivated algebraically, and all of Chapter 3 maintains an algebraic flavor. The convergence of the algorithm is proved inductively in Section 3.8. There are some geometrical considerations throughout the chapter, however, to assist in the understanding of the development, and the last section of Chapter 3 is about convexity. The only material from this last section used later in the text is the definition of convexity, and that is not used again until Chapter 9.

The Duality Theorem is proved in Section 4.3 of Chapter 4, but only after some indication of its importance is given by means of the examples in Section 4.2. In Chapter 4 matrix notation is used for the first time in the representation of the linear programming problem.

Sensitivity analysis is presented at two levels in Chapter 5. In Section 5.1 two examples are developed to show some of the problems of sensitivity analysis and to show how the theory of duality can be used to resolve some of these problems. The more general study is begun in Section 5.2 with the development of the matrix representation of the simplex tableau. Here it is assumed that the reader is familiar with matrix multiplication and the inverse of a matrix. In Section 5.6 the Dual Simplex Algorithm is presented. Although the algorithm is motivated by problems raised in Section 5.5, Section 5.6 is independent of the theory of these preceding sections and could, in fact, be read directly after Chapter 4. The Dual Simplex Algorithm is used in Sections 5.7, 6.3, and 6.4.

Chapter 6 provides an introduction to integer programming. Two algorithms that can be used to solve integer programming problems are presented. Except for the fact that both these algorithms use the Dual Simplex Algorithm as a tool, this chapter could be read after Chapter 3.

The first three sections of Chapter 7 deal with the transportation problem. A special algorithm (Ford–Fulkerson) is developed for the resolution of these problems, and in Section 7.3 the material is related to the assignment problem. Section 7.4 is independent of these first three sections. In this section one approach to a nondeterministic model is demonstrated by means of an example. As a supplementary benefit, the resulting optimization problem has many upper bound constraints, and so special solution techniques for such problems are illustrated. The entire chapter can be read after Chapter 4.

Two-person, zero-sum games are the subject of Chapter 8. First the axioms that form the foundation of the theory are discussed at some length to help the reader understand not only the concept of a solution to a game, but also the limitations on the applicability of the theory. Then, using the Duality Theorem of linear programming, the existence of solutions to two-person, zero-sum games is demonstrated. Computational techniques and examples conclude the chapter.

Utility theory is introduced in the first section of Chapter 9. The remainder of the chapter is devoted to two-person, non-zero-sum games. These games provide excellent examples of some of the difficulties that can be encountered when attempting to formulate mathematical models of complicated situations that involve human behavior. In discussing these games, factors not relevant in the theory of two-person, zero-sum games, such as the possibility of cooperation between the participants, are considered, and then various approaches and solution concepts are explored, primarily by means of examples.

Appendix A, as stated, contains a brief summary of the topics from linear algebra used in the text. Appendix B contains flow charts and possible test problems for use by students who are developing their own computer programs of the simplex algorithm.

EXERCISES

Problem sets containing computational exercises, problems testing understanding, and examples motivating new material conclude each section of the text. Because the problems are placed in each section and not simply at the conclusion of each chapter, the reader is constantly encouraged to test and develop his or her understanding of the material. An instructor's manual containing the worked-out solutions to many of the problems is available.

ACKNOWLEDGMENTS

First, I would like to thank Professors Joseph G. Ecker, James A. Murtha, and Edward J. Smerek for the many valuable suggestions and constructive criticisms they have provided. I am also grateful to the many students at Boston College who have used preliminary versions of the text. They, with their questions and answers, frowns and comments, have contributed greatly to the development of the material. Finally, I wish to acknowledge the professional expertise that the staff at Wiley has provided in the production of the text.

Paul R. Thie

Contents

1 Mathematical Models

SECTION 1.1 APPLYING MATHEMATICS

Recent history has shown us that many problems of our technically oriented society yield to mathematical descriptions and solutions. Problems as complex as sending people into space or maximizing the profit of a giant industrial conglomerate or problems as simple as balancing our own monthly budget or winning at the game of Nim are susceptible to mathematical formulations. This book is concerned with two specific fields of mathematics, linear programming and game theory, that offer insights into certain problems of the real world and techniques for solutions to some of these problems.

In order best to understand how one goes about applying a mathematical theory to the solution of some mundane problem, consider the stages that a problem passes through from organization to conclusion. We list four: (1) recognition of the problem; (2) formulation of a mathematical model; (3) solution of the mathematical problem; and (4) translation of the results back into the context of the original problem. These four stages are by no means exclusive or that well-defined. Certainly other authors have broken down the problem-solving operation in other ways, but the four steps listed indicate the framework that the applied mathematician works in.

The meaning of the first stage, recognition of the problem, is self-explanatory. The meaning of the second stage, formulation of a mathematical model, can be much more mysterious, conjuring visions of a precisely built representation of a small, snow-covered village at a scale of $\frac{1}{87}$. Actually, although the meaning of this step can be made quite clear, it is usually the most critical and difficult step to implement in the entire operation. The development of the mathematical model consists of translating the problem into mathematical terms, that is, into the language and concepts of mathematics. As an example of this process, consider what is called the "word problems" of high school algebra. Here the mathematics is trivial and the problems unrealistic, but many students stumble over the difficulties inherent in translating some concocted word problem into an algebraic equation, that is, in formulating the mathematical model. It was not always easy to determine how much 40% antifreeze solution to drain from the 20-qt cooling system in order to attain a 75% solution by adding a 90% antifreeze mixture.

In the development of a mathematical model of a complex situation, two basic and opposing elements are encountered. On the one hand, one seeks simplifying assumptions and overlooks minor details so that the resulting mathematical problem yields to a successful analysis. On the other hand, the model must adequately reflect reality so that the knowledge gained from the study of the model can be applied to the original problem. The ability to select those elements of a problem that are of major importance and disregard those of minor importance probably comes best from experience. Throughout the text and, in particular, in the next two sections, examples and problems requiring the development of a mathematical model are given. Although in many instances problems from a text may immediately single out the important elements and may seem somewhat artificial, much skill is to be gained by attempting them; practice model building and problem solving whenever possible.

Once the mathematical model has been formulated, one comes to the third stage in the process, the solution of the mathematical problem. It should be emphasized that this can entail much more than just computing the difference of a function at the end points of an interval or finding the solution to a system of equations. Even if the known theory does provide a complete theoretical solution to the problem, the specific answer to the problem at hand must still be calculated. It could very well be that further analysis does not provide any simplification of the problem, and only through involved computations can an estimate of the solution be made. Thus, finding a solution to a problem could mean determining a technique to approximate a solution that is financially feasible to implement within a given computer's capabilities and provides error estimates within given tolerance limits.

The meaning of the fourth step of the operation, the translation of the results back into the context of the original problem, is clear. Of course, more than a simple numerical answer is called for. The simplifying assumptions on which the solution is based must be understood, and the changes in the problem that would invalidate these assumptions should be considered.

We now give two examples of specific and well-known problems and begin the development of the associated mathematical models.

SECTION 1.2 THE DIET PROBLEM

The diet problem is one of the classical illustrations of a problem that leads to a linear programming model. The problem is concerned with providing at minimal cost a diet adequate for a person to sustain himself or herself. Simply stated, what is the least expensive way of combining various amounts of available foods in a diet that satisfies a person's nutritional requirements?

In order to develop a mathematical model of this problem, first the various aspects of the problem must be considered. Here the two competing needs for simplification and realism come into play as one attempts to state in precise terms the different components of the problem. For example, just how does one determine the basic nutritional requirements? We must consider the age, sex, size and activity of our subject. We must determine what nutrients, among the many known nutrients such as calories, proteins, and the multitude of vitamins and minerals, are essential. Can a need for one be met by a combination of others? Is it the case that too much of a certain nutrient is harmful and therefore forces an upper bound on the intake of that quantity? Should we provide for some variety in the diet, hopefully to meet nutritional requirements unknown to us at the present time?

Another component of the problem requiring study is the foods to be used for consideration in the diet. What foods can we assume are available? For example, can we assume that fresh fish, fruits or vegetables, or frozen foods are available? Once the foods to be used in the problem are established, the nutrient values of these foods must be determined. Here again only approximations can be made, since the nutrient value of a certain type of food, say apples or hamburger, not only varies from sample to sample because of lack of uniformity, but is also contingent on conditions and duration of storage and the method of preparation for consumption. The cost of a food can also fluctuate due to seasonal and geographical variances.

Once suitable approximations for the nutritional requirements of man and the nutrient values and cost of the available foods have been determined, a mathematical problem involving finding the minimum of a linear function can be formulated. In order to demonstrate this, we will consider a much simplified version of the diet problem.

Suppose we wish to minimize the cost of meeting our daily requirements of proteins, vitamin C, and iron with a diet restricted to apples, bananas, carrots, dates, and eggs. The nutrient values and cost of a unit of each of these five foods, along with the meaning of a unit of each, is given in the following table.

Food	*Measure of a Unit*	*Protein (g/unit)*	*Vitamin C (mg/unit)*	*Iron (mg/unit)*	*Cost (cents/unit)*
Apples	1 med.	0.4	6	0.4	8
Bananas	1 med.	1.2	10	0.6	10
Carrots	1 med.	0.6	3	0.4	3
Dates	$\frac{1}{2}$ cup	0.6	1	0.2	20
Eggs	2 med.	12.2	0	2.6	15

Our daily diet requires at least 70 g of protein, 50 mg of vitamin C, and 12 mg of iron. Since we are assuming that our supply of these foods is unlimited, it is obvious that we can find a diet that meets our needs, for example, a diet consisting of 6 units of eggs and 5 units of bananas would be more than adequate, as you can easily verify.

Our problem then is to determine the least expensive way of combining various amounts of the five foods to meet our three daily requirements. Hence the decision to be made involves the number of units of each of the five foods to consume daily. To translate this question into a mathematical problem, introduce five variables A, B, C, D, and E, where A is defined as the number of units of apples to be used in the daily diet, B the number of units of bananas, C the number of units of carrots, D the number of units of dates, and E the number of units of eggs. The cost in cents of such a diet is given by the function $f(A,B,C,D,E) = 8A + 10B + 3C + 20D + 15E$, found by using the cost column in the above table. It is this function that we wish to minimize.

However, there are clearly restrictions imposed by the problem on the possible values of the variables A, B, C, D, and E, that is, restrictions on the domain of the function f. First, all the variables must be nonnegative. And to guarantee that the daily nutritional requirements are fulfilled, the following three inequalities must be satisfied.

$$\begin{aligned} 0.4A + 1.2B + 0.6C + 0.6D + 12.2E &\geq 70 \\ 6A + 10B + 3C + 1D &\geq 50 \\ 0.4A + 0.6B + 0.4C + 0.2D + 2.6E &\geq 12 \end{aligned}$$

These inequalities are determined by considering the total input of the three required nutrients in a diet consisting of A units apples, B units of bananas, and so on. For example, since 1 unit of apples contains 0.4 g of protein, A units contain $0.4A$ g. Similarly, B units of bananas contain $1.2B$ g of protein, C units of carrots contain $0.6C$ units of protein, D units of dates $0.6D$ units, and E units of eggs $12.2E$ units. Adding these five terms gives the total intake of protein. Since our daily requirement of 70 g of protein is a minimal requirement and more is allowable, we have the first inequality. Similarly, the other two inequalities follow.

In sum, the resulting mathematical problem is to determine the minimum value of the function

$$f(A,B,C,D,E) = 8A + 10B + 3C + 20D + 15E$$

with the possible values of A, B, C, D, and E restricted by the inequalities

$$\begin{aligned} 0.4A + 1.2B + 0.6C + 0.6D + 12.2E &\geq 70 \\ 6A + 10B + 3C + 1D + &\geq 50 \\ 0.4A + 0.6B + 0.4C + 0.2D + 2.6E &\geq 12 \\ A,B,C,D,E &\geq 0 \end{aligned}$$

In 1944 George Stigler [30] considered the general diet problem. Stigler discussed the questions we raised and others, and he justified modifications and simplifications. For human nutritional requirements, Stigler decided on nine common nutrients (calories, protein, calcium, iron, vitamins A, B_1, B_2, C, and niacin), and estimated their needs from data supplied by the National Research Council. Stigler initially considered 77 types of foods and determined average nutrient value and costs. From this he was able to construct a diet that satisfied all the basic nutritional requirements and cost only $39.93 a year (less than 11 cents/day) for the year 1939. The diet consisted solely of wheat flour, cabbage, and dried navy beans.

SECTION 1.3 THE PRISONER'S DILEMMA

In the context of game theory, the word "game" in general refers to a situation or contest involving two or more players with conflicting interests, with each player having partial but not total control over the outcome of the conflict. The following is an example of such a situation. However, at this stage we are not yet able to translate the conflicting interests represented in the example into a precise mathematical problem, as contrasted to the example developed in the last section. Indeed, one of the major contributions of game theory is the resulting study of the question of what it means to "solve" a game.

The situation we consider is as follows. A certain democratic republic has a unicameral legislature with a membership drawn primarily from two major political parties. Before the assembly is a bill sponsored by a citizens' group designed to restrict the power and influence of the senior members of each political party. On this issue the legislators can be divided into three approximately equal groups—two groups whose members will follow the directives of their respective party leaders, and a third group of responsible representatives who consider passage of the bill more important than the maintenance of party loyalties and will support the bill regardless of circumstances.

Consider now this situation from the viewpoint of the leaders of the two parties. Due to the nature of things they would like to see the bill defeated, but their constituents overwhelmingly support the bill. However, an impending general election complicates matters. Because they are fairly adaptable people, the leaders know that they could, in fact, work moderately well within the limits set by the bill, so each group believes that the most beneficial outcome of the vote on the bill would be for their party to profess support for the bill while the opposition party opposes the bill. Of course, this would mean that the bill would pass, but the wave of public support generated for the one party voting for the bill would be a prevailing factor in the impending election. Thus the problem is how should each group of leaders direct their respective faithful party members to vote on the bill.

To answer this question, the leaders of one of the parties gather to consider the various possible outcomes of the vote on the bill. The most favorable outcome as far as they are concerned is for their party to support the measure and the opposition to oppose it. They denote this outcome by the ordered pair (Y,N) (they vote "yea" and the opposition votes "nay"). The least favorable outcome is the reverse of this situation, with their party members opposing but the opposition favoring passage of the bill [the (N,Y) outcome]. The two remaining possible outcomes are for both parties to support the bill [outcome (Y,Y)] and for both parties to oppose the bill [outcome (N,N)]. Neither of these outcomes would be a factor in the election, since the public reaction, either good or bad, would be balanced evenly between the two parties. However, outcome (N,N) is preferred over outcome (Y,Y), on the grounds that if both parties oppose the bill, it would be defeated and so the power of the party leaders would remain unaffected. Thus the leaders of the party linearly order the four possible outcomes, from most to least favorable, as follows.

$$(Y,N) > (N,N) > (Y,Y) > (N,Y)$$

Wishing to make this analysis even more precise and, hopefully, instructive, some of the leaders propose to assign numerical weights to each of these outcomes. They claim that such an assignment could reflect not only the above linear ordering, but also could measure how much more one outcome is preferred over another. They point out, for example, that a consideration in some contest of the three outcomes win \$3, win \$2, and win \$1 would not be identical to a consideration of the three outcomes win \$100, win \$2, and win \$1. Seeing the merits of this proposition, the leaders continue their deliberations on the four possible outcomes of the vote on the bill. Since outcomes (Y,N), (Y,Y), and (N,Y) all result in passage of the bill, their relative merits can be measured only by their effects in the impending election. Moreover, because of the equivalent strengths across the country of the two parties, the leaders believe that the advantage of (Y,N) over (Y,Y) is equal to the advantage of (Y,Y) over (N,Y). In fact, they argue that public reaction to support of the bill by only one party could be the determining factor in the election contests in up to

12 representative districts. Accepting this as a general unit and arbitrarily assigning the value 0 to outcome (Y,Y), they set (Y,N) worth 12 units and (N,Y) worth -12 units. There remains to be considered outcome (N,N), which lies between (Y,N) and (Y,Y) in the linear ordering. The assigning of a weight to this outcome is not immediate but, after a subcommittee review, prolonged debate, and various trade-offs in other matters, the political leaders accept the value of 6 units for this outcome.

Suppose that the leaders of the other party conduct similar deliberations and, since the positions of the two parties are comparable, reach the same conclusions. Then, to each possible outcome is attached two numerical weights, the value of that outcome to each party. Let us denote this pair of weights by an ordered pair of numbers, with the first component being the value of that particular outcome to one fixed party, called say Party D, and the second component being the value to the other party, say Party R. Then this situation can be represented by the following tableau.

		Party R	
		Vote "yea"	*Vote "nay"*
Party D	*Vote "yea"*	(0,0)	(12, −12)
	Vote "nay"	(−12,12)	(6,6)

Thus, for example, the outcome of a "nay" vote by Party D and a "yea" vote by Party R is $(-12,12)$; that is, that outcome is worth -12 units for Party D and 12 units for Party R.

This completes our analysis of this situation for the time being. It will be resumed in Chapter 9. We have formulated a two-person, non-zero-sum game in which each player has two possible moves, but we do not yet have a precisely stated mathematical problem to be solved. In fact, one of the major contributions of game theory is the analysis accompanying an attempt to define exactly what one would mean by a "solution" to the game or a "resolution" of the conflict. Such an analysis for a certain type of game is made in Chapter 8, where a complete mathematical model is formulated for finite, two-person, zero-sum games and the resulting mathematical problems are resolved (terms such as zero-sum are defined there).

The assigning of meaningful weights to the various possible outcomes is not properly a part of game theory, but is the function of utility theory (see Section 9.1). In the example of this section the use of game theory actually begins with the above tableau. Moreover, it is assumed in the theory that the information contained in that tableau is known to both parties. However, the theory does distinguish various interpretations of the conflict situation, such as whether or not the players can communicate with each other before the event, whether or not they can cooperate with each other, and whether or not agreements made are actually binding.

A word of explanation as to the meaning of the title of this section is in order. The game that has been developed in this section is an example of a certain type of two-person game. The archetype of game in this category, and the game that lends its name to the category, is the following example of a "prisoner's dilemma."

Two men are arrested on suspicion of armed robbery. The district attorney is convinced of their guilt, but lacks sufficient evidence for conviction at a trial. He points out to each prisoner, separately, that he can either confess or not confess. If one confesses and other does not, the district attorney promises immunity for the confessor and a 2-year jail sentence for the convicted partner. If both confess, he promises leniency and the probable result of a 1-year jail sentence for each prisoner. If neither confess, he promises to throw the book at them on a concealed weapon charge, with a 6-month jail sentence resulting for each.

The possible actions and the corresponding outcomes for the two prisoners are given by the following tableau. The outcomes are stated in terms of ordered pairs, with the first component representing the length of a prison term in months for Prisoner 1, and the second component the length for Prisoner 2.

		Prisoner 2	
		Confess	*Not Confess*
Prisoner 1	*Confess*	$(-12, -12)$	$(0, -24)$
	Not Confess	$(-24, 0)$	$(-6, -6)$

The negative signs indicate the undesirable nature of the outcomes (certainly a 12-month sentence is more favorable than a 24-month sentence; that is, $-12 > -24$). The similarity between this tableau and the previous one should be apparent, since the positions of the numbers in the linear ordering of the preferences and in the tableaux correspond. In fact, in this particular case, all the corresponding entries in the two tableaux differ by a fixed amount, 12.

SECTION 1.4 THE ROLES OF LINEAR PROGRAMMING AND GAME THEORY

Using as a base the four-step description of the operation of applying mathematics given in Section 1.1, an outline of how the fields of linear programming and game theory fit into this general scheme can be given.

In Section 1.2 an example of a linear programming problem was given. There occur many problems from business, industry, warfare, economics, and so on, that can be reduced to problems of this type, problems of finding the optimal value of some given linear function while the domain of the function is restricted by a system of linear equations or inequalities. The major concern here is not to determine whether or not an optimal value exists, but to develop a technique to determine quickly and easily the optimal value and where it occurs. Thus, from a mathematical point of view, we wish to develop for linear programming problems a method to use in the third stage of the process, finding the solution of the mathematical problem. In that realistic problems arising from a complex situation may have many variables and many constraints, it is clear that we need a method adaptable for use on modern computers. Moreover, since this is a mathematics text, we will prove that our solution is complete.

In Section 1.3 an example of a game theory problem was given. Our first concern with games will be with what we called the two-person, zero-sum games. Although the extent of our assumptions may seem to limit the applicability of the theory, this theory still serves as the foundation for the study of more complex games. Moreover, two-person, zero-sum games provide the opportunity to consider at a theoretical level the second stage in the process of applying mathematics, the formulation of the mathematical model. What one means by "the solution to a game" is not at all apparent, and axioms must be established that define this concept precisely and, hopefully, adequately reflect the economic or social situations to which game theory might be applied. This is in contrast to linear programming problems, where the desire to maximize profits or minimize costs translates immediately into a problem of optimizing a particular function.

From our discussion so far the problems of game theory and linear programming may seem to be totally unrelated, but this is not the case. Once our mathematical model for two-person, zero-sum games is developed, the problems of existence and calculation of a solution to a game will be related to our theory of linear programming. Here the unifying concept will be the notion of duality. Duality will be introduced in Chapter 4, and the main theorem of that chapter, the Duality Theorem, will provide the answer to the principal question of our study of games, that is, the question of existence of a solution.

2 The Linear Programming Model

SECTION 2.1 HISTORY

The basic problem of linear programming, determining the optimal value of a linear function subject to linear constraints, arises in a wide variety of situations, but theory that we will develop is of recent origin.

In 1939, the Russian mathematician L. V. Kantorovich published a monograph entitled "Mathematical Methods in the Organization and Planning of Production" [20]. Kantorovich recognized that a broad class of production problems led to the same mathematical problem, and that this problem was susceptible to solution by numerical methods. However, Kantorovich's work went unrecognized.

In 1941 Frank Hitchcock [18] formulated the transportation problem, and in 1945 George Stigler [30] considered the problem referred to in Section 1.2 of determining an adequate diet for an individual at minimal cost. Through these problems and others, especially problems related to the World War II effort, it became clear that a feasible method for solving linear programming problems was needed. Then in 1947 George Dantzig [4] developed the simplex method. This technique is the basis of the next chapter. John von Neumann recognized the importance of the concept of duality, the mathematical thread uniting linear programming and game theory, and the first published proof of the Duality Theorem is that of Gale, Kuhn, and Tucker [12].

Since the late 1940s, many other computational techniques and variations have been devised, usually for certain types of problems or for use with certain types of computing hardware. The theory has been applied extensively in industry. On the one hand, management has been forced to define explicitly its desired objectives and given constraints. This has brought about a much greater understanding of the decision-making process. On the other hand, the actual techniques of linear programming have been successfully applied in the petroleum industry, the food processing industry, the iron and steel industry, and many more.

SECTION 2.2 THE BLENDING MODEL

The diet problem described in Section 1.2 is an example of a general type of linear programming problem that involves blending or combining various ingredients. The cost and composition or characteristics of the various ingredients are known, and the problem is to determine how much of each of the ingredients to blend together so that the total cost of the mixture is minimized while the composition of the mixture satisfies specified requirements. In the diet problem, foods were combined to form a diet minimizing costs and meeting basic nutritional requirements.

The construction of the mathematical model for problems of this type follows quickly once the usually more difficult task of defining the characteristics and cost of the ingredients and required composition of the blend has been accomplished. Assuming that all this information is at hand, what amounts of each of the ingredients to blend together must be decided. Thus, assign variables to represent these amounts. The cost function, the function to be optimized, can then be constructed by considering the cost of each of the ingredients and assuming that the total cost is the sum of the individual costs. The system of constraints, that is, the set of restrictions of the variables, follows by considering the requirements specified for the final blend.

Example

In order to feed his stock a farmer can purchase two kinds of feed. The farmer has determined that his herd requires 60, 84, and 72 units of the nutritional elements A, B, and C, respectively, per day. The contents and cost of a pound of each of the two feeds are given in the following table.

	Nutritional Elements (units/lb)			Cost (cents/lb)
	A	*B*	*C*	
Feed 1	3	7	3	10
Feed 2	2	2	6	4

Obviously the farmer could use only one feed to meet the daily nutritional requirements. For example, it can be easily seen that 24 lb of the first feed would provide an adequate diet at a daily cost of \$2.40. However, the farmer wants to determine the least expensive way of providing an adequate diet by combining the two feeds. To do this thc farmer should consider all possible diets that satisfy the specified requirements, and then select from this set that diet of minimal cost.

To translate this into a mathematical problem, let x be the number of pounds of Feed 1 and y the number of pounds of Feed 2 to be used in the daily diet. Then by definition, x and y must be nonnegative. Moreover, a diet consisting of x lb of Feed 1 and y lb of Feed 2 would contain $3x + 2y$ units of nutritional element A. Since 60 units of elements A are required daily, we must have $3x + 2y \geq 60$. We are assuming that providing more than the minimal requirements of any of the nutritional elements will have no harmful effects, and so any diet providing at least 60 units of element A will satisfy this requirement. Thus the inequality and not an equality.

In order to provide insight into the nature of linear programming, this particular problem will be solved geometrically. Now the set of diets satisfying the above requirement can be illustrated graphically. All the points (x,y) in the first quadrant satisfying the inequality are shown in Figure 2.1.

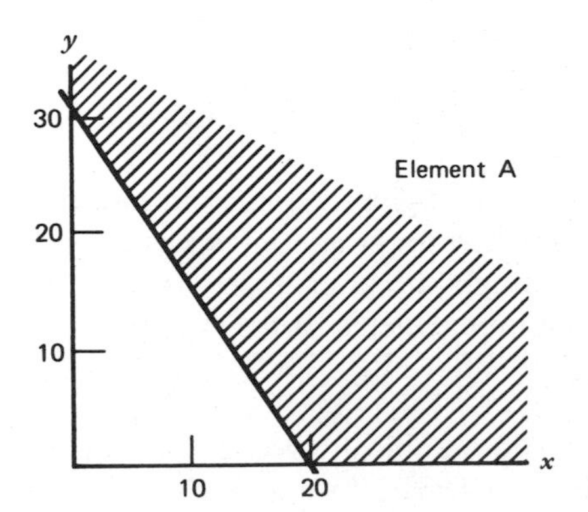

Figure 2.1

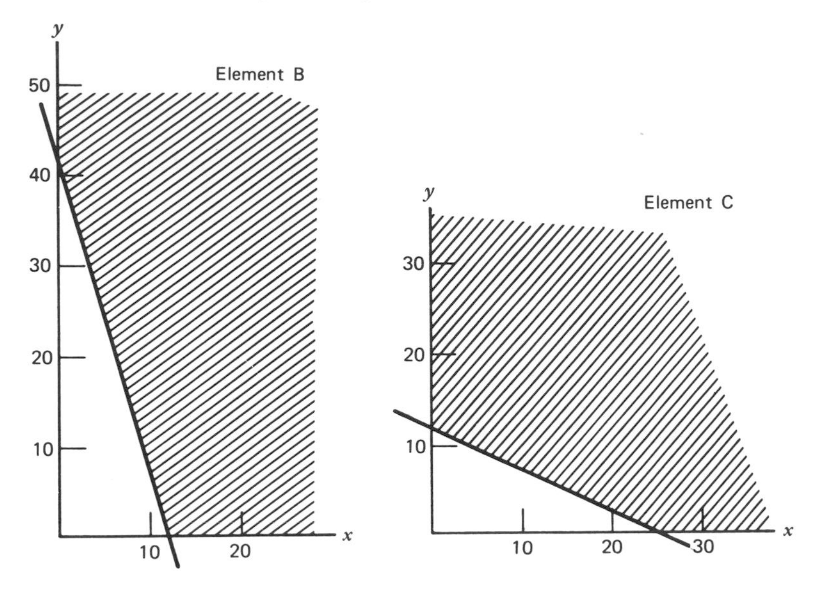

Figure 2.2

The other two nutritional requirements demand that

$$7x + 2y \geq 84 \qquad \text{and} \qquad 3x + 6y \geq 72$$

The corresponding regions in the first quadrant are sketched in Figure 2.2.

We must consider all diets that satisfy all three requirements. They are given graphically by the hatched region in Figure 2.3.

The cost in cents of a diet of x lb of Feed 1 and y lb of Feed 2 is $10x + 4y$. Thus we must determine the minimum of the function $f(x,y) = 10x + 4y$, while the x and y are restricted to the hatched region in Figure 2.3.

Consider the graphs of the family of lines determined by the equations $10x + 4y = c$, where c is constant. In Figure 2.4, some of these lines are graphed for various values of c. Note that all the lines have the same slope, and that the lines move to the left as c decreases.

Each of the parallel lines consists of points that give the same value for the cost function $10x + 4y$. Thus we seek that line furthest to the left that still intersects the hatched region of Figure 2.3. The line through the point (6,21) is that line, as illustrated in Figure 2.5. Thus the cost of a minimal diet is $10 \cdot 6 + 4 \cdot 21 = 144$ cents, and this diet consists of 6 lb of Feed 1 and 21 lb of Feed 2.

Notice that this problem could be solved graphically because we sought the optimal combination of only two ingredients. Clearly, if we were choosing from a

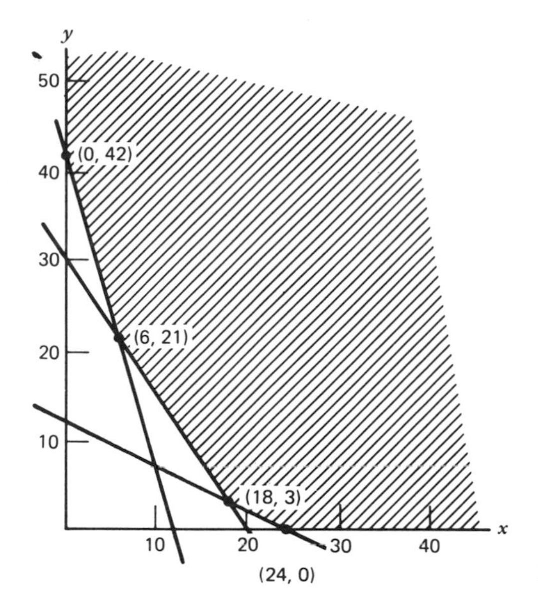

Figure 2.3

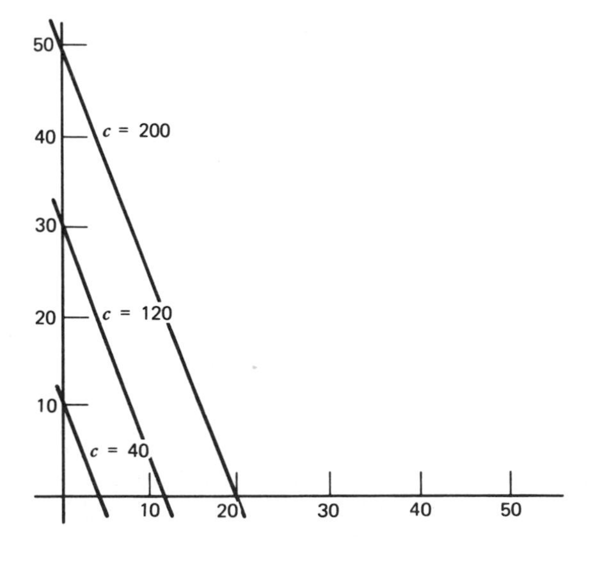

Figure 2.4

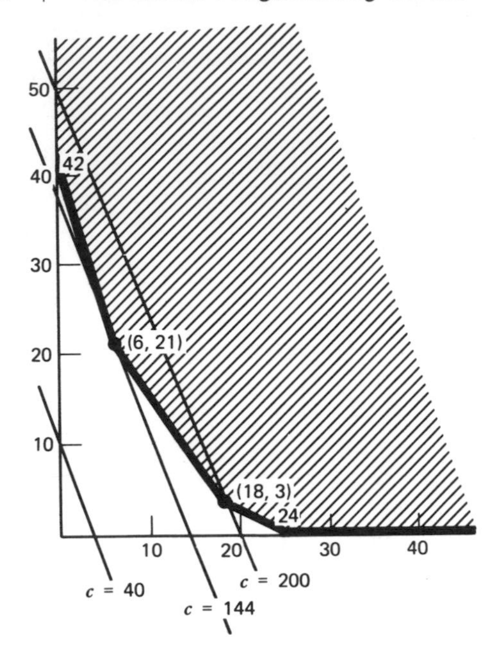

Figure 2.5

multitude of ingredients, the mathematical problem, while essentially the same, would be intractable to a graphical approach. The goal of Chapter 3 is to develop an efficient method of solving the general problem, regardless of size.

PROBLEM SET 2.2

Problems 1–6 refer to the example in this section.

1. A salesperson offers the farmer a new feed for his stock. One pound of this feed contains 2, 4, and 4 units of the nutritional elements A, B, and C, respectively, and costs 7 cents. By considering a blend that consists of equal parts of Feeds 1 and 2, show that the use of this new feed cannot reduce the minimal cost of an adequate diet.

2. Suppose the cost of Feed 1 is increased to 14 cents/lb. Solve the problem now by using Figure 2.3. Has the cost of the minimal diet changed? Has the diet necessary to achieve this minimal cost changed? Is this diet unique?

3. Suppose the costs of the two feed fluctuate because of availability and demand. However, the farmer has determined that as long as the ratio of the cost of Feed 1 to the cost of Feed 2 is between $\frac{3}{2}$ and $\frac{7}{2}$, an adequate diet of minimal cost can be achieved by using 6 lb of Feed 1 and 21 lb of Feed 2. Explain.

4. What should the ratio of the costs of the feeds be in order to warrant the use of a diet consisting of 18 lb of Feed 1 and 3 lb of Feed 2? When should the farmer use only Feed 1 for his stock?

5. After reviewing her father's mathematical formulation of the feed problem, the farmer's daughter claims that in general the constraining inequalities should be equalities. She reasons that money must be wasted if some of the nutritional elements are fed the stock at a level above the minimal requirements. Is this true?

6. After some study, the farmer has decided that 40 units of nutritional element D is also critical for the daily feeding of his stock. One pound of Feeds 1 and 2 contain 4 and 2 units of element D, respectively. How does this change the analysis of the original problem?

7. Solve the following.
(a) A poultry producer must feed his stock daily at least 124 units of nutritional element A and 60 units of nutritional element B. He has available two feeds. One pound of Feed 1 costs 16 cents and contains 10 units of A and 3 units of B. One pound of Feed 2 cost 14 cents and contains 4 units of A and 5 units of B. Determine the least expensive adequate feeding diet.
(b) As above, but suppose that now the cost of Feed 2 has doubled to 28 cents/lb.

Formulate mathematical models for the following problems.

8. A paint manufacturer must produce a base for its line of indoor domestic paints. Four chemicals, A, B, C, and D, are critical in its manufacture. The final composition of the base by weight must be at least 5% of Chemical A, 3% of Chemical B, 26% of Chemical C, and no more that 15% of Chemical D. The manufacturer can produce this base by combining three crude minerals. The compositions by weight and the costs of these minerals are given in the following table.

	% Chemical A	*% Chemical B*	*% Chemical C*	*% Chemical D*	*Cost (dollars/lb)*
Mineral 1	0	5	30	20	4.00
Mineral 2	6	8	30	10	7.50
Mineral 3	7	0	25	16	3.00

The manufacturer could use just Mineral 2. However, he wants to know if some combination of the three minerals will provide a base with the desired characteristics at a lower cost. (In working with percentages, ambiguities can arise. To avoid these, determine the least expensive way of producing a fixed amount of the base, say 1 lb. One pound of the base must contain at least 0.05 lb of Chemical A, 0.03 lb of Chemical B, and so on. Do not forget that the sum of the weights of the minerals used must also equal this fixed amount.)

9. A firm wants to market bags of lawn fertilizer that contain 23% nitrogen, 7% phosphoric acid, and 7% soluble potash. They have available Chemicals A, B, C, D, and E that can be combined for their product. The contents in pounds and cost in dollars of 100 lb of each are:

	A	*B*	*C*	*D*	*E*
Nitrogen	18	28	0	30	16
Phosphoric Acid	12	5	6	7	3
Potash	0	5	18	8	2
Cost	10	23	10	30	15

How much of each chemical should be used to minimize costs?

10. A coin is to be minted containing 40% silver, 50% copper, and 10% nickel. The mint has available Alloys A, B, C, and D, with the following compositions and costs.

	A	*B*	*C*	*D*
% Silver	30	35	50	40
% Copper	60	35	50	45
% Nickel	10	30	0	15
Costs/lb	$11	12	16	14

What blend of these alloys provides the desired composition at minimal costs?

11. The manager of a fleet of trucks needs an antifreeze solution containing at least 50% pure antifreeze and 5% anticorrosion additives. He has available three commercial products A, B, and C, with characteristics and costs given in the following table. What blend will provide a suitable solution at minimal costs?

	A	*B*	*C*
% Antifreeze	60	18	75
% Additives	10	3	0
Cost (dollars/gal)	1.6	0.5	1.4

SECTION 2.3 THE PRODUCTION MODEL

Production models and their variations occur frequently in linear programming applications. Central to these problems is an operation or production system, say a factory or a refinery. Commodities such as raw materials, capital, and labor are input into the system and are acted on by various productive processes. The results are the output or goods produced, and the basic problem is to operate the system in a way that maximizes profit using limited resources or minimizes costs while meeting specified production requirements or some combination of these goals.

Example 1

Suppose a boat manufacturer produces two types of boats for the sports and camping trade, a family rowboat and a sports canoe. The boats are molded from aluminum by means of a large pressing machine and are finished by hand labor. A rowboat requires 50 lb of aluminum, 6 min of machine time, and 3 h of finishing labor; a canoe requires 30 lb of aluminum, 5 min of machine time, and 5 hr of finishing labor. For the next 3 months the company can commit up to 1 ton of aluminum, 5 hr of machine time, and 200 hr of labor for the manufacture of the small boats. The company realizes \$50 profit on the sale of a rowboat and \$60 profit on the sale of a canoe. Assuming that all boats made can be sold, how many of each type should be manufactured in the next 3 months in order to maximize profits?

Here the decision to be made involves the number of rowboats and the number of canoes to be produced in the next 3 months. Thus, let R and C denote these numbers, with R the number of rowboats and C the number of canoes. Then the profit for the company, measured in dollars, from its small boat line will be $50R + 60C$, and this is the function to be maximized.

Now the quantities R and C cannot be negative. Moreover, they are limited by the amount of resources available for the production of the boats. Specifically, at most 1 ton of aluminum can be used, and so we must have $50R + 30C \leq 2000$. Similarly, consideration of available machine time and finishing labor leads to the inequalities

$$6R + 5C \leq 300 \qquad \text{and} \qquad 3R + 5C \leq 200$$

Thus the mathematical problem is to determine R and C that maximize the function $50R + 60C$ and satisfy the constraints $R \geq 0$, $C \geq 0$,

$$\begin{aligned} 50R + 30C &\leq 2000 \\ 6R + 5C &\leq 300 \\ 3R + 5C &\leq 200 \end{aligned}$$

Example 2

Consider the operation of one division in a large plant. The division is responsible for manufacturing two parts of the plant's final product. The division manager has available four different processes to produce these two parts; each process uses various amounts of labor and two raw materials. The inputs and outputs for 1 hr of each of the four processes is given in the following table.

	Input			Output	
Process	*Labor (man-hr)*	*Raw Material A (lb)*	*Raw Material B (lb)*	*Units of Part 1*	*Units of Part 2*
1	20	160	30	35	55
2	30	100	35	45	42
3	10	200	60	70	0
4	25	75	80	0	90

The division is responsible for producing each week 2100 units of Part 1 and 1800 units of Part 2. The division manager has at her disposal each week 4 tons of Raw Material A and 2 tons of Raw Material B, and 1000 man-hours of labor. One pound of Raw Material A costs the firm \$3 and 1 lb of Raw Material B costs \$7. Because of labor contracts, the plant must pay its employees a full weeks' salary, regardless of whether or not the employees are used that week, and so the cost of the 1000 man-hr of labor is fixed. However, the division manager can request her workers to work up to an extra 200 man-hr per week in overtime, at a cost of \$8/hr to the firm. The plant vice-president in charge of production wants to know if the division can meet its weekly production requirements with the material on hand without using overtime and, if so, the minimal cost of this operation. Second because the decision to allow overtime must be made at the plant level, the vice-president wants some estimate on how much money if any the division can save by using overtime.

In order to respond to her supervisor's questions, the division manager must consider the problem in two stages and at each stage must determine the optimal use of her facilities. In the first stage overtime is not available, and thus the manager must decide only on how to best utilize the four available processes. Let x_i denote the number of hours a week that Process i is used, for $i = 1,2,3,4$. The constraints imposed by the limited amounts of labor and raw materials are the following.

$$\begin{aligned} 20x_1 + 30x_2 + 10x_3 + 25x_4 &\leq 1000 \\ 160x_1 + 100x_2 + 200x_3 + 75x_4 &\leq 8000 \\ 30x_1 + 35x_2 + 60x_3 + 80x_4 &\leq 4000 \end{aligned}$$

The output requirements give

$$\begin{aligned} 35x_1 + 45x_2 + 70x_3 \qquad &\geq 2100 \\ 55x_1 + 42x_2 \qquad + 90x_4 &\geq 1800 \end{aligned}$$

Thus the initial question of determining whether or not the weekly production requirements can be met with the available materials is translated into the mathematical problem of determining if there exists four nonnegative numbers x_1, x_2, x_3, x_4 that satisfy these five inequalities. A solution to this problem would provide a suitable or feasible way of operating the division, and this suggests that any nonnegative solution to the system of constraints in a linear programming problem be called a *feasible solution.* In this particular problem, the existence of a feasible solution is easy to verify. The weekly use of 30 hr of Process 3 and 20 hr of Process 4 will produce the exact number of needed parts and will not even exhaust any of the supplies of labor and raw materials.

The cost of the operation when overtime is not employed depends only on the amounts of the raw materials used and is given in dollars by the function

$$\begin{aligned} f(x_1,x_2,x_3,x_4) &= 3(160x_1 + 100x_2 + 200x_3 + 75x_4) \\ &\quad + 7(30x_1 + 35x_2 + 60x_3 + 80x_4 \\ &= 690x_1 + 545x_2 + 1020x_3 + 785x_4 \end{aligned}$$

Thus the first optimization problem is to determine x_1, x_2, x_3, x_4 that minimize

$$f(x_1,x_2,x_3,x_4) = 690x_1 + 545x_2 + 1020x_3 + 785x_4$$

subject to

$$\begin{aligned} 20x_1 + 30x_2 + 10x_3 + 25x_4 &\le 1000 \\ 160x_1 + 100x_2 + 200x_3 + 75x_4 &\le 8000 \\ 30x_1 + 35x_2 + 60x_3 + 80x_4 &\le 4000 \\ 35x_1 + 45x_2 + 70x_3 \qquad\quad &\ge 2100 \\ 55x_1 + 42x_2 \qquad\qquad + 90x_4 &\ge 1800 \\ x_1, x_2, x_3, x_4 \ge 0 \qquad\qquad\qquad & \end{aligned} \tag{1}$$

The possibility of using overtime introduces one more decision the division manager can make: how much if any overtime to use to reduce the total production cost. Let x_5 denote the number of hours of overtime employed. Then $0 \le x_5 \le 200$. As before, x_1, x_2, x_3, x_4 represent the hours of use of the four processes. Note, however, that this is a different problem, so the optimal schedule here may employ amounts of the processes different from the previous optimal schedule. The first constraint, the restriction on available man-hours, is the only inequality that needs to be changed. Now the total man-hours used cannot exceed $1000 + x_5$, so this inequality becomes

$$20x_1 + 30x_2 + 10x_3 + 25x_4 \le 1000 + x_5$$

The total cost function must also reflect the cost of the overtime, but the cost of the raw materials is measured as before. Thus the new total cost function, say g, can be defined by simply adding the cost of overtime to the original cost function f. Thus we must minimize

$$g(x_1,x_2,x_3,x_4,x_5) = 8x_5 + f(x_1,x_2,x_3,x_4)$$

Hence the second problem is to minimize the function

$$g(x_1,x_2 x_3,x_4,x_5) = 8x_5 + 690x_1 + 545x_2 + 1020x_3 + 785x_4$$

subject to

$$\begin{aligned}
20x_1 + 30x_2 + 10x_3 + 25x_4 &\leq 1000 + x_5 \\
160x_1 + 100x_2 + 200x_3 + 75x_4 &\leq 8000 \\
30x_1 + 35x_2 + 60x_3 + 80x_4 &\leq 4000 \\
35x_1 + 45x_2 + 70x_3 \qquad &\geq 2100 \\
55x_1 + 42x_2 \qquad + 90x_4 &\geq 1800 \\
x_1,x_2,x_3,x_4,x_5 \geq 0,\ x_5 &\leq 200
\end{aligned} \tag{2}$$

One final question. Suppose the vice-president in charge of profits wants to make some estimate on the production costs of the firm's products and, in order to do this, requests the division manager to estimate the costs of manufacturing 1 unit of Parts 1 and 2. It would be easy to determine the cost of 1 unit of Part 1, for example, if the division produced only this type of part and Process 3 was used in its production. Then the total cost of 1 hr of operation of this process, and here the cost of the man-hours involved would need to be included, divided by the number of units of Part 1 produced, would give a unit cost. However, this is not the situation. Not only could several processes be involved in the production of Part 1, but also the output of the processes can be mixed. Moreover, how can we measure the true costs of the labor and raw materials? It could be, for example, that a minimal cost production schedule leaves a surplus of Raw Material A but exhausts the available supply of Raw Material B, and thus Raw Material B is more precious. Should this fact also be included in the costs of the raw materials? We will consider such questions later, in Chapters 4 and 5, after the concept of duality has been introduced. (See in particular Problem 12 of Section 5.1.)

PROBLEM SET 2.3

The first three problems refer to the second example of this section.

1. As long as all workers in the division are interchangeable, there arises a restriction on the use of overtime; that the total number of regular man-hours available be exhausted before overtime is used. In this problem that would mean that, for an optimal schedule considering the use of overtime, if the quantity $20x_1 + 30x_2 + 10x_3 + 25x_4$ is less than or equal to 1000, x_5 must be 0. Prove that any optimal solution to (2) must have this property.

2. Suppose the firm can sell excess units of Part 1 on the market for \$3 and excess units of Part 2 for \$2. Change the first mathematical problem not involving overtime to incorporate this new fact.

3. Suppose the firm can supplement its supply of Raw Material A by purchasing the material from an outside source. Assume that an unlimited amount can be purchased, and that the cost would be \$4/lb. Construct the associated mathematical problem, assuming that the use of overtime is also possible.

4. Solve the first example of this section graphically.

5. Extremum problems for functions of several variables are discussed in multivariable calculus. The standard technique involves taking first partials and setting them equal to zero. What information does this method provide for the first example of this section?

Formulate the mathematical models for Problems 6–10.

6. An electronics firm manufactures integral circuits for radios, televisions, and stereos. For the next month it has available 1500 units of materials and 920 units of labor. The requirements and selling price of one of each of the above products is given in the following table.

	Units of Material	*Units of Labor*	*Selling Price (dollars)*
Radio	2	1	8
TV	12	8	60
Stereo	15	6	45

Determine a production schedule that maximizes income.

7. An oil refinery has available three different processes to produce gasoline. Each process produces varying amounts of three grades of gasoline: regular, low-lead, and premium. These amounts, in hundreds of gallons per hour of operation, are given in the following table, along with the cost in dollars of an hour's operation of each of the processes.

	Regular	*Low-Lead*	*Premium*	*Cost*
Process 1	3	4	2	160
Process 2	6	6	8	400
Process 3	6	3	4	300

The refinery must meet the weekly demands of 3600 gal of regular, 2000 gal of low-lead, and 3000 gal of premium. Determine that operation of the refinery that satisfies these demands and minimizes costs.

8. A fruit grower has two systems for picking crops. In the first system, the help works individually and, because of the selectivity and care of the picker, this method yields more choice produce than regular produce. In the second system four pickers work with a machine; although this method has a greater harvest, it yields less choice produce than regular produce and costs proportionately more because of the operating expense of the machine. The exact amounts of input and output for 1 hr of operation of each system are contained in the following table.

	Input (man-hr)	*Output (bu)*		*Costs (dollars)*
		Choice	*Regular*	
System 1	1	4	2	2
System 2	4	20	40	11

Weekly the grower must supply the retail outlet with 480 bu of choice produce and deliver 800 bu of regular produce to the local cannery. The grower has available 100 man-hr of labor per week. Determine an operating schedule that meets these demands and minimizes costs.

9. A small steel plant has available three processes for the production of steel. The processes require varying amounts of labor, ore, and coal and produce not only steel, but also one side product with a limited saleability. The relevant data for an hour's operation of each process is as follows.

	Input			*Output*	
	Labor (man-hr)	*Ore (lb)*	*Coal (lb)*	*Steel (lb)*	*Side Product (lb)*
Process 1	8	200	145	150	35
Process 2	11	140	120	125	15
Process 3	7	300	225	185	75

For a week's operation the plant has available up to 350 man-hr of labor at $5.75/hr, up to 5 tons of ore at $43/ton, and an unlimited amount of coal at $12/ton. All the steel produced can be sold for $150/ton, and up to 1 ton of the side product can be sold for $37/ton (any amount above 1 ton is considered worthless). Because of operational restrictions, no one process can be employed for more than 40 hr in any week. Determine an operating schedule that maximizes net income. (*Suggestion.* To incorporate the value of the production of the side product into the function to be optimized, divide its total amount produced into two increments, say S_1 and S_2, where S_1, $0 \leq S_1 \leq 2000$ lb represents the first ton produced.)

10. A farmer has 100 acres of tillable land on which corn, tomatoes, beans, peas, or carrots can be planted. The labor, costs, and gross income for an acre of each of these crops is as follows.

	Labor (man-hr)		*Gross Income*
Corn	5	\$ 20	\$ 60
Tomatoes	120	200	1000
Beans	25	55	200
Peas	35	40	225
Carrots	40	75	325

The costs include fertilizer, seeds, insecticides, and so on. The farmer has available up to 3600 man-hr of labor at \$3.25/hr, and \$3000 in capital to invest and can borrow up to another \$12,000 if desired. Any such loan would have a 9% annual interest rate, but would be repaid within 4 months. Land unused for any of these vegetables must be maintained by planting ground cover. Cover crops require 2 man-hr and cost \$9/acre. Determine a planting schedule that maximizes net income.

11. A shop with three furniture makers produces uniquely designed chairs and sofas from fabric and wood. The requirements for each piece are as follows.

	Fabric	*Wood*	*Labor*
Chair	3 yd	6 units	9 man-hr
Sofa	8 yd	5 units	4 man-hr

For a week's operation, the shop has available 96 yd of fabric, 90 units of wood, and 120 man-hr of labor. A profit of \$70 is realized from the sale of a chair and a profit of \$60 from a sofa. Determine a weekly production schedule that maximizes profit.

(a) Formulate a mathematical model for this example.

(b) Show graphically that the solution to the problem calls for the manufacture of $7\frac{3}{11}$ chairs and $9\frac{3}{11}$ sofas.

(c) How would one implement such a schedule?

(d) Answer to part c? Show that the weekly production of 8 chairs and 10 sofas, or of 8 chairs and 9 sofas, or of 7 chairs and 10 sofas are all impossible within the given restrictions, but that 7 chairs and 9 sofas can be produced. Thus this feasible schedule, with integral components, appears to be the desired schedule.

(e) Show that it is also possible for the shop to produce weekly, within the given limitations, 10 chairs and 6 sofas. Compare the profit associated with this production schedule with the profit of the (7,9) production schedule of part d.

(f) *Conclusion.* Problems requiring integral answers may require special techniques.

SECTION 2.4 THE TRANSPORTATION MODEL

Transportation problems were one of the first types of problems analyzed in the early history of linear programming. The general problem arises when goods available at several sources, say warehouses, must be shipped to various destinations, say retail outlets. With fixed amounts available at the sources and fixed demands to be met at the destinations, the problem is to determine a shipping schedule that minimizes transportation costs. It is assumed that the costs of shipping goods from a source to a destination are directly proportional to the amount of goods shipped.

Example

A paper manufacturer having two mills must supply weekly three printing plants with newsprint. Mill 1 produces 350 tons of newsprint a week and Mill 2, 550 tons. Plant 1 requires 300 tons/week, Plant 2, 400 tons, and Plant 3, 200 tons. The shipping costs, in dollars per ton, are given in the following table.

		Plants		
		1	2	3
Mills	1	17	22	15
	2	18	16	12

The problem is to determine how many tons each mill should ship to each plant so that the total transportation cost is minimal.

To formulate the mathematical model, let x_{ij} denote the amount in tons to be shipped weekly from Mill i to Plant j, for $i = 1,2$ and $j = 1,2,3$. Then each x_{ij} must be nonnegative. Moreover, the amount shipped from each mill cannot exceed the supply. Thus we must have

$$
\begin{aligned}
x_{11} + x_{12} + x_{13} &= 350 \\
x_{21} + x_{22} + x_{23} &= 550
\end{aligned}
$$

We have equalities here, since the total weekly supply is equal to the total demand, and thus all the available newsprint at both mills must be shipped, leaving no surplus. So that the demands at each printing plant are met, the following equalities must be satisfied.

$$
\begin{aligned}
x_{11} + x_{21} &= 300 \\
x_{12} + x_{22} &= 400 \\
x_{13} + x_{23} &= 200
\end{aligned}
$$

The total shipping cost is $17x_{11} + 22x_{12} + 15x_{13} + 18x_{21} + 16x_{22} + 12x_{23}$, and it is this function of the six variables x_{ij}, $i = 1,2$, $j = 1,2,3$, that we wish to minimize.

One complication in problems of this type can occur when the commodity to be shipped is not divisible. For example, problems involving the shipment of automobiles or colored television sets would require an integral solution. (See also Problem 11 of Section 2.3.) Considerable study has been done in this field of integer programming, and the theory has been found to have many applications. We will consider the topic in Chapter 6.

PROBLEM SET 2.4

1. Even though the example in this section has six variables, the following elementary analysis does lead to the solution of the problem. It costs \$1/ton less to supply Plant 1 from Mill 1 instead of Mill 2. However, supplying Plant 2 from Mill 2 saves \$6/ton, and supplying Plant 3 from Mill 2 saves \$3/ton. The greatest relative savings comes from supplying Mill 2 as much as possible from Plant 2. Continuing this argument, show that the optimal shipping schedule has $x_{11} = 300$, $x_{12} = 0$, $x_{13} = 50$, $x_{21} = 0$, $x_{22} = 400$, $x_{23} = 150$.

Formulate mathematical models for the following problems.

2. A canned goods supplier has two warehouses serving four outlets. The East Coast Warehouse has 600 cases on hand and the West Coast Warehouse has 1000 cases on hand. The shipping costs, in cents per case, and the requirements for the four outlets, all located east of the Mississippi, are given in the following table.

		Outlet 1	*Outlet 2*	*Outlet 3*	*Outlet 4*
Shipping	*East Coast Warehouse*	20	16	30	20
costs	*West Coast Warehouse*	45	40	50	44
Requirements (cases)		300	350	400	450

Determine a shipping schedule that minimizes transportation costs.

3. Three beverage plants supply five wholesale outlets with cases of soft drinks. The weekly demands and transportation costs (in cents per case) are as follows.

		Outlets				
		1	*2*	*3*	*4*	*5*
	1	6.2	—	5.1	10.1	8.0
Plants	2	6.5	10.5	4.3	11.3	6.5
	3	6.3	9.0	—	10.8	—
Requirements (cases)		1000	1200	3000	400	2200

The dashes in the table indicate the impossibility of shipping cases between the corresponding plants and outlets. The weekly production of Plant 1 is 4000 cases, of Plant 2, 2000 cases, and of Plant 3, 3000 cases. Suppose also that the weekly surplus at each plant can be sold locally for $1.20/case at Plant 1, $1.10/case at Plant 2, and $1.14/case at Plant 3. Determine a shipping schedule that minimizes transportation costs and that takes into account the amount accrued from the sale of the surplus.

4. A commodity is to be shipped from three warehouses to four outlets, each outlet receiving 120 units. The shipping costs in dollars per unit are:

		Outlets			
		1	*2*	*3*	*4*
	1	12	15	10	25
Warehouses	2	10	19	11	30
	3	21	30	18	40

Warehouse 1 has available 100 units, Warehouse 2, 150 units, and Warehouse 3, 300 units. For any unit not shipped, there is a storage charge of $6/unit at Warehouses 1 and 2 and $12/unit at Warehouse 3. Moreover, because of labor contracts, Outlet 2 must receive more units from Warehouse 2 than from Warehouse 1, and also more than from Warehouse 3. Determine a minimal cost shipping and storing schedule.

5. The following is an example of a transshipment problem. Consider the network of Cities 1–7.

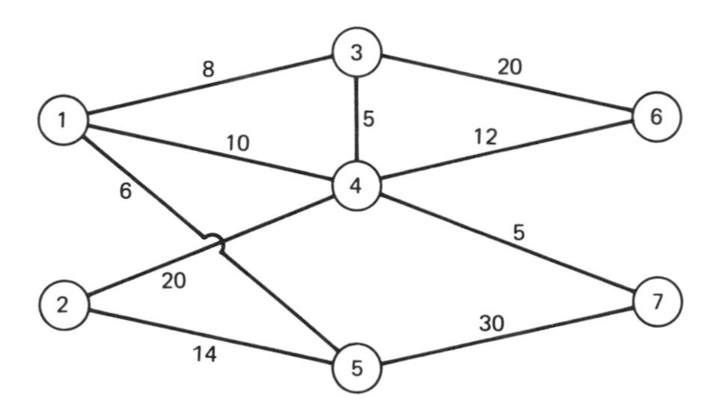

At City 1, 20 units of a commodity are produced, and at City 2, 30 units. Cities 3–5 have no need for the commodity, but Cities 6 and 7 require 25 units each. The lines in the above network indicate the available transportation links between the cities; the numbers indicate the capacity in units of the links. Is it possible to satisfy the given demands with the given supplies using this transportation network?

SECTION 2.5 THE DYNAMIC PLANNING MODEL

The operation of many systems or processes can be divided into distinct time periods that allow for the flexibility of activities during each period and such that decisions for one period affect not only that period but also subsequent periods. For example, the yearly operation of a giant steel plant can be divided into 12 monthly time periods. In each period labor, capital, and raw materials are combined to produce steel, with varying monthly demands on the product. At the beginning of each month, the amount of steel to be produced must be decided on. If demands are low, should employees be laid off or left idle or should surplus steel be produced and stored? Should future high demands be met by immediate increased production and storage or by hiring and training personnel? If the cost and availability of raw materials varies due to, say weather factors and mining conditions, then the decision on the amount of raw materials to purchase in any one month is influenced by present and future needs, storage capacity, and available capital.

Example 1

Consider the operation of a dealer of home heating oil. Suppose the dealer owns a storage tank with a capacity for 10,000 gal of oil, and that initially has 3000 gal in it. Each month for the next 3 months the dealer can sell up to 8000 gal of oil per month, charging 47 cents/gal during the first month, 50 cents/gal the second, and 52 cents/gal the third. Furthermore, the dealer can purchase up to 5000 gal of oil each month either for distribution during the month or for storage for later use. The cost to the dealer of this oil is 40 cents/gal during the first month, 42 cents/gal during the second month, and 45 cents/gal during the third. How much oil should the dealer purchase, sell, and store during each month in order to maximize profits? Assume that any oil left in the storage tank after the third month has a value of only 38 cents/gal.

This example is a bit more involved than those we have seen so far. To formulate the mathematical model, we must, as before, first assign variables to represent the amounts of each activity that the dealer can perform. Since, at the beginning of each month, the dealer must decide on how much oil to buy, distribute, and store during that month, three variables will be needed for each period. In particular, let P_i denote the number of gallons of oil purchased by the dealer during Month i, where $i = 1,2,3$. Similarly, let D_i represent the number of gallons of oil distributed during Month i and S_i the number of gallons stored.

There are many restrictions on these nine variables. Obviously they must be nonnegative, and they all have fixed upper bounds, with $S_i \leq 10{,}000$, $P_i \leq 5000$, and $D_i \leq 8000$, for $i = 1,2,3$. However the quantities are also interrelated. For each month, the oil purchased during that month plus the oil stored from the previous month must equal the total amount of oil delivered and stored during that month. Thus, for the first month, $3000 + P_1 = S_1 + D_1$. Similarly, for the next 2 months we must have

$$S_1 + P_2 = S_2 + D_2 \qquad \text{and} \qquad S_2 + P_3 = S_3 + D_3$$

The total profit for the operation is equal to the income from the oil sold plus the value of the oil left in the storage tank less the cost of the oil purchased. Thus the function to be maximized is

$$47D_1 + 50D_2 + 52D_3 + 38S_3 - (40P_1 + 42P_2 + 45P_3)$$

Example 2

A firm is given a short-term government contract to produce a total of 3700 units of some commodity over a period of 4 weeks. Producing the commidity can involve three new cost factors to the firm: the hiring of new workers, the purchasing of material for the commodity from an outside source, and the imposing of a penalty for late deliveries. The operation of the plant that keeps the total cost of these factors at a minimum is to be determined.

Specifically, the delivery schedule for the 3700 units is as follows.

End of Week	1	2	3	4
Number of Units	700	1200	1000	800

The plant has a stable working force of 35. New workers can be hired, but require a week of training, with one experienced worker capable of training five new workers each week. A worker can produce 25 units/week, but the trainees and trainers are not involved in production. All new workers hired for work on this contract cannot be retained by the firm, and so must be laid off by the end of the fourth week, if not sooner, at a cost of \$125/worker. All workers receive \$250/week.

One raw material is required in the production of this commodity, with each unit of the commodity needing 2 lb of the material. A subsidiary of the firm produces 1 ton of this material each week, but the material is perishable and can be used only during the week it is produced. However, the firm can also purchase an unlimited amount of the material on the market at a cost of \$3/lb above its own production costs.

In order to insure considerations for future contracts, the plant must deliver the 3700 units by the end of the fourth week. There is, however, a \$5/unit/week penalty for all units not delivered on schedule. On the other hand, we will assume that there is no penalty or storage charge for any units delivered early.

To construct a mathematical model of this situation, the contingencies involving labor, raw materials, and the delivery schedule must all be considered. Moreover, these elements can and do vary from week to week, and so decisions on the operation of the plant must be made each week. Thus we have in essence four time intervals: the beginning of the first, second, third, and fourth weeks.

Let us first consider labor. Each week the firm must decide how to employ its labor force. The activities to be established at the beginning of each week are as follows.

Activity	*Denoted by, for Week i, i = 1,2,3,4*
1. New workers to be hired	H_i
2. Workers to be fired	F_i
3. Workers to train and be trained	T_i
4. Workers to be idle	I_i
5. Workers to produce	P_i

Let M_i denote the number of pounds of raw material the firm should purchase during Week i from the outside source. Let D_i denote the number of units of the commodity produced and delivered during Week i. In order to measure the penalty costs for late deliveries, let U_i denote the accumulated number of units required but not delivered during Week i. Thus, if for some week the number of units delivered is less than the scheduled number required for that week plus any deficit accrued from previous weeks, the associated U_i would measure that difference. If the number delivered is greater than or equal to that sum, the U_i would be 0.

The constraints imposed by labor during the first week come from considering the employment of the total labor force and the training of the new workers. Note that F_1 must be 0 because, at the beginning of the first week, there are no new workers capable of being fired. We have

$$35 + H_1 = T_1 + I_1 + P_1 \qquad \text{and} \qquad H_1 + \frac{H_1}{5} = T_1$$

Constraints imposed on the number of units produced are as follows.

$$\begin{aligned} 25P_1 &= D_1 \\ 2000 + M_1 &\geq 2D_1 \\ D_1 + U_1 &\geq 700 \end{aligned}$$

The last inequality may need some clarification. If the optimal production schedule calls for D_1 to be less than 700, then U_1 will be the difference between 700 and D_1, and the quantity $D_1 + U_1$ will equal 700. On the other hand, if the optimal schedule calls for D_1 to be greater than 700, then U_1 will be zero and $D_1 + U_1$ greater than 700.

The production costs in dollars for the first week's operation are

$$250(35 + H_1) + 3M_1 + 5U_1$$

The constraints and costs for the next three weeks follow.

Second week:

$$35 + H_1 + H_2 - F_2 = T_2 + I_2 + P_2$$

$$H_2 + \frac{H_2}{5} = T_2$$

$$F_2 \leq H_1$$

$$25P_2 = D_2$$

$$2000 + M_2 \geq 2D_2$$

$$D_2 + U_2 \geq 1200 + (700 - D_1)$$

Costs: $250(35 + H_1 + H_2 - F_2) + 125F_2 + 3M_2 + 5U_2$

Third week:

$$35 + H_1 + H_2 - F_2 + H_3 - F_3 = T_3 + I_3 + P_3$$

$$H_3 + \frac{H_3}{5} = T_3$$

$$F_3 \leq H_1 + H_2 - F_2$$

$$25P_3 = D_3$$

$$2000 + M_3 \geq 2D_3$$

$$D_3 + U_3 \geq 1000 + (1200 + 700 - D_1 - D_2)$$

Costs: $250(35 + H_1 + H_2 - F_2 + H_3 - F_3) + 125F_3 + 3M_3 + 5U_3$

Fourth week:

$$35 + H_1 + H_2 - F_2 + H_3 - F_3 + H_4 - F_4 = T_4 + I_4 + P_4$$

$$H_4 + \frac{H_4}{5} = T_4$$

$$F_4 \leq H_1 + H_2 - F_2 + H_3 - F_3$$

$$25P_4 = D_4$$

$$2000 + M_4 \geq 2D_4$$

$$D_1 + D_2 + D_3 + D_4 = 3700$$ (all units must be delivered by the end of the fourth week)

Costs: $250(35 + H_1 + H_2 - F_2 + H_3 - F_3 + H_4 - F_4) + 125F_4 + 3M_4$

At the end of the fourth week any worker hired for this project and not yet laid off must be fired. Let F_5 denote this number. Then $125F_5$ is the cost of this activity, while F_5 is given by the equation

$$F_5 = H_1 + H_2 + H_3 + H_4 - F_2 - F_3 - F_4$$

The total cost, the function to be minimized, is given by the sum of these five costs. The nonnegative variables are restricted by all the above equalities and inequalities.

PROBLEM SET 2.5

1. Show that any optimal solution to the mathematical problem of the first example of this section must have $S_3 = 0$.

Problems 2 and 3 refer to the second example of this section.

2. The constraints relating the amount of raw material used and the number of units produced, $2000 + M_i \geq 2D_i, i = 1,2,3,4$, are all inequalities. The use of an equality here would prevent the consideration of what flexibility in the firm's operation?

3. Intuitively F_2 should be 0. Prove that any optimal solution to the resulting mathematical problem must have $F_2 = 0$. (*Hint.* Show that if $F_2 = k > 0$, then another solution can be found by letting $F_2 = 0$ and reducing H_1 by k. How does this change affect the value of the objective function?)

Formulate mathematical models for the following problems.

4. For the next 3 months a dealer in Commodity A can buy from the producer and sell to the retailer units of A at the following prices per unit.

A	*Buy*	*Sell*
Month 1	\$31	\$40
Month 2	33	44
Month 3	36	48

During any particular month the dealer can buy at most 450 units and sell at most 600 units and, moreover, can rent storage space from a local warehouse for up to 300 units at any one time at \$2/month/unit. Determine an optimal buying, selling, and storing program, assuming the dealer has no units of *A* on hand initially and wants none on hand at the termination of the 3-month period.

5. (a) Suppose the agent of Problem 4 can also buy and sell Commodity B, with the following prices per unit.

B	*Buy*	*Sell*
Month 1	$80	$ 95
Month 2	85	110
Month 3	95	125

The dealer can buy at most 200 units of *B* and sell at most 250 units during any one month and can also store *B* at the local warehouse, but space is limited. Assume that the warehouse has 300 yd^3 of space available, at $2/$yd^3$, and that a unit of A requires 1 yd^3 and a unit of B requires 2 yd^3. Again the dealer has no stock on hand and wants none at the end of the 3 months. Determine an optimal buying, selling, and storing program utilizing both commodities.

(b) In the above, any units stored represent an investment of capital. Reconsider the problem, assuming that a maximum of $10,000 can be borrowed each month for this purpose, with an accompanying 2%/month interest rate.

6. A subsidiary division of an automobile plant produces automobile engines. For the next four quarters, the demands of the plant are

Quarter	1	2	3	4
Number of Engines	500	625	650	550

There is an initial inventory of 100 engines. The division can produce 475 engines in a quarter using its normal facilities. By the use of overtime, up to an additional 100 engines can be produced in any quarter, at a cost of $26/engine above the normal costs. Any engines on hand at the end of a quarter can be stored at a cost of $5/engine each quarter. Any quarterly demand not met by the division cost the main plant in underutilization $33/engine for each quarter of the deficiency. By the end of the fourth quarter, all the demands must be met. Determine an operating schedule that minimizes costs.

7. Using Material C, a firm produces Commodities A and B. The requirements for the manufacture of a unit of each is as follows.

	Labor (man-hr)	*C (units)*
A	3	5
B	2	8

The firm has available each month 400 man-hr of labor at a cost of \$4/hr, and up to an additional 100 man-hr of overtime at \$6/hr. The firm has 3500 units of C in stock, but can obtain no more. For the next 3 months up to 100 units each of A and B can be sold, at the following prices per unit.

	A	*B*
Month 1	\$22	\$16
Month 2	23	20
Month 3	24	25

The firms' warehouse, with a 3000 ft^3 capacity, can store units of A and B produced during 1 month for sale during a later month, but it must also store all the unused units of C. One unit of A requires 8 ft^3 of space, one unit of B, 9 ft^3 of space, and one unit of C, 0.8 ft^3 of space. At the end of the 3-month period, any units of C left in stock must be disposed of at a cost of \$1.50/unit. Determine an operating schedule that maximizes profit.

SECTION 2.6 SUMMARY

Now that we have seen some examples of problems leading to linear programming models, it should be emphasized that our list of categories and examples is not by any means comprehensive nor are our categories mutually exclusive. Many types of problems occurring in the world and amenable to linear programming techniques are not confined to the categories we described. Moreover, our examples were somewhat straightforward. Real-life problems, usually with a multitude of interrelated components to be considered, tend to be much more complicated and may lend themselves to several different approaches.

However, all the examples we have considered led to the same mathematical problem, that of finding the maximum or minimum of a linear function on a set determined by a family of linear equations or inequalities. This is the basic problem of linear programming, the optimization of a linear function subject to a system of linear constraints.

In order to be able to establish a linear programming model for some real-life optimization problem, the system or operation under study must be amenable to some basic assumptions. First, the system must be decomposable into a number of elementary operations called activities. An activity usually can be considered to be a transformation process that converts input such as labor or raw materials into the product of the operation, such as the manufactured goods. For example, in the feed problem of Section 2.2, an activity is the process of converting feed into three nutritional elements. In the example of the paper manufacturer of Section 2.4, an activity is the transportation of newsprint from a mill to a printing plant. It is these activities that are combined in varying amounts to attain the stated objective. The amount or rate at which an activity operates or functions is called its activity level.

Second, the objective of the entire operation, when measured in terms of the activity levels, must be a linear function. This means simply that if x_j measures the level of activity j, there are constants c_j such that the product $c_j x_j$ measures the attainment of the objective from the operation of activity j, and such that the total output of the operation, if there are n activities, is given by the sum $c_1 x_1 + c_2 x_2 + \cdots + c_n x_n$. Thus the objective function is a linear function of the x_j's.

Third, the restrictions on the various input and output items and the requirements on the relationships between these items must be linear in form, that is, given by either linear equations or inequalities in the x_j's. Thus the component processes or activities of the problem must be linear, and so, for example, doubling the quantities of all the items input into an activity must have the effect of doubling the output of the activity. In the first example in Section 2.3, dealing with the production of small boats, the activity of producing a boat depends linearly on the input of raw material, machine time, and labor.

Certainly real-life situations can fail to satisfy the requirements listed above, but can still be open to a linear programming model that provides accurate and useful information. One simplifying principle is to neglect incidental details. For example, the paper manufacturer of Section 2.4 can probably realistically ignore the quantity of newsprint lost in shipment from mill to plant. Also, the linearity restrictions may only be an approximation of the situation at hand, but at least would still lead to a good first estimate of the desired solution. For example, in transportation problems, the cost of shipping a unit of goods may decrease if the volume of goods shipped is increased, due to a more economical use of available equipment or due to the accessibility of other means of transportation suitable for large shipments. The problems that we have considered belong to what is called the "deterministic class"; that is, they have involved no uncertainty. For example, the output of newsprint at the two paper mills and requirements of newsprint at the three printing plants are stated precisely. However, few real-life situations can be so predicted with certainty. One approach here is to work with average values for the quantities under study. Another approach is to develop a probabilistic or "stochastic" model. An elementary example of such a model is given in Section 7.4.

The actual construction of a mathematical model of a real-life problem involves several steps, steps that you may have already recognized in doing the examples and exercises of the preceding sections. First, the entire operation under study must be decomposed into its component activities. Then the items and units used to measure the activity levels must be determined. It is the rate of these activity levels that is subject to our control and that is represented by the variables in the problem. Finally, the objective function, the function to be optimized, and the linear constraints must be identified. The constraints on the system usually come about by consideration of the input and output items in the system and the restrictions and relationships between them. And although it may seem that the execution of these steps may become somewhat straightforward, that is not the case. Experience with programming techniques and understanding of the real-life problem under study are necessary in order to be able to distinguish the significant elements of the problem from the inconsequential and to interpret and employ any solution found for the problem properly.

The goal of the next chapter will be to develop a mechanical technique for solving the general mathematic problem resulting from the execution of the above steps. This general problem is called the basic problem of linear programming.

PROBLEM SET 2.6

Formulate mathematical models for the following problems.

1. A farmer must determine a plan to feed his stock through the coming winter. He has two types of stock, each with distinct nutritional requirements. To feed the stock the farmer has available 1000 lb of grain harvested over the summer. However, this supply of grain is not adequate to meet the nutritional demands of the entire stock over the winter, and so the farmer must supplement this supply with feeds purchased from the local co-op. Determine a feeding plan that utilizes all the available grain, satisfies the nutritional demands, and minimizes the amount spent on the supplementary feeds. The data follow.

Nutritional demands (minimal number of units required/winter):

	Element A	*Element B*	*Element C*
Stock 1	150	360	650
Stock 2	90	700	450

Nutritional contents (units per pound) and costs (cents per pound) of the grain and the two available feeds:

	Element A	*Element B*	*Element C*	*Cost*
Grain	0.2	0.9	0.8	0
Feed 1	1	5	10	15
Feed 2	3	7	13	23

2. A pet food manufacturer produces weekly 600 lb of dog food and 250 lb of cat food. The foods must contain minimal percentages of four nutritional elements, A, B, C, and D, as follows.

	A	*B*	*C*	*D*
Dog Food	10	5	8	8
Cat Food	8	16	5	12

The manufacturer has available six different meat by-products to combine to meet these demands, but only limited quantities of some are available. The available weekly supply, costs, and contents of the by-products are as follows.

			Contents (% nutritional element)			
By-product	*Supply (lb/wk)*	*Cost (cents/lb)*	*A*	*B*	*C*	*D*
1	Unlimited	33	10	0	5	8
2	300	29	8	20	6	10
3	500	30	15	14	8	10
4	400	28	6	12	10	15
5	Unlimited	37	0	18	13	20
6	200	23	12	10	4	6

Determine how much of each by-product should be combined in the production of each food so that total costs are minimized.

3. A firm combines Raw Materials A, B, and C in the production of two products. The requirements (in pounds) for the manufacture of a unit of each product is as follows.

	Raw material		
	A	*B*	*C*
Product 1	4	12	8
Product 2	7	9	10

The firm has available 1 ton of A, 2 tons of B, and $1\frac{1}{2}$ tons of C. All units of the products made can be sold. The firm realizes a profit of \$1.20/unit on the first 200 units of Product 1 sold, and \$1/unit on the remainder sold; and a profit of \$1.40/unit on the first 150 units of Product 2 sold, and \$1.05/unit on the remainder. Determine a production schedule that maximizes profit.

4. The firm of Problem 3 has changed owners. Because of a subsequent expanded market, the firm can now sell all units produced at the fixed prices of \$5.55/unit for Product 1 and \$6.30/unit for Product 2. However, now the firm must purchase the necessary raw materials from outside sources, and the costs, which must be considered in determining profits, vary. The firm can purchase up to 1 ton of Raw Material A at 20 cents/lb, and any amount over 1 ton at 35 cents/lb; up to 2 tons of B at 10 cents/lb, and any amount over at 20 cents/lb; and up to $1\frac{1}{2}$ tons of C at 15 cents/lb and any amount over at 25 cents/lb. Determine a production schedule that maximizes net profit, assuming that there is an additional overhead cost to the firm of \$1/unit for each unit of Product 1 and of Product 2 produced.

5. A poultry producer has available 112 rods^2 of land in which he can raise during the next 12-week period chickens (in units of 28), ducks (in units of 18), or turkeys (in units of 8). The space and labor requirements, and the profit—excluding labor costs—from the sale after the 12-week breeding period are as follows.

	Space (rods^2/unit)	*Labor (man-hr/week/unit)*	*Profit (excluding labor costs) (dollars/unit)*
Chickens	1.2	3	130
Ducks	1	2	86
Turkeys	0.8	1	44

The producer has available each week 200 man-hr of labor at \$3/hr, and up to 45 man-hr of overtime at \$3.50/hr. What stock should the producer raise over the 12-week period in order to maximize his net income (profits less labor costs)?

6. A machine shop assembles transuniversals for sale to the local automobile plant. Because of high demands all units assembled can be sold for \$188/unit. Three major components, C_1, C_2, and C_3 are required in the assembly of each transuniversal, and the shop can either purchase these components from outside sources or manufacture the components themselves. In the internal manufacture of the components, and also in the final assembly of the transuniversals, labor and machine time on two machines M_1 and M_2 are required. The requirements follow.

	Labor (man-hr/unit)	*Machine time (min/unit)*	
		M_1	M_2
Transuniversal assembly	7	35	25
Manufacture of C_1	0.3	10	8
Manufacture of C_2	0.5	15	20
Manufacture of C_3	1	13	12

The shop has available each week 800 man-hr of labor at $4/hr and 200 man-hr of overtime at $6/hr, and 80 hr of machine time for M_1 and 100 hr for M_2. The costs per unit of the components, if purchased externally or produced internally (internal costs exclude labor costs), are as follows.

	C_1	C_2	C_3
Purchase Price	$37	$53	$45
Production Cost	$29	$40	$30

Determine a production schedule that maximizes net income.

7. Problem 14 of Section 7.2.

8. A firm must meet the demands of seven markets for a commodity. The firm has three plants at which limited amounts of the commodity can be produced, and the firm can also buy unlimited amounts of the finished commodity from an outside source. The production capabilities of the plants, the demands of the markets, and the total costs (production or purchase costs plus transportation costs) of supplying the markets from the varying sources are given in the following tables. Determine a supply schedule that minimizes overall costs.

Supply

Plant	1	2	3
Number of Units	700	600	400

Demand

Market	1	2	3	4	5	6	7
Number of Units	150	300	425	325	200	250	250

Costs (dollars/unit)	***Markets*** 1	2	3	4	5	6	7
Plant 1	5	7	8	6.5	9	12	3
Plant 2	13	16.5	10	12	14	18	9
Plant 3	10	12	11.5	9	12.5	14	7
Outside Source	21	25	35	24	27	38	20

9. A firm supplies six outlets with two commodities, A and B, produced at three plants. The transportation costs vary from plant to market and are also dependent on the commodity being shipped. Moreover, there are upper bound capacities on the combined total number

of units that can be shipped from each plant to each market. Determine if it is possible to meet the demands with the supplies given the restrictions on the shipping capacities and, if so, determine a minimal cost shipping schedule. The data are:

Supplies	*Plant*	1	2	3
	Units of A	1000	2500	1500
	Units of B	1400	1500	1100

Demands	*Outlet*	1	2	3	4	5	6
	Units of A	1200	600	1100	1000	500	600
	Units of B	800	800	1000	500	300	600

Costs (cents/unit)

Plant		*Outlet* 1	2	3	4	5	6
1	*A*	26	35	27	32	23	40
	B	17	19	13	20	12	25
2	*A*	48	56	70	45	55	60
	B	20	32	45	25	30	32
3	*A*	30	39	38	40	35	32
	B	21	32	30	33	29	25

Shipping Capacities

Plant	*Outlet* 1	2	3	4	5	6
1	600	500	1000	500	300	550
2	400	500	2000	2000	350	450
3	400	500	0	2000	250	400

3
The Simplex Method

SECTION 3.1 THE GENERAL PROBLEM

In the last chapter all the examples led to one basic mathematical problem: the optimization of a linear function subject to a system of linear constraints. In this chapter we will develop a technique for solving this basic problem.

One minor complication to studying the problem is that the optimization problem can take various forms. For example, we have seen both maximization and minimization problems and constraint sets that have consisted of equalities and inequalities in both directions. However, this difficulty is easily resolved as all linear programming problems can be transformed into equivalent problems that are in what we call *standard form.*

Definition The *standard form* of the linear programming problem is to determine a solution of a set of equations

$$\begin{aligned} a_{11}x_1 + a_{12}x_2 + \cdots + a_{1n}x_n &= b_1 \\ a_{21}x_1 + a_{22}x_2 + \cdots + a_{2n}x_n &= b_2 \\ &\vdots \\ a_{m1}x_1 + a_{m2}x_2 + \cdots + a_{mn}x_n &= b_m, \end{aligned} \tag{1}$$

with

$$x_j \geq 0, j = 1, \ldots, n,$$

that minimizes the linear form

$$z = c_1x_1 + c_2x_2 + \cdots + c_nx_n$$

It is this standard form of the linear programming problem, a minimization problem involving only equalities, that we will solve. Thus our first task is to show that any linear programming problem can be formulated as a problem in standard form, where the number of equalities, m, and the number of variables, n, are determined by the problem.

Consider first a linear programming problem with a system of constraints that contains inequalities. For example, suppose a particular diet problem reduces to the mathematical problem of minimizing $3x_1 + 2x_2 + 4x_3$ subject to the constraints

$$\begin{aligned} 30x_1 + 100x_2 + 85x_3 &\leq 2500 \\ 6x_1 + 2x_2 + 3x_3 &\geq 90 \\ x_1, x_2, x_3 &\geq 0 \end{aligned}$$

Such a problem could result from seeking a diet of minimal costs that places an upper bound on calorie intake and a lower bound on protein intake. We will show that this problem is equivalent to the following problem derived from the original problem by the addition of two new nonnegative variables, x_4 and x_5.

$$\text{Minimize} \quad 3x_1 + 2x_2 + 4x_3$$

subject to

$$\begin{aligned} 30x_1 + 100x_2 + 85x_3 + x_4 &= 2500 \\ 6x_1 + 2x_2 + 3x_3 - x_5 &= 90 \\ x_1, x_2, x_3, x_4, x_5 &\geq 0 \end{aligned}$$

Notice that if $(x_1^*, x_2^*, x_3^*, x_4^*, x_5^*)$ is a solution to the second constraint set then, since x_4^* and x_5^* are restricted to nonnegative values, $30x_1^* + 100x_2^* + 85x_3^* = 2500 - x_4^* \leq 2500$ and $6x_1^* + 2x_2^* + 3x_3^* = 90 + x_5^* \geq 90$. Therefore (x_1^*, x_2^*, x_3^*) is a solution to the first constraint set. Similarly, if (x_1^*, x_2^*, x_3^*) is a solution to the first constraint set, there exist x_4^* and x_5^* [let $x_4^* = 2500 - (30x_1^* + 100x_2^* + 85x_3^*)$ and $x_5^* = 6x_1^* + 2x_2^* + 3x_3^* - 90$] that are nonnegative and such that $(x_1^*, x_2^*, x_3^*, x_4^*, x_5^*)$ is a solution to the second constraint set. Thus solutions of the two constraint sets correspond, with corresponding solutions having the same first three coordinates. However, the form to be minimized, $3x_1 + 2x_2 + 4x_3$, depends only on the first three coordinates. Hence the minimal value of the linear function for both problems will be the same, and points where this minimum is achieved for one problem will correspond to points with this same property for the other problem.

Clearly this technique generalizes. Given any problem with a system of constraints containing inequalities, by adding additional nonnegative variables, an equivalent problem can be formulated with a constraint system consisting only of equalities. The number of variables added would equal the number of inequalities in the system of constraints. These variables added are called *slack variables.* In fact, they usually can be interpreted as measuring the slack or surplus of the items or requirements of the problem. For example, in the preceding diet problem, suppose the first restriction does come from consideration of the calorie intake and the second from the protein intake. Then, for a fixed diet, the slack variable x_4 measures the number of calories below the maximum calorie requirement, and x_5 measures the number of units of protein above the minimum protein requirement for that diet.

Second, suppose a linear programming problem seeks to maximize the linear function $c_1x_1 + c_2x_2 + \cdots + c_nx_n$. But the problem of maximizing this function is equivalent to the problem of minimizing its negative: $-c_1x_1 - c_2x_2 - \cdots - c_nx_n$. Thus a maximization problem can be easily formulated as a minimization problem by multiplying the function to be optimized by -1.

The last restriction on the standard form of the linear programming problem is that all the variables be nonnegative. For most problems this restriction comes naturally from the physical interpretation of the variables. In all the examples we have considered the variables could assume only nonnegative values. However, for some complicated production system involving various processes and options, it could be that some commodity that is input for some process is output for another, and it is not clear whether this commodity will be input or output in the optimal operation of the system. Thus we may wish to formulate the problem with a variable not restricted in sign. Suppose that x_1 is a variable unrestricted in sign for a linear optimization problem. However, any number can be written as the difference of two (not unique) nonnegative numbers. Hence we can introduce into the problem two nonnegative variables, say x_1' and x_1'', and replace x_1 everywhere in the problem with the difference $x_1' - x_1''$. This will give an equivalent problem with the unrestricted variable replaced by two nonnegative variables.

As a result of these methods, for any linear programming problem, an equivalent problem can be constructed that is in standard form.

Example

The problem of maximizing $3x_1 - 2x_2 - x_3 + x_4$

subject to

$$\begin{aligned} 4x_1 - x_2 \qquad\quad + x_4 &\le 6 \\ -7x_1 + 8x_2 + x_3 \qquad &\ge 7 \\ x_1 + x_2 \qquad\quad + 4x_4 &= 12 \\ x_1, x_2, x_3 \ge 0, \quad x_4 &\text{ arbitrary} \end{aligned}$$

is equivalent to:

Minimize $-3x_1 + 2x_2 + x_3 - (x_4' - x_4'')$

subject to

$$
\begin{aligned}
4x_1 - x_2 \qquad + x_4' - x_4'' + x_5 \qquad &= 6 \\
-7x_1 + 8x_2 + x_3 \qquad\qquad - x_6 &= 7 \\
x_1 + x_2 \qquad + 4x_4' - 4x_4'' \qquad &= 12 \\
x_1, x_2, x_3, x_4', x_4'', x_5, x_6 &\geq 0
\end{aligned}
$$

In a linear programming problem, the linear function to be optimized is called the *objective function*. Any point $(x_1, x_2, \ldots, x_n)$ with nonnegative coordinates that satisfies the system of constraints is called a *feasible solution* to the problem. For a particular problem, a feasible solution can be interpreted as a way of operating the system under study so that all of the requirements are fulfilled, that is, as a feasible way of operation.

Thus our basic problem is to determine, from among the set of all feasible solutions, a point that minimizes the objective function. Moreover, in order to be able to handle involved real-life problems, we need a technique for determining the solution that is basically computational and adaptable to computer use. Existence theorems derived from, say, the theory of continuous functions on compact sets or the theory of linear functions on convex sets, although mathematically quite attractive, do not provide an efficient means for actually finding a desired solution.

The method that will be developed in this chapter for solving the basic linear programming problem is called the simplex method. It is due to George Dantzig [4], and this method and its various modifications are the primary means used today to solve linear optimization problems. One additional feature of this method that is useful for practical application and also very attractive mathematically is that the method can handle the exceptional cases. For example, the method can determine if a problem has, in fact, any feasible solutions and, if so, if the objective function actually assumes a minimum value.

The basic step in the simplex method is derived from the familiar pivot operation used to solve linear equations. In the next section we pause briefly from our consideration of the standard linear programming problem to consider linear equations.

PROBLEM SET 3.1

1. (a) In the example of this section, $x_1 = 4$, $x_2 = 12$, $x_3 = 0$, $x_4' = 21$, $x_4'' = 22$, $x_5 = 3$, $x_6 = 61$ is a solution to the second constraint set. Find the corresponding solution to the first constraint set.

(b) Conversely, $x_1 = 1$, $x_2 = 3$, $x_3 = 5$, $x_4 = 2$ is a solution to the first constraint set. Find a corresponding solution to the second. In this case is your answer unique?

2. Explain why the following constraint sets are not equivalent.

Set A:

$$\begin{aligned} x_1 + x_2 &\le 6 \\ x_1 + 2x_2 &\le 10 \\ x_1, x_2 &\ge 0 \end{aligned}$$

Set B:

$$\begin{aligned} x_1 + x_2 + x_3 &= 6 \\ x_1 + 2x_2 + x_3 &= 10 \\ x_1, x_2, x_3 &\ge 0 \end{aligned}$$

Hint. $x_1 = 3$ and $x_2 = 3$ satisfy the inequalities of Set A.

Can you find an x_3 such that $(3,3,x_3)$ satisfies the equalities of Set B?

This shows that when introducing slack variables, the same variable cannot be used for different inequalities.

3. Put the following problems into standard form.

(a) Maximize $3x_1 - 2x_2$

subject to

$$\begin{aligned} 5x_1 + 2x_2 - 3x_3 + x_4 &\le 7 \\ 3x_2 - 4x_3 &\le 6 \\ x_1 + x_3 - x_4 &\ge 12 \\ x_1, x_2, x_3, x_4 &\ge 0 \end{aligned}$$

(b) Minimize $x_2 + x_3 + x_4$

subject to

$$\begin{aligned} x_1 + x_2 &\ge 6 \\ x_2 + x_3 - x_4 &\le 1 \\ 5x_1 - 6x_2 + 7x_3 - 8x_4 &\ge 2 \end{aligned}$$

$x_1 \ge 0$, $x_2 \le 0$, x_3, x_4 unrestricted

(c) Maximize $x_1 + 6x_2 + 12x_3$

subject to

$-x_1 - x_2 + x_4 \ge$ Maximum of $7x_1 + 2x_2$ and $5x_2 + x_3 + x_4$

$x_1, x_2, x_3, x_4 \ge 0$

(d) Maximize $-x_1 - x_2 + 2x_3 + x_5$

subject to

$$
\begin{aligned}
x_1 + 7x_2 + 16x_3 &\le 4x_4 + x_5 \\
x_3 + 12x_4 &\ge x_1 + 6x_2 \\
9x_5 &\le x_2 + 3x_4 \\
x_1, x_2, x_3, x_4, x_5 &\ge 0
\end{aligned}
$$

4. Many times the amount of slack or surplus of a commodity enters into the initial formulation of the problem; it is a factor in the function to be optimized. For example, in a production problem, there could be a cost associated with the storage of the surplus production of a commodity. For another example, formulate the mathematical model for the following.

Two warehouses supply two retail outlets with 100-lb bags of lime. Warehouse A has 1000 bags, and Warehouse B has 2000 bags. Both outlets need 1200 bags. The transportation costs in cents per bag are given in the following table.

		Outlet	
		1	2
Warehouse	*A*	5	4
	B	12	9

However, there is a storage charge of 2 cents/bag for all bags left at Warehouse A and 8 cents/bag for those left at Warehouse B. Determine a shipping schedule that minimizes the total cost.

5. In the text it was suggested that when putting a linear programming problem with unrestricted variables into canonical form, each unrestricted variable is to be replaced by a pair of nonnegative variables. Actually, this method is inefficient if the problem has more than one unrestricted variable; we need to introduce only one additional variable to handle all the unrestricted variables. For example, if a problem has unrestricted variables x_1 and x_2, show that replacing x_1 with $x_1' - x_0$ and x_2 with $x_2' - x_0$ where x_1', x_2', and x_0 are new nonnegative variables leads to an equivalent problem.

6. Show that the following problems are equivalent.

Problem A: Minimize $x_1 + 2x_2 - 3x_3 + 4x_4$

subject to

$$
\begin{aligned}
3x_1 - 2x_2 + 5x_3 - 6x_4 &= 20 \\
x_1 + 7x_2 - 6x_3 + 9x_4 &= 30 \\
x_1 \ge 0,\ x_2, x_3, x_4 &\text{ unrestricted}
\end{aligned}
$$

Problem B: Minimize $x_1 + 2x_2' - 3x_3' + 4x_4' - 3x_0$

subject to

$$3x_1 - 2x_2' + 5x_3' - 6x_4' + 3x_0 = 20$$
$$x_1 + 7x_2' - 6x_3' + 9x_4' - 10x_0 = 30$$
$$x_1, x_2', x_3', x_4', x_0 \geq 0$$

7. Using the technique suggested in Problem 5, determine a linear programming problem in standard form and with only five variables and equivalent to the linear programming problem of Problem 3b.

SECTION 3.2 LINEAR EQUATIONS AND BASIC FEASIBLE SOLUTIONS

The pivot operation used in solving linear equations consists of replacing a system of equations with an equivalent system in which a selected variable is eliminated from all but one of the equations. The operation revolves around what is called the *pivot term*. The pivot term can be the term in any one of the equations that contains the selected variable with a non-zero coefficient. In the first step of the pivot operation the equation containing the pivot term is divided by the coefficient in that term, thus producing an equation in which the selected variable has coefficient one. Multiples of this equation are added to the remaining equations in such a way that the selected variable is eliminated from these remaining equations.

It is easy to show that the solution set of the system of equations resulting from the pivot operation is identical to the solution set of the original system, that is, that the systems are equivalent (Problem 9). In general, repeated use of this pivot operation can lead to a system of equations whose solution set is obvious.

Example 1

Solve

$$\begin{aligned} x_1 + 4x_2 + 2x_3 &= 6 \\ 3x_1 + 14x_2 + 8x_3 &= 16 \\ 4x_1 + 21x_2 + 10x_3 &= 28 \end{aligned}$$

We arbitrarily select x_1 as the first variable to be eliminated from two of the equations, and the $1x_1$ term of the first equation as the pivot term. Notice that we could have also selected the $3x_1$ term of the second equation or the $4x_1$ term of the third equation for the pivot term. However, the arithmetic associated with the selection of the $1x_1$ term is less involved because of the unit coefficient. The pivot operation at this term consists of dividing the first equation by one, subtracting three times the first equation from the second, and subtracting four times the first equation from the third. The resulting equivalent system is

$$\begin{aligned} x_1 + 4x_2 + 2x_3 &= 6 \\ 2x_2 + 2x_3 &= -2 \\ 5x_2 + 2x_3 &= 4 \end{aligned}$$

Continuing, we arbitrarily select x_2 as the next variable to be eliminated from two of the equations. Since we are striving to simplify the system, the next pivot term should not be the $4x_2$ term of the first equation; pivoting here would reinstate into the last two equations the x_1 variable. Pivoting at the x_2 term of either of the other two equations, however, will isolate the x_2 variable to that pivoting equation without destroying the isolated status of the x_1 variable. Using the $2x_2$ term of the second equation as the pivot term gives

$$\begin{aligned} x_1 \qquad - 2x_3 &= 10 \\ x_2 + x_3 &= -1 \\ - 3x_3 &= 9 \end{aligned}$$

At this stage one might solve the third equation for x_3 and use this value and the first two equations to compute the associated values for x_1 and x_2. Actually, that operation is essentially equivalent to the pivot operation with the $-3x_3$ term of the third equation as pivot term. Pivoting at this term gives

$$\begin{aligned} x_1 \qquad\qquad &= 4 \\ x_2 \qquad &= 2 \\ x_3 &= -3 \end{aligned}$$

Now this system of equations is equivalent to the original system. However, the solution set for this system obviously consists only of the point $(4, 2, -3)$, so we have proven that this point is the unique solution to the original problem.

As we have seen with this example, repeated use of the pivot operation led to a system of three equations with three unknowns in a special form, where each variable appeared in one and only one equation, and in that equation had coefficient one. This form, called the *canonical form*, is crucial to the simplex method. We now define it, along with the associated term of *basic variable*.

Definition A system of m equations and n unknowns, with $m \leq n$, is in *canonical form* with a distinguished set of m *basic variables* if each basic variable has coefficient one in one equation and zero in the others, and each equation has exactly one basic variable with coefficient one.

Thus, in the example, the three pivoting operations led to a system of equations in canonical form with basic variables x_1, x_2, and x_3. However, not every system of equations is equivalent to a system in canonical form. See, for example, Problem 3. Moreover, a system of equations with $m < n$ may be equivalent to different systems in canonical form, with the difference resulting from the choice of basic variables.

Example 2

Consider the system

$$\begin{aligned} 2x_1 + 3x_2 - x_3 &= 6 \\ 2x_1 - x_2 + 7x_3 &= 5 \end{aligned} \tag{1}$$

Pivoting at the $2x_1$ term of the first equation gives

$$\begin{aligned} x_1 + (\tfrac{3}{2})x_2 - (\tfrac{1}{2})x_3 &= 3 \\ -4x_2 + 8x_3 &= -1 \end{aligned} \tag{2}$$

Now pivoting at the $-4x_2$ term of the second equation gives

$$\begin{aligned} x_1 \qquad + (\tfrac{5}{2})x_3 &= \tfrac{21}{8} \\ x_2 - 2x_3 &= \tfrac{1}{4} \end{aligned} \tag{3}$$

Thus the original system is equivalent to this new system, which is in canonical form with basic variables x_1 and x_2. The solution set can be obviously given by

$$\{(x_1,x_2,x_3) \mid x_3 \in R, x_1 = \tfrac{21}{8} - \tfrac{5}{2}x_3, x_2 = \tfrac{1}{4} + 2x_3\}$$

One particular solution can be determined from (3) by setting x_3, the nonbasic variable, equal to 0. This gives the solution $(\frac{21}{8},\frac{1}{4},0)$, with the basic variables equal to the constant terms on the right-hand side of the equations.

If, however, we use the system of equations in (2) and pivot at the $8x_3$ term of the second equation, the result is

$$\begin{aligned} x_1 + \tfrac{5}{4}x_2 \qquad &= \tfrac{47}{16} \\ -\tfrac{1}{2}x_2 + x_3 &= -\tfrac{1}{8} \end{aligned}$$

This system is in canonical form with basic variables x_1 and x_3. Letting the nonbasic variable equal zero here gives the solution $(\frac{47}{16},0,-\frac{1}{8})$.

The solutions described above found by setting the nonbasic variable equal to zero when the system was in canonical form are called *basic solutions*. The term is defined precisely as follows.

Definition Assume a system of equations is in canonical form with a specified set of basic variables. The associated *basic solution* is that solution to the system with the basic variables given by the constant terms in the equations and the nonbasic variables equal to zero.

Consider now the standard form of the linear programming problem.

$$\text{Minimize} \quad c_1x_1 + c_2x_2 + \cdots + c_nx_n = z$$

subject to

$$\begin{aligned}
a_{11}x_1 + a_{12}x_2 + \cdots + a_{1n}x_n &= b_1 \\
a_{21}x_1 + a_{22}x_2 + \cdots + a_{2n}x_n &= b_2 \\
&\vdots \\
a_{m1}x_1 + a_{m2}x_2 + \cdots + a_{mn}x_n &= b_m \\
x_1, x_2, \ldots, x_n &\geq 0
\end{aligned}$$

The system of constraints consists of m equations with n unknowns. Armed with the technology of the pivot operation, one possible way of simplifying the problem would be to determine a system of equations in canonical form equivalent to the given system of constraints. However, we are interested only in the solutions to the system of constraints that have nonnegative coordinates, that is, that are feasible. In attempting to put the constraint system into canonical form, an arbitrary selection of basic variables could easily lead to a system with some negative constant terms and thus an associated basic solution that is not even feasible. Those basic solutions to the constraint set, however, that are also feasible will prove to be the critical points when using the simplex method to determine the minimum of the objective function. We call these points the *basic feasible solutions*; the basic variables are determined by the basic variables of the associated canonical form.

Example 3

Consider the problem of

$$\text{Minimizing} \quad x_1 - x_2 + 2x_3 - 5x_4 = z$$

subject to

$$\begin{aligned}
x_1 + x_2 + 2x_3 + x_4 &= 6 \\
3x_2 + x_3 + 8x_4 &= 3 \qquad (4) \\
x_1, x_2, x_3, x_4 &\geq 0
\end{aligned}$$

The system of constraints consists of two equations in four unknowns. Pivoting at the $3x_2$ term of the second equation gives the equivalent system

$$\begin{aligned}
x_1 \quad + \tfrac{5}{3}x_3 - \tfrac{5}{3}x_4 &= 5 \\
x_2 + \tfrac{1}{3}x_3 + \tfrac{8}{3}x_4 &= 1
\end{aligned} \qquad (5)$$

The associated basic solution is feasible, and so the point (5,1,0,0) is a basic feasible solution with basic variables x_1 and x_2.

From the system in (5), pivoting at the $\frac{1}{3}x_3$ term of the second equation results in the system

$$\begin{aligned} x_1 - 5x_2 \qquad - 15x_4 &= 0 \\ 3x_2 + x_3 + 8x_4 &= 3 \end{aligned}$$

Thus (0,0,3,0) is a basic feasible solution with basic variables x_1 and x_3. Note that a basic variable, here x_1, can equal zero. If this happens, the basic feasible solution is said to be *degenerate*. As we will see later when considering the simplex method, theoretical complications arise from the possibility of degeneracy. Returning to the system in (5), if we pivot at the $-\frac{5}{3}x_4$ term of the first equation, we get

$$\begin{aligned} -\tfrac{3}{5}x_1 \qquad - x_3 + x_4 &= -3 \\ \tfrac{8}{5}x_1 + x_2 + 3x_3 \qquad &= 9 \end{aligned}$$

Here the associated basic solution has $x_4 = -3$, and so it is not a basic feasible solution.

Verify that the other basic feasible solutions are $(\frac{45}{8},0,0,\frac{3}{8})$, with basic variables x_1 and x_4, and (0,0,3,0) with basic variables either x_2 and x_3, or x_3 and x_4. The distinction on account of basic variables between the three basic feasible solutions with coordinates (0,0,3,0) is a minor one, but will prove reasonable later on.

Using the column vectors associated with the coefficient matrix of any system of equations, the basic feasible solutions as, indeed, all solutions to the system, have a simple geometric interpretation. We demonstrate using the above example.

Example 3 (continued)

The system of constraints for the problem can be expressed in vector form as follows.

$$x_1\begin{bmatrix}1\\0\end{bmatrix} + x_2\begin{bmatrix}1\\3\end{bmatrix} + x_3\begin{bmatrix}2\\1\end{bmatrix} + x_4\begin{bmatrix}1\\8\end{bmatrix} = \begin{bmatrix}6\\3\end{bmatrix}$$

Thus the system of two equations and four variables is equivalent to the problem of expressing the vector $\begin{bmatrix}6\\3\end{bmatrix}$ as a linear combination of the vectors $\begin{bmatrix}1\\0\end{bmatrix}$, $\begin{bmatrix}1\\3\end{bmatrix}$, $\begin{bmatrix}2\\1\end{bmatrix}$, and $\begin{bmatrix}1\\8\end{bmatrix}$. Moreover, for our purposes, we are restricted to solutions with nonnegative coordinates.

Suppose now we wish to determine geometrically if x_1 and x_2 can serve as basic variables for a basic feasible solution. If so, the nonbasic variables x_3 and x_4 will equal zero, and the resulting vector equation reduces to

$$x_1\begin{bmatrix}1\\0\end{bmatrix} + x_2\begin{bmatrix}1\\3\end{bmatrix} = \begin{bmatrix}6\\3\end{bmatrix}, \qquad x_1 \text{ and } x_2 \geq 0$$

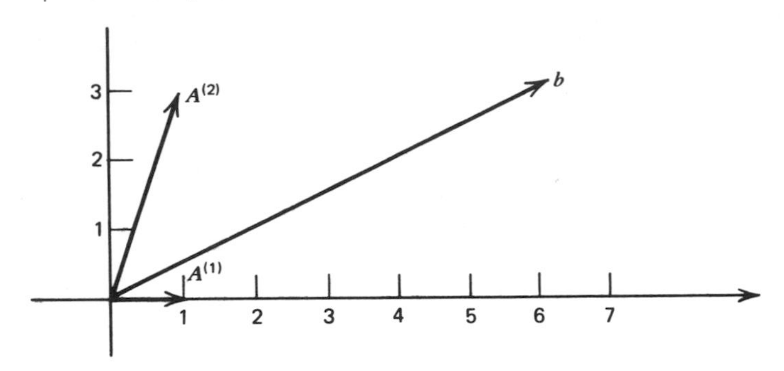

Figure 3.1

Using the notation

$$A^{(1)} = \begin{bmatrix} 1 \\ 0 \end{bmatrix},\ A^{(2)} = \begin{bmatrix} 1 \\ 3 \end{bmatrix},\ \text{and } b = \begin{bmatrix} 6 \\ 3 \end{bmatrix}$$

these vectors in R^2 are sketched in Figure 3.1. Now the set of points of the form $x_1 A^{(1)}$ for $x_1 \geq 0$ is the line ray emanating from the origin in R^2 in the direction of $A^{(1)}$, and similarly for the points $x_2 A^{(2)}$ with $x_2 \geq 0$. The set of points of the form $x_1 A^{(1)} + x_2 A^{(2)}$, x_1 and $x_2 \geq 0$, can be determined using the usual rule for addition of vectors. This region [the *convex cone* of $A^{(1)}$ and $A^{(2)}$] is illustrated in Figure 3.2. Since b lies in this region, a solution to the system of equations with x_1 and x_2 nonnegative and x_3 and x_4 equal to 0 must exist. This solution is the point (5,1,0,0) found previously.

To extend these ideas, let $A^{(3)} = \begin{bmatrix} 2 \\ 1 \end{bmatrix}$ and $A^{(4)} = \begin{bmatrix} 1 \\ 8 \end{bmatrix}$. From the graph in Figure 3.3 we see that b cannot be expressed as a sum of the form $x_2 A^{(2)} + x_4 A^{(4)}$ with x_2 and $x_4 \geq 0$. Thus x_2 and x_4 cannot serve as basic variables for a basic feasible solution. Furthermore, it can be seen that any other pair of variables can

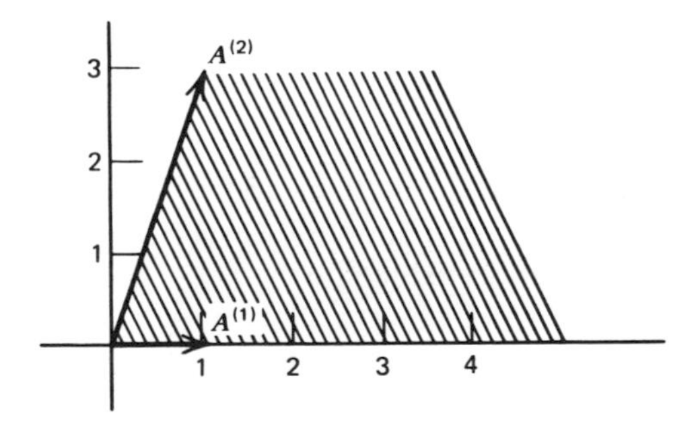

Figure 3.2

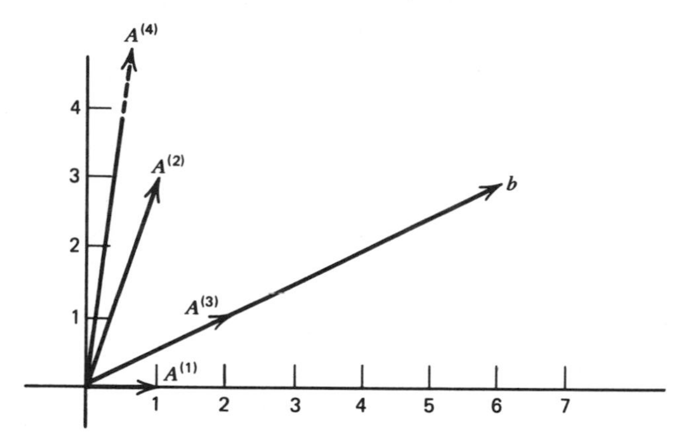

Figure 3.3

serve as basic variables for a basic feasible solution. Note also that b is a multiple of $A^{(3)}$ alone. Thus any basic feasible solution with x_3 as a basic variable will be degenerate. These results all correspond to the solutions previously found algebraically.

One final step on the standard linear programming problem will put the problem into the form to which the simplex method can be applied. This step can be considered as an extension of the problem of putting the system of constraints into canonical form, and we demonstrate it first by means of the problem of Example 3.

The function to be minimized was $f(x_1,x_2,x_3,x_4) = x_1 - x_2 + 2x_3 - 5x_4$, and the system of constraints, using the system of (5) with basic variables x_1 and x_2, was

$$\begin{aligned} x_1 \quad + \tfrac{5}{3}x_3 - \tfrac{5}{3}x_4 &= 5 \\ x_2 + \tfrac{1}{3}x_3 + \tfrac{8}{3}x_4 &= 1 \end{aligned}$$

From these equations it is obvious that the value of the objective function f at any point (x_1,x_2,x_3,x_4) satisfying the constraints can be given by

$$\begin{aligned} &x_1 - x_2 + 2x_3 - 5x_4 \\ &= [5 - \tfrac{5}{3}x_3 + \tfrac{5}{3}x_4] - [1 - \tfrac{1}{3}x_3 - \tfrac{8}{3}x_4] + 2x_3 - 5x_4 \\ &= \tfrac{2}{3}x_3 - \tfrac{2}{3}x_4 + 4 \end{aligned}$$

Thus on this system of constraints, the problem of minimizing f is equivalent to the problem of minimizing the function $\tfrac{2}{3}x_3 - \tfrac{2}{3}x_4 + 4$. With this new function our goal of expressing the function to be optimized in terms of only the nonbasic variables is attained.

Definition The standard linear programming problem will be said to be in *canonical form* with a distinguished set of basic variables if:

(a) The system of constraints is in canonical form with this distinguished set of basic variables.
(b) The associated basic solution is feasible.
(c) The objective function is expressed in terms of only the nonbasic variables.

If the first two conditions of this definition are satisfied for a linear programming problem, the system of constraints can be used, as in the above example, to eliminate the basic variables from the objective function. In putting and maintaining a problem in canonical form, we will abuse the language somewhat and always speak of one fixed objective function. Certainly in the above example the function $x_1 - x_2 + 2x_3 - 5x_4$ does not equal the function $\frac{2}{3}x_3 - \frac{2}{3}x_4 + 4$. However, the problems of optimizing these functions on the given constraint set are equivalent, and it is this equivalency that we have in mind when we say, for example, that the objective function is now given by $\frac{2}{3}x_3 - \frac{2}{3}x_4 + 4$.

You may be somewhat puzzled by our earlier remark that, when determining the minimum of the objective function of a linear programming problem, the basic feasible solutions are the critical points to be considered. Why, when trying to minimize a function, should we wish to restrict our attention to only those feasible solutions of the constraint set that are basic and therefore have at least $n - m$ zero coordinates? For example, in a diet problem with five nutritional requirements and 15 foods to choose from, is it possible to find a diet of minimal cost that uses at most only five of the foods? As we will show in this chapter, the answer to this question is "yes." In fact, we will show by an algebraic argument that if the objective function does have a minimum value, that value is assumed by at least one basic feasible solution.

The significance of the basic feasible solutions can also be demonstrated by considering the geometry of the linear programming problem. The set of feasible solutions to the constraint set of a linear programming problem form a convex set, and the optimal value of a linear function on a convex set is attained at a "corner" or vertex of the set. Moreover, the basic feasible solutions are vertices of this convex set of solution points. Since we do not need the concept of convexity and these related results in the algebraic development that follows, we postpone this discussion until the last section of this chapter.

Actually, the role played by the basic feasible solutions can be seen from the elementary geometric approach used to solve the blending problem developed in Section 2.2. The problem there was to determine a blend of two feeds that minimized costs and met three nutritional requirements. Letting x_1 denote the amount of Feed 1 and x_2 the amount of Feed 2 in a diet, the associated mathematical problem was to

$$\text{Minimize} \quad 10x_1 + 4x_2$$

$$\text{subject to}$$

$$\begin{aligned}
3x_1 + 2x_2 &\geq 60 \\
7x_1 + 2x_2 &\geq 84 \\
3x_1 + 6x_2 &\geq 72 \\
x_1, x_2 &\geq 0
\end{aligned}$$

Putting this into standard form gives the following.

$$\text{Minimize} \quad 10x_1 + 4x_2$$

$$\text{subject to}$$

$$\begin{array}{lllll}
3x_1 + 2x_2 & - x_3 & & & = 60 \\
7x_1 + 2x_2 & & - x_4 & & = 84 \\
3x_1 + 6x_2 & & & - x_5 & = 72 \\
\multicolumn{5}{l}{x_1, x_2, x_3, x_4, x_5 \geq 0}
\end{array}$$

The slack variables x_3, x_4, and x_5 measure the surplus amounts of the nutritional elements A, B, and C in a given diet. Now the geometric argument based on Figure 2.5 of that section showed that if the linear function had a minimal value, the function would assume that value at a corner or vertex of the region shaded in Figure 2.3. The four vertices of the shaded region in Figure 2.3 are the points (0,42), (6,21), (18,3), and (24,0). They occur on the boundaries of the regions defined by the original three inequalities, that is, when some of the inequalities are actually equalities and the corresponding slack variables therefore equal zero. In fact, the solutions to the constraint set in standard form corresponding to these four points are:

$$\begin{aligned}
(0,42) &\leftrightarrow (0,42,24,0,180) \\
(6,21) &\leftrightarrow (6,21,0,0,72) \\
(18,3) &\leftrightarrow (18,3,0,48,0 \\
(24,0) &\leftrightarrow (24,0,12,84,0)
\end{aligned}$$

Notice that each of the four points in the right-hand column has two coordinates at zero level. These four points are basic feasible solutions to the constraint set in standard form, and so, if the objective function is bounded below, the minimal value must occur at a basic feasible solution.

PROBLEM SET 3.2

1. Solve the following using the pivot operation.

(a)
$$\begin{aligned} 3x_2 - 3x_3 &= 15 \\ x_1 + x_2 + x_3 &= 0 \\ 3x_1 + 5x_2 + 3x_3 &= 4 \end{aligned}$$

(b)
$$\begin{aligned} 3x_1 + 2x_2 - 7x_3 &= 1 \\ x_1 - 5x_2 - 6x_3 &= -4 \end{aligned}$$

(c)
$$\begin{aligned} x_1 + x_2 \qquad - 2x_4 &= 5 \\ - 3x_2 + x_3 + 4x_4 &= 2 \end{aligned}$$

2. A system of equations is redundant if one of the equations in the system is a linear combination of the other equations. Show by using the pivot operation that the following system is redundant. Is this system equivalent to a system of equations in canonical form?

$$\begin{aligned} x_1 + x_2 - 3x_3 &= 7 \\ -2x_1 + x_2 + 5x_3 &= 2 \\ 3x_2 - x_3 &= 16 \end{aligned}$$

3. A system of equations is inconsistent if the system has no solution. Show by using the pivot operation that the following systems are inconsistent. Are either of these systems equivalent to a system in canonical form?

(a)
$$\begin{aligned} x_1 + 2x_2 &= 3 \\ x_1 + 2x_2 &= 4 \end{aligned}$$

(b)
$$\begin{aligned} x_1 + x_2 - 3x_3 &= 7 \\ -2x_1 + x_2 + 5x_3 &= 2 \\ 3x_2 - x_3 &= 15 \end{aligned}$$

4. (a) Solve the following system of equations by finding an equivalent system in canonical form with basic variables x_1 and x_2.

$$\begin{aligned} 2x_1 + x_2 - 2x_3 &= 17 \\ x_1 \qquad - x_3 &= 4 \end{aligned}$$

(b) Is this system equivalent to a system in canonical form with basic variables x_1 and x_3?

(c) Interpret these results geometrically.

5. Suppose a system of equations contains the following terms:

$$\begin{aligned} &ax_1 + bx_2 \\ &cx_1 + dx_2 \end{aligned}$$

where the a, b, c, and d are constants, $a \neq 0$.

The system is then replaced with an equivalent system by pivoting at the ax_1 term. Show that these four terms become

$$x_1 + \frac{b}{a}x_2$$

$$0x_1 + \left(d - \frac{bc}{a}\right)x_2$$

The expression $d - bc/a$ provides a way of remembering the effect of the pivot operation on any term not in the row or column of the pivot term.

6. For the linear programming problem of

$$\text{Minimizing} \quad 5x_1 + 2x_2 + 3x_3 + x_4$$

subject to

$$\begin{aligned} x_1 + x_2 - 2x_3 + 3x_4 &= 2 \\ -2x_1 \qquad + x_3 \qquad &= 2 \\ x_1, x_2, x_3, x_4 &\geq 0 \end{aligned}$$

(a) Show geometrically that there can be only two basic feasible solutions to the problem.
(b) Compute these two basic feasible solutions.
(c) Show that the objective function is bounded below.
(d) Assume that the minimal value of the objective function is attained at a basic feasible solution, and determine this minimal value.

7. Following the outline in Problem 6, complete the problem of Example 3.

8. (a) Put the constraint set from the standard form of blending problem considered in this section into canonical form with basic variables x_1, x_2, and x_5. The associated basic feasible solution is (6,21,0,0,72).
(b) The objective function for this problem is $10x_1 + 4x_2$. By eliminating the x_1 and x_2 variables by using the equations found in part a, this function can be expressed in terms of only x_3 and x_4. Verify that the form reduces to $144 + x_3 + x_4$.
(c) In that we are considering only feasible solutions to the constraint set, using part b, give another proof that the minimal value of the objective function is 144.

9. Prove that the system of equations resulting from a given system by applying the pivot operation is equivalent to (has the same solution set as) the original system.

10. Prove that although there may be different ways of driving a system of equations into canonical form with a specified set of basic variables, there is a unique basic solution associated with this specified set of basic variables.

11. True or false: A system of equations is equivalent to a system of equations in canonical form if and only if the original system has at least one solution?

12. Construct a linear programming problem with four variables and three equations for which there exists degenerate feasible solutions with exactly two non-zero coordinates.

SECTION 3.3 INTRODUCTION TO THE SIMPLEX METHOD

In this section the simplex method for solving linear programming problems will be introduced. The basic ideas behind the technique will be demonstrated by means of a specific example. The goal of this section is to develop motivation and understanding, and the theorems related to the simplex method will be proven only in the subsequent sections of this chapter.

Let us consider the following problem in standard form.

$$\text{Minimize} \quad -4x_1 + x_2 + x_3 + 6x_4 + 3x_5 = z$$

subject to

$$\begin{aligned} -6x_1 \qquad\quad + x_3 - 2x_4 + 2x_5 &= 6 \\ 3x_1 + x_2 - x_3 + 7x_4 + x_5 &= 9 \\ x_1, x_2, x_3, x_4, x_5 \geq 0 & \end{aligned} \tag{1}$$

The simplex method can begin only with the problem in canonical form. To put the problem into canonical form, we could first arbitrarily select two variables to be basic variables and then, by pivoting, attempt to put the system of constraints into canonical form with these variables as basic variables, with the hope that the associated basic solution would be feasible. Or, because here we have a problem with only two constraints, we could determine, using elementary vector geometry, a pair of variables that would serve as basic variables for a feasible solution. In general, however, finding an initial basic feasible solution to a problem can be a major difficulty. This problem will be resolved in Section 3.6. For now, assume that we know that for the problem at hand the variables x_2 and x_3 can serve as basic variables for a feasible solution. Pivoting at the $1x_3$ term of the first equation will put the system of constraints into canonical form. This gives

$$\begin{aligned} -6x_1 \qquad\quad + x_3 - 2x_4 + 2x_5 &= 6 \\ -3x_1 + x_2 \qquad\quad + 5x_4 + 3x_5 &= 15 \end{aligned} \tag{2}$$

The associated basic solution, (0,15,6,0,0), is feasible, as promised. Now these two equations can be used to eliminate the basic variables x_2 and x_3 from the expression for the objective function z, given by

$$-4x_1 + x_2 + x_3 + 6x_4 + 3x_5 = z \tag{3}$$

In fact, simply subtracting the two equations in (2) from the equation in (3) gives

$$5x_1 + 0x_2 + 0x_3 + 3x_4 - 2x_5 = z - 21$$

Hence the objective function can be given by the form

$$5x_1 + 3x_4 - 2x_5 + 21 = z$$

Thus the problem in canonical form with basic variables x_2 and x_3 is to

Minimize z with

$$\begin{aligned}
-6x_1 \qquad + x_3 - 2x_4 + 2x_5 &= 6 \\
-3x_1 + x_2 \qquad + 5x_4 + 3x_5 &= 15 \\
5x_1 \qquad + 3x_4 - 2x_5 &= -21 + z \\
x_1, x_2, x_3, x_4, x_5 &\geq 0
\end{aligned} \tag{4}$$

The value of the objective function at the associated basic feasible solution (0,15,6,0,0) is 21. Now the key idea behind the simplex method is to move to another basic feasible solution that gives a smaller value for z by replacing exactly one basic variable from the present set. As we will see, the mechanics for this replacement will be provided by the simple pivot operation. However, what variable from the set of nonbasic variables x_1, x_4, and x_5 to insert into the basis, and what basic variable, x_2 or x_3, to replace in order to reduce the value of z is not obvious.

These questions are answered first by considering the objective function $z = 5x_1 + 3x_4 - 2x_5 + 21$. In this expression for z, the x_5 variable has a negative coefficient. Thus a feasible solution to the constraint set with x_1 and x_4 still equal to zero, but with x_5 greater than zero, will give a smaller value to z. This suggests that we move x_5 into the set of basic variables and attempt to make x_5 as large as possible.

But what basic variable, x_2 or x_3, should we replace? To answer this, consider the constraint set with the conditions imposed by this situation, that the nonbasic variables x_1 and x_4 equal 0. From (4) we have

$$\begin{aligned}
x_3 + 2x_5 &= 6 \\
x_2 + 3x_5 &= 15
\end{aligned}$$

Solving for x_3 and x_2 gives

$$\begin{aligned}
x_3 &= 6 - 2x_5 \\
x_2 &= 15 - 3x_5
\end{aligned} \tag{5}$$

Clearly, x_5 cannot be made arbitrarily large because in order to have a solution to the constraint set with $x_1 = x_4 = 0$, x_2 and x_3 must satisfy these equations and will therefore become negative. In fact, since x_2 and x_3 must be nonnegative, x_5 is restricted by the two inequalities

$$0 \leq 6 - 2x_5 \qquad \text{and} \qquad 0 \leq 15 - 3x_5$$

that is, $x_5 \leq 3 = \frac{6}{2}$ and $x_5 \leq 5 = \frac{15}{3}$. Now, since x_5 must satisfy both these inequalities, the maximum possible value for x_5 is 3. Letting $x_5 = 3$ and using (5) to calculate x_3 and x_2, we have the feasible solution $x_1 = x_4 = 0$, $x_5 = 3$, $x_3 = 0$, and $x_2 = 6$. The value of z at this point is 15, six less than the value at the first basic feasible solution. The basic variables for this solution point (0,6,0,0,3) should be

x_2 and x_5, since these two variables assume positive values. Thus x_5 should replace x_3 in the basis. In fact, by letting x_5 equal the minimum of 3 and 5, we were guaranteed that x_3 would assume the value 0, because this minimum value, 3, was the bound coming from the x_3 equation in (5). To determine the variable to extract from the basis then, we need only determine the basic variable of that equation in the modified constraint set (5) that leads to the minimal bound. And each of these bounds of $3 = \frac{6}{2}$ and $5 = \frac{15}{3}$ is the ratio of the constant term in the equation to the coefficient of the x_5 variable. This suggests a simple procedure for determining the variable to extract from the basis, a procedure that will be spelled out in detail in the next section.

The simplex method consists of the continuation of this process. In order to proceed, however, the problem must be in canonical form with basic variables x_2 and x_5. To do this we use the pivot operation. With the system of constraints expressed as in (4), the first equation contains the basic variable x_3, which is to be replaced with the variable x_5. Hence pivoting at the $2x_5$ term of this equation will put the system of constraints into canonical form with basic variables x_2 and x_5. Moreover, the effect of this pivot operation on the third equation in (4) would be to eliminate the variable x_5 from that equation also. Then the objective function z would be expressed in terms of only the variables x_1, x_3, and x_4. Thus the effect of the pivot operation at the $2x_5$ term of the first equation in (4) applied to all three equations would be to transform the entire problem into the desired canonical form. Pivoting here gives

$$\begin{aligned} -3x_1 + \tfrac{1}{2}x_3 - x_4 + x_5 &= 3 \\ 6x_1 + x_2 - \tfrac{3}{2}x_3 + 8x_4 &= 6 \\ -x_1 + x_3 + x_4 &= -15 + z \end{aligned} \tag{6}$$

Now we proceed exactly as before. The variable x_1 has a negative coefficient in the expression for the objective function, and so should be inserted into the basis. Letting $x_3 = x_4 = 0$, the constraint set of (6) becomes

$$\begin{aligned} -3x_1 + x_5 &= 3 \\ 6x_1 + x_2 &= 6 \end{aligned} \quad \text{or} \quad \begin{aligned} x_5 &= 3 + 3x_1 \\ x_2 &= 6 - 6x_1 \end{aligned} \tag{7}$$

Since x_2 and x_5 must be nonnegative, we have

$$\begin{aligned} 0 &\le 3 + 3x_1 \\ 0 &\le 6 - 6x_1 \end{aligned} \quad \text{or} \quad \begin{aligned} -1 &\le x_1 \\ x_1 &\le 1 \end{aligned}$$

The first inequality places no upper bound on x_1, so the upper limit for x_1 is determined solely by the second inequality, the inequality resulting from the x_2 equation in (7). Thus x_1 should replace x_2 in the basis. Letting $x_1 = 1$ gives the basic feasible solution (1,0,0,0,6), and the value of the objective function at this point is 14.

One lingering question that we have so far assiduously avoided is the following. When do we know that the minimal value of the objective function has been achieved and the process can terminate? Our example will now provide the answer to this question.

We have seen that a reduced value for z can be determined by using x_1 and x_5 as basic variables instead of x_2 and x_5. Accordingly, we put the system into canonical form with these as basic variables by pivoting at the $6x_1$ term of the second equation in (6). This gives

$$\begin{array}{rcl} \frac{1}{2}x_2 - \frac{1}{4}x_3 + 3x_4 + x_5 &=& 6 \\ x_1 + \frac{1}{6}x_2 - \frac{1}{4}x_3 + \frac{4}{3}x_4 \quad &=& 1 \\ \frac{1}{6}x_2 + \frac{3}{4}x_3 + \frac{7}{3}x_4 \quad &=& -14 + z \end{array}$$

The objective function is given by $z = \frac{1}{6}x_2 + \frac{3}{4}x_3 + \frac{7}{3}x_4 + 14$. In contrast to the two previous situations, here the coefficients of the nonbasic variables are all positive. This means in fact that the value of the objective function at *any* feasible solution to the constraint set must be at least 14, since all the coordinates of a feasible solution are nonnegative. Thus our process is terminated. The minimal value of the objective function can be no less than 14, and this value is attained at the point (1,0,0,0,6).

To summarize, the simplex method begins with the problem in canonical form. We move from one basic feasible solution to another by replacing exactly one basic variable at each step, with the new basic feasible solution providing a reduced value of the objective function (except possibly when there is degeneracy, a complication to be discussed later). Consideration of the coefficients of the objective function tells us if the minimal value has been achieved and, if not, what variable to insert into the basis. Consideration of the modified constraint set tells us what variable to extract from the basis. And one simple pivot operation at each step keeps the entire system in proper form.

In the next section we will make precise the simplex method for the general problem and will consider the case when the objective function is not bounded below. (See also Problem 3.) In Section 3.6 a method for determining an initial basic feasible solution based on the simplex method will be discussed.

PROBLEM SET 3.3

1. Minimize $x_1 + x_2 + 4x_3 + 7x_4$

subject to

$$\begin{array}{rcl} x_1 + x_2 + 5x_3 + 2x_4 &=& 8 \\ 2x_1 + x_2 + 8x_3 \quad &=& 14 \\ x_1,x_2,x_3,x_4 \geq 0 && \end{array}$$

(a) The variables x_1 and x_2 can serve as basic variables for a basic feasible solution. Show that the problem expressed with these as basic variables is:

$$\begin{aligned} x_1 \quad\quad + 3x_3 - 2x_4 &= 6 \\ x_2 + 2x_3 + 4x_4 &= 2 \\ -x_3 + 5x_4 &= -8 + z \end{aligned}$$

(b) Entering x_3 into the basis will reduce the value of z. Why? Show that the variable to be replaced is x_2.

(c) Perform the pivot operation. Show that the minimal value of the objective function is 7 and is achieved at (3,0,1,0).

2. Use the simplex method to do the following problem. The problem is stated in canonical form with basic variables x_2 and x_3. Notice that at the first step in the simplex method, either x_1 or x_4 can enter the basis.

Minimize $\quad -x_1 - 2x_4 + x_5$

subject to

$$\begin{aligned} x_1 \quad\quad + x_3 + 6x_4 + 3x_5 &= 2 \\ -3x_1 + x_2 \quad\quad + 3x_4 + x_5 &= 3 \\ x_1, x_2, x_3, x_4, x_5 &\geq 0 \end{aligned}$$

3. In the following problem the objective function does not have a minimum. However, the problem is stated in canonical form with basic variables x_1 and x_2, and the simplex method can be initiated. What occurs after the first pivot operation that makes this problem different from our other examples? Can you prove, using the resulting equations, that the objective function is in fact not bounded below?

Minimize $\quad 4x_3 - 6x_4$

subject to

$$\begin{aligned} x_2 - 6x_3 + 2x_4 &= 6 \\ x_1 \quad\quad + 2x_3 - x_4 &= 5 \\ x_1, x_2, x_3, x_4 &\geq 0 \end{aligned}$$

SECTION 3.4 THEORY OF THE SIMPLEX METHOD

In this section we develop the simplex method for a general linear programming problem. To initiate the algorithm, the problem must be in canonical form. In Section 3.1 we showed that any linear programming problem is equivalent to a problem in standard form, and in Section 3.6 we will show how to drive a problem in standard form into canonical form. In fact, the technique developed in Section

3.6 will make use of the ideas developed in this section. Thus, for the time being, we assume that our general problem is in canonical form. Suppose the problem has m constraints and n variables, with the first m variables as basic variables. The problem is then:

Minimize z where

$$\begin{aligned}
x_1 + \quad \cdots + a_{1,m+1}x_{m+1} + \cdots + a_{1n}x_n &= b_1 \\
x_2 + \cdots + a_{2,m+1}x_{m+1} + \cdots + a_{2n}x_n &= b_2 \\
\ddots \qquad & \\
x_m + a_{m,m+1}x_{m+1} + \cdots + a_{mn}x_n &= b_m \\
c_{m+1}x_{m+1} + \cdots + c_nx_n &= z_0 + z
\end{aligned}$$

$$x_1, x_2, \ldots, x_n \geq 0 \tag{1}$$

The a_{ij}'s, b_i's, c_j's, and z_0 are constants and, since the associated basic solution is feasible, $b_i \geq 0$, $i = 1, \ldots, m$.

Example 1

Minimize z with

$$\begin{aligned}
x_1 \quad + 2x_3 - \quad x_4 &= 10 \\
x_2 - \quad x_3 - 5x_4 &= 20 \\
2x_3 + 3x_4 &= 60 + z \\
x_1, x_2, x_3, x_4 &\geq 0
\end{aligned}$$

Here we have a problem with $m = 2$ constraints, $n = 4$ variables, and in canonical form. The associated basic feasible solution is (10,20,0,0), and the value of the objective function z at this point is -60. Note that in this particular problem the coefficients $c_3 = 2$ and $c_4 = 3$ are nonnegative. Since x_3 and x_4 are restricted to be nonnegative, the smallest value $z = 2x_3 + 3x_4 - 60$ can possibly attain is -60, the value of the objective function at the (10,20,0,0) solution point. This suggests our first theorem.

Theorem 1 (optimality criterion) *For the linear programming problem of* (1), *if* $c_j \geq 0, j = m + 1, \ldots, n$, *then the minimal value of the objective function is* $-z_0$ *and is attained at the point* $(b_1, b_2, \ldots, b_m, 0, 0, \ldots, 0)$.

Proof For any point satisfying the set of constraints, the value of the objective function is given by $z = c_{m+1}x_{m+1} + \cdots + c_nx_n - z_0$. Since any feasible solution to the constraints has nonnegative coordinates, the smallest possible value for the sum $c_{m+1}x_{m+1} + \cdots + c_nx_n$ is zero. Thus the minimal possible value for z is $-z_0$, and this value is assumed at the point $(b_1, b_2, \ldots, b_m, 0, \ldots, 0)$. ###

Hence the problem is resolved if all the c_j's are nonnegative. Assume now that at least one c_j, say c_s, is negative. Then we attempt to enter the variable x_s into the basis. In order to determine what basic variable to replace, we consider the constraint set with all the nonbasic variables except x_s equal to zero. This gives

$$\begin{aligned} x_1 + a_{1s}x_s &= b_1 \qquad \text{or} \qquad & x_1 &= b_1 - a_{1s}x_s \\ x_2 + a_{2s}x_s &= b_2 & x_2 &= b_2 - a_{2s}x_s \\ &\vdots & &\vdots \\ x_m + a_{ms}x_s &= b_m & x_m &= b_m - a_{ms}x_s \end{aligned} \tag{2}$$

Example 2

Minimize z with

$$\begin{aligned} x_1 \quad + 2x_3 - x_4 &= 10 \\ x_2 - x_3 - 5x_4 &= 20 \\ 2x_3 - 3x_4 &= 60 + z \\ x_1,x_2,x_3,x_4 \geq 0 \end{aligned}$$

Except for a change in sign in c_4, this is the problem of Example 1. As before, (10,20,0,0) is a feasible solution, and the value of the objective function $z = 2x_3 - 3x_4 - 60$ at this point is -60. However, here we could reduce the value of z if we could find feasible solutions to the constraint set with x_4 positive and x_3 equal to zero, since $c_4 = -3$ is negative. Setting $x_3 = 0$, the constraints reduce to

$$\begin{aligned} x_1 - x_4 &= 10 \qquad \text{or} \qquad & x_1 &= 10 + x_4 \\ x_2 - 5x_4 &= 20 & x_2 &= 20 + 5x_4 \end{aligned}$$

Note that if we fix x_4 at any positive number and then use these two equations to solve for x_1 and x_2, the resulting values will be positive. Thus all points in the set

$$\{(x_1,x_2,0,x_4) \mid x_4 \geq 0, x_1 = 10 + x_4, x_2 = 20 + 5x_4\}$$

are feasible solutions to the system of constraints. But the function $z = 2x_3 - 3x_4 - 60$ is unbounded below on this set. This suggests our next theorem.

Theorem 2 *For the linear programming problem of* (1), *if there is an index* s, $m + 1 \leq s \leq n$, *such that* $c_s < 0$ *and* $a_{is} \leq 0$ *for all* $i = 1,2,\ldots,m$. *then the objective function is not bounded below.*

Proof Assume there is an index s satisfying the conditions of the theorem. Since the coefficients a_{is} are all nonpositive, the m equations of (2) can be used to find a set S of feasible solutions to the constraints with x_s assuming arbitrarily large values, the original basic variables x_1 to x_m positive values, and the remaining variables value zero. But the objective function is given by the form

$$z = c_m x_{m+1} + \cdots + c_s x_s + \cdots + c_n x_n - z_0,$$

and on S, this reduces to $z = c_s x_s - z_0$. Since $c_s < 0$, z is unbounded below on S.
###

Assume now that $c_s < 0$ and that at least one $a_{is} > 0$. Then the argument above breaks down, because if $a_{is} > 0$, the equation $x_i = b_i - a_{is}x_s$ places a limit on how large x_s can become. In fact, in order for x_i to remain nonnegative, we must have $0 \le b_i - a_{is}x_s$, that is, $x_s \le b_i/a_{is}$ for $a_{is} > 0$. Thus our goal now is simply to replace in the basis one of the basic variables $x_1,\ldots,x_m$ with the variable x_s. Because of the term $c_s x_s$ in the expression for the objective function, the value of z at this new basic feasible solution hopefully will be reduced. Our one demand on this new basis is that the associated basic solution be feasible. Hence the equations of (2) for which $a_{is} > 0$ restrict our choice of the variable to extract from the basis. Since we must have $x_s \le b_i/a_{is}$ for all i with $a_{is} > 0$, the largest possible value for x_s is $\text{Min}\{b_i/a_{is} \mid a_{is} > 0\}$. Suppose this minimum value is attained when $i = r$. Then letting $x_s = b_r/a_{rs}$ will give $x_i \ge 0$ for $i = 1,\ldots,m$ and, in particular, $x_r = b_r - a_{rs}x_s = 0$. Since x_r takes on the value zero here, it appears that x_r is the variable to be replaced in the basis. And since in (1) the rth equation of the constraints isolates x_r, the problem can be put into canonical form with basic variables $x_1,\ldots,x_{r-1},x_{r+1},\ldots,x_m,x_s$ by a single pivot operation at the $a_{rs}x_s$ term of the rth equation. Before formally stating and proving these results, we give an example.

Example 3

Minimize z with

$$\begin{aligned}
x_1 \qquad\qquad\quad + 2x_4 - x_5 &= 10\\
x_2 \qquad\; - x_4 - 5x_5 &= 20\\
x_3 + 6x_4 - 12x_5 &= 18\\
- 2x_4 + 3x_5 &= 60 + z\\
x_1,x_2,x_3,x_4,x_5 &\ge 0
\end{aligned}$$

The problem is in canonical form with basic variables x_1, x_2, and x_3. The associated basic feasible solution is (10,20,18,0,0), and the value of the objective function at this point is -60. However, $c_4 = -2$ is negative, and so we attempt to reduce the value of z by inserting x_4 into the basis. Letting $x_5 = 0$, the constraints reduce to

$$\begin{aligned}
x_1 + 2x_4 &= 10 &\quad \text{or} \quad x_1 &= 10 - 2x_4\\
x_2 - x_4 &= 20 & x_2 &= 20 + x_4\\
x_3 + 6x_4 &= 18 & x_3 &= 18 - 6x_4
\end{aligned}$$

The second equation places no restriction on x_4. However, the first requires that $x_4 \leq \frac{10}{2} = 5$, and the third that $x_4 \leq \frac{18}{6} = 3$. The largest possible value for x_4 with $x_5 = 0$, is the minimum of 3 and 5, that is, 3. Letting $x_4 = 3$ gives $x_3 = 0$. Thus x_4 should replace x_3 in the basis and, since the third equation of the constraints isolates x_3, pivoting at the $6x_4$ term of this equation should keep the problem in canonical form, but with basic variables x_1, x_2, and x_4. In fact, pivoting here yields the following equivalent problem.

Minimize z with

$$\begin{aligned}
x_1 \quad - \tfrac{1}{3}x_3 \quad\quad + 3x_5 &= 4 \\
x_2 + \tfrac{1}{6}x_3 \quad\quad - 7x_5 &= 23 \\
\tfrac{1}{6}x_3 + x_4 - 2x_5 &= 3 \\
\tfrac{1}{3}x_3 \quad\quad - x_5 &= 66 + z
\end{aligned}$$

The problem remains in canonical form, but with basic variables x_1, x_2, and x_4. The associated basic solution (4,23,0,3,0) is feasible, and the value of the objective function at this point is -66. Although the optimal value of z has not yet been attained, we have as promised moved to a basic feasible solution yielding a smaller value for z while maintaining the problem in canonical form.

Theorem 3 *In the problem of* (1), *assume that there is an index s such that* $c_s < 0$ *and that at least one* $a_{is} > 0, i = 1, \ldots, m$. *Suppose*

$$\frac{b_r}{a_{rs}} = \text{Min}\left\{\frac{b_i}{a_{is}} \middle| 1 \leq i \leq m \quad \textit{and} \quad a_{is} > 0\right\}.$$

Then the problem can be put into canonical form with basic variables

$$x_1, x_2, \ldots, x_{r-1}, x_{r+1}, \ldots, x_m, x_s.$$

The value of the objective function at the associated basic feasible solution is

$$-z_0 + \frac{c_s b_r}{a_{rs}}$$

Proof Consider the problem of (1) under the assumptions of the theorem. The coefficient $a_{rs} \neq 0$ (it is, in fact, >0), and so the term $a_{rs}x_s$ of the rth equation can be used as the pivot term in the pivot operation applied to the $m + 1$ equations. By pivoting here the system of constraints will be expressed in canonical form with basic variables $x_1, \ldots, x_{r-1}, x_{r+1}, \ldots, x_m, x_s$. The constant terms, b_i^* say, $i = 1, \ldots, m$, on the right-hand side of the equations, become

$$b_i^* = b_i - \frac{a_{is}b_r}{a_{rs}}, \qquad i = 1, \ldots, m; \quad i \neq r$$

$$b_r^* = \frac{b_r}{a_{rs}} \tag{3}$$

Clearly $b_r^* \geq 0$. If $a_{is} \leq 0$ then, since $b_r \geq 0$ and $a_{rs} > 0$, $b_i^* \geq b_i \geq 0$. If $a_{is} > 0$ and $i \neq r$, by the choice of r, $b_i/a_{is} \geq b_r/a_{rs}$, and so $b_i \geq a_{is}b_r/a_{rs}$. Hence $b_i^* \geq 0$. Therefore the basic solution associated with these basic variables is feasible.

Now the objective function is given in (1) by the form $c_{m+1}x_{m+1} + \cdots + c_s x_s + \cdots + c_n x_n = z_0 + z$. The effect of the pivot operation on this equation will be to eliminate the x_s term from the equation, producing the equation

$$c_r^* x_r + c_{m+1}^* x_{m+1} + \cdots + c_{s-1}^* x_{s-1} + c_{s+1}^* x_{s+1} + \cdots + c_n^* x_n = z_0^* + z \quad (4)$$

with $z_0^* = z_0 - c_s b_r/a_{rs}$.

Thus the objective function is expressed in terms of only the new nonbasic variables and the value of this function at the new basic feasible solution is $-z_0 + c_s b_r/a_{rs}$.

Notice the result of this pivot operation applied to the system of constraints and the objective function. The problem remains in canonical form with the original basic variable x_r replaced with the variable x_s. The value of the objective function at this new basic feasible solution is equal to the value $-z_0$ at the original basic feasible solution plus the quantity $c_s b_r/a_{rs}$. Since we have assumed that $c_s < 0$ and $a_{rs} > 0$, $c_s b_r/a_{rs}$ is less than or equal to zero, and is strictly less than zero if b_r is strictly positive. Thus, if $b_r > 0$, the pivot operation has left the system in canonical form at a basic feasible solution with a smaller value for the objective function. Let us assume for the time being that this is always the case, that any basic feasible solution to the system of constraints has no basic variable equal to zero. A basic solution with some basic variables equal to zero is called a *degenerate solution*, so we are assuming that all basic feasible solutions are nondegenerate.

Under this nondegeneracy hypothesis, Theorem 3 states that if at least one of the coefficients c_j, $m + 1 \leq j \leq n$, is negative, say c_s, and if at least one of the coefficients a_{is}, $1 \leq i \leq m$, is positive, then a specific pivot operation leaves the problem in canonical form at a basic feasible solution that gives a reduced value for the objective function. Now we can continue. If the new coefficients of the objective function are all nonnegative, we are at the minimal value for the objective function, as Theorem 1 applies. If one of these coefficients is negative and if all of the coefficients of the associated variable are nonpositive in the constraint set, the objective function is unbounded below, as Theorem 2 applies. Otherwise we can apply Theorem 3 again, driving to another basic feasible solution with an even smaller value for the objective function. Since at each step the value of the objective function is reduced (due to the nondegeneracy assumption), there can be no repetition of basic feasible solutions. The different values for the objective function guarantee that a particular basic feasible solution can appear at most once in the process (see Problem 10 of Section 3.2). Now there are at most a finite number of basic solutions, as there are only $\binom{n}{m} = n!/[n!(n - m)!]$ ways of selecting m basic variables from a set of n variables. Thus this process must eventually terminate. Either the minimum value of the objective function will be reached or the function will be proven to be unbounded.

This is the simplex method, with a proof, using the nondegeneracy hypothesis, that the process must terminate after a finite number of steps with either Theorems 1 or 2 applying. The nondegeneracy assumption is quite critical. If some basic feasible solutions were degenerate, the pivot operation of Theorem 3 applied in a row with $b_i = 0$ would leave the value of the objective function unchanged. After several steps of this, we would have no assurance that basic feasible solutions would not reappear, possibly causing the process to cycle indefinitely. In fact, examples of cycling have been constructed (see, for example, Beale [1]). Thus, from a mathematical point of view, our proof of convergence of the process is inadequate. In Section 3.8 of this chapter we will provide a complete proof that, for any linear programming problem, there exists a sequence of pivot operations that will drive the problem to completion.

From a practical point of view, however, a pleasant phenomenon occurs. The cliché "whatever can go wrong will go wrong" does not apply. Although degeneracy certainly occurs quite frequently in linear programming applications, simple rules such as those described below for deciding what variables to interchange at those steps of the pivot operation where the choice is not unique have been sufficient to prevent cycling in real-life problems. So far the only problems known to cycle are the examples constructed solely for this purpose.

We now summarize the steps of the simplex method, starting with the problem in canonical form.

1. If all the $c_j \geq 0$, the minimum value of the objective function has been achieved (Theorem 1).
2. If there exists an s such that $c_s < 0$ and $a_{is} \leq 0$ for all i, the objective function is not bounded below (Theorem 2).
3. Otherwise pivot (Theorem 3). To determine the pivot term:
   (a) Pivot in any column with a negative c_j term. If there are several negative c_j's, pivoting in the column with the smallest c_j may reduce the total number of steps necessary to complete the problem. Assume we pivot in column s.
   (b) To determine the row of the pivot term, find that row, say row r, such that

$$\frac{b_r}{a_{rs}} = \text{Min}\left\{\frac{b_i}{a_{is}} \,\middle|\, a_{is} > 0\right\}$$

   Notice that here only those ratios b_i/a_{is} with $a_{is} > 0$ are considered. If the minimum of these ratios is attained in several rows, a simple rule such as choosing that row with the smallest index can be used to determine the pivoting row.
4. After pivoting, the problem remains in canonical form at a different basic feasible solution. Now return to step 1.

If the problem contains degenerate basic feasible solutions, proceed as above. These steps should still be adequate to drive the problem to completion.

PROBLEM SET 3.4

1. Complete the problem of Example 3.

2. Solve the following using the ideas developed in this section.

(a) Minimize $x_3 + x_4$
subject to

$$\begin{aligned} x_1 \qquad\qquad\quad - x_4 &= 5 \\ x_2 + 2x_3 - 3x_4 &= 10 \\ x_1, x_2, x_3, x_4 &\geq 0 \end{aligned}$$

(b) Minimize x_3 subject to the constraints of part a.
(c) Minimize $x_3 - x_4$ subject to the constraints of part a.
(d) Minimize $x_3 - x_4$

subject to

$$\begin{aligned} x_1 \qquad\qquad\quad - x_4 &= 5 \\ x_2 + 2x_3 \qquad\quad &= 10 \\ x_1, x_2, x_3, x_4 &\geq 0 \end{aligned}$$

(e) Minimize $-x_3 + x_4$ subject to the constraints of part d.
(f) Minimize $-x_3 + x_4$

subject to

$$\begin{aligned} x_1 \qquad + x_3 - x_4 &= 0 \\ x_2 + 2x_3 \qquad\quad &= 10 \\ x_1, x_2, x_3, x_4 &\geq 0 \end{aligned}$$

(g) Minimize $-x_3 - x_4$ subject to the constraints of part f.

3. Calculate the coefficient c_r^* in (4). Can the variable removed from the basis at one step of the pivot operation return to the basis on the next step?

4. Using the form for the objective function given in (1) and the coordinates of the new basic feasible solution given in (3), by direct calculation show that the value of the objective function at the new basic feasible solution is as stated in Theorem 3.

5. Using (3), determine when the pivot operation will go from a nondegenerate basic feasible solution to a degenerate basic feasible solution.

6. Suppose a problem is in canonical form and the associated basic feasible solution degenerate, and that x_1 is a basic variable with the value zero. The pivot operation is performed with the x_1 variable extracted from the basis. Describe the new basic feasible solution.

7. Although the objective function of a linear programming problem can certainly have only one minimal value, for some problems this value may be attained at more than one point. Show this. (*Example.* Minimize x_2 subject to $x_1 + x_2 + x_3 = 1$, $x_1, x_2, x_3 \geq 0$. See also Problem 4 of Section 3.5.)

8. Suppose a problem is driven to the canonical form in (1), and $c_j > 0$ for $m + 1 \leq j \leq n$. Prove that the minimal value $-z_0$ of the objective function is attained *only* at the point $(b_1, \ldots, b_m, 0, \ldots, 0)$.

9. Extend the formulas in the proof of Theorem 3 expressing the results of the pivot operation at the a_{rs} term. Show that for any $j \neq s$,

$$a_{ij}^* = a_{ij} - \frac{a_{is}a_{rj}}{a_{rs}}, \qquad i \neq r$$

$$a_{rj}^* = \frac{a_{rj}}{a_{rs}}$$

$$c_j^* = c_j - \frac{c_s a_{rj}}{a_{rs}}$$

10. Consider the linear programming problem of (1). Suppose that the value of the function

$$z' = c'_{m+1}x_{m+1} + \cdots + c'_n x_n - z'_0$$

equals the value of the objective function

$$z = c_{m+1}x_{m+1} + \cdots + c_n x_n - z_0$$

at all solutions to the system of constraints of (1). Prove that

$$z'_0 = z_0 \qquad \text{and} \qquad c'_j = c_j \qquad \text{for all } j,\ m + 1 \leq j \leq n$$

Conclusion. Given a linear programming problem in canonical form with a specified set of basic variables, the coefficients in the expression for the objective function are unique.

SECTION 3.5 THE SIMPLEX TABLEAU AND EXAMPLES

At each step of the simplex method it is crucial to know only the basic variables and the values of the coefficients in the system of equations. To facilitate computation of a solution, at each step all we need do is record this information. This suggests a notation similar to the "detached coefficient" notation used for solving linear equations. We illustrate with the example of Section 3.3. The problem, expressed in canonical form with basic variables x_2 and x_3 was, as in (4) of Section 3.3, to minimize z with

$$\begin{aligned} -6x_1 \quad\quad + x_3 - 2x_4 + 2x_5 &= 6 \\ -3x_1 + x_2 \quad\quad + 5x_4 + 3x_5 &= 15 \\ 5x_1 \quad\quad\quad + 3x_4 - 2x_5 &= -21 + z \end{aligned}$$

$$x_1, x_2, x_3, x_4, x_5 \geq 0$$

This information is recorded in tableau form in Table 3.1.

Table 3.1

	x_1	x_2	x_3	x_4	x_5	
x_3	-6	0	1	-2	2	6
x_2	-3	1	0	5	3	15
	5	0	0	3	-2	-21

The first column gives the basic variables. The first two rows correspond to the system of constraints, with the constant terms given in the last column. The last row corresponds to the equation defining the objective function, with the constant term on the right-hand side of that equation in the last column and the z term suppressed from the tableau as it remains fixed throughout the simplex method.

We now apply the simplex method. As noted in Section 3.3, the -2 in the x_5 column of the last row indicates that we should pivot in that column. To determine the pivoting row, we compare the ratios b_i/a_{is}, for $a_{is} > 0$, as in Theorem 3 of

Table 3.2

	x_1	x_2	x_3	x_4	x_5	
x_3	-6	0	1	-2	(2)	6
x_2	3	1	0	5	3	15
	5	0	0	3	-2	-21
x_5	-3	0	$\frac{1}{2}$	-1	1	3
x_2	(6)	1	$-\frac{3}{2}$	8	0	6
	-1	0	1	1	0	-15

Section 3.4, and find that row in which the minimum is attained. In this case $\frac{6}{2}$ is less than $\frac{15}{3}$ and, therefore, we should pivot at the 2 in the first row, replacing the basic variable x_3 with the variable x_5. The tableau representing the result of this pivot operation can be constructed from the present tableau by dividing the first row by 2 and then adding multiples of this row to the remaining rows in such a way as to generate zeros in the x_5 column. We illustrate in Table 3.2, placing this new tableau directly below the original tableau.

The second tableau represents the problem as stated in (6) of Section 3.3. The associated basic feasible solution is (0,6,0,0,3), and the value of the objective function at this point is the negative of the constant in the lower right-hand corner of the tableau, $-(-15) = 15$.

Pivoting now at the 6 in the x_1 column of the second row gives the tableau of Table 3.3.

Table 3.3

x_5	0	$\frac{1}{2}$	$-\frac{1}{4}$	3	1	6
x_1	1	$\frac{1}{6}$	$-\frac{1}{4}$	$\frac{4}{3}$	0	1
	0	$\frac{1}{6}$	$\frac{3}{4}$	$\frac{7}{3}$	0	-14

Since all the constants in the last row, excluding the -14, are nonnegative, the minimum value of the objective function has been attained. This value, $-(-14) = 14$, is attained at the basic feasible solution (1,0,0,0,6), as can be read from the final tableau.

Hereafter the steps of the simplex method for any example will be recorded using this tableau notation. We emphasize that if at any time you are confused or bewildered by a statement based on the tableau presentation of a problem, simply translate the information in the tableau back into a clearly stated problem with the system of constraints and the objective function defined as usual, that is, "attach back" the variables. The tableau remains just a notation for a linear programming problem and the associated equations.

Example 1

$$\text{Maximize} \quad 2x_1 + 3x_2 + 3x_3$$

subject to

$$\begin{aligned} 3x_1 + 2x_2 \qquad\quad &\leq 60 \\ -x_1 + x_2 + 4x_3 &\leq 10 \\ 2x_1 - 2x_2 + 5x_3 &\leq 50 \\ x_1, x_2, x_3 &\geq 0 \end{aligned}$$

Introducing three slack variables and putting the problem into standard form gives the following.

$$\text{Minimize} \quad -2x_1 - 3x_2 - 3x_3$$

subject to

$$\begin{aligned} 3x_1 + 2x_2 \qquad\quad + x_4 \qquad\qquad &= 60 \\ -x_1 + x_2 + 4x_3 \qquad + x_5 \qquad &= 10 \\ 2x_1 - 2x_2 + 5x_3 \qquad\qquad + x_6 &= 50 \\ x_j \geq 0, j = 1,\ldots,6 & \end{aligned}$$

The system of constraints for this problem is in canonical form with basic variables x_4, x_5, and x_6, the associated basic solution, (0,0,0,60,10,50), is feasible, and the objective function is written in terms of the nonbasic variables. Thus the simplex method can be initiated. Table 3.4 gives the resulting tableaux.

Notice that the first pivot could have been made in either the first, second, or third column. From the last tableau we see that, for the problem as stated in standard form, the minimal value of the objective function is -70, and this value is

Table 3.4

	x_1	x_2	x_3	x_4	x_5	x_6	
x_4	3	2	0	1	0	0	60
x_5	-1	(1)	4	0	1	0	10
x_6	2	-2	5	0	0	1	50
	-2	-3	-3	0	0	0	0
x_4	(5)	0	-8	1	-2	0	40
x_2	-1	1	4	0	1	0	10
x_6	0	0	13	0	2	1	70
	-5	0	9	0	3	0	30
x_1	1	0	$-\frac{8}{5}$	$\frac{1}{5}$	$-\frac{2}{5}$	0	8
x_2	0	1	$\frac{12}{5}$	$\frac{1}{5}$	$\frac{3}{5}$	0	18
x_6	0	0	13	0	2	1	70
	0	0	1	1	1	0	70

attained at the point (8,18,0,0,0,70). Since the original problem was a maximization problem with no slack variables, the optimal value for the original objective function is 70 and is attained at the point (8,18,0).

Example 2

$$\text{Maximize} \quad 2x_2 + x_3$$

subject to

$$\begin{aligned} x_1 + x_2 - 2x_3 &\leq 7 \\ -3x_1 + x_2 + 2x_3 &\leq 3 \\ x_1, x_2, x_3 &\geq 0 \end{aligned}$$

The standard form of the problem is

$$\text{Minimize} \quad -2x_2 - x_3$$

subject to

$$\begin{aligned} x_1 + x_2 - 2x_3 + x_4 \qquad &= 7 \\ -3x_1 + x_2 + 2x_3 \qquad + x_5 &= 3 \\ x_j \geq 0, j = 1, \ldots, 5 \end{aligned}$$

This problem is in canonical form with basic variables x_4 and x_5, and the steps of the simplex algorithm are displayed in Table 3.5.

Table 3.5

	x_1	x_2	x_3	x_4	x_5	
x_4	1	1	-2	1	0	7
x_5	-3	①	2	0	1	3
	0	-2	-1	0	0	0
x_4	④	0	-4	1	-1	4
x_2	-3	1	2	0	1	3
	-6	0	3	0	2	6
x_1	1	0	-1	$\frac{1}{4}$	$-\frac{1}{4}$	1
x_2	0	1	-1	$\frac{3}{4}$	$\frac{1}{4}$	6
	0	0	-3	$\frac{3}{2}$	$\frac{1}{2}$	12

The three negative entries in the third column of the last tableau indicate that that objective function is unbounded below.

Example 3

Minimize $-4x_1 + x_2 + 30x_3 - 11x_4 - 2x_5 + 3x_6$

subject to

$$\begin{aligned} -2x_1 \quad\quad + 6x_3 + 2x_4 \quad\quad - 3x_6 + x_7 &= 20 \\ -4x_1 + x_2 + 7x_3 + x_4 \quad\quad - x_6 \quad\quad &= 10 \\ - 5x_3 + 3x_4 + x_5 - x_6 \quad\quad &= 60 \end{aligned}$$

$x_j \geq 0, j = 1, \ldots, 7$

Table 3.6

	x_1	x_2	x_3	x_4	x_5	x_6	x_7	
x_7	-2	0	6	2	0	-3	1	20
x_2	-4	1	7	(1)	0	-1	0	10
x_5	0	0	-5	3	1	-1	0	60
	0	0	13	-6	0	2	0	110
x_7	(6)	-2	-8	0	0	-1	1	0
x_4	-4	1	7	1	0	-1	0	10
x_5	12	-3	-26	0	1	2	0	30
	-24	6	55	0	0	-4	0	170
x_1	1	$-\frac{1}{3}$	$-\frac{4}{3}$	0	0	$-\frac{1}{6}$	$\frac{1}{6}$	0
x_4	0	$-\frac{1}{3}$	$\frac{5}{3}$	1	0	$-\frac{5}{3}$	$\frac{2}{3}$	10
x_5	0	1	-10	0	1	(4)	-2	30
	0	-2	23	0	0	-8	4	170
x_1	1	$-\frac{7}{24}$	$-\frac{7}{4}$	0	$\frac{1}{24}$	0	$\frac{1}{12}$	$\frac{5}{4}$
x_4	0	$\frac{1}{12}$	$-\frac{5}{2}$	1	$\frac{5}{12}$	0	$-\frac{1}{6}$	$\frac{45}{2}$
x_6	0	$\frac{1}{4}$	$-\frac{5}{2}$	0	$\frac{1}{4}$	1	$-\frac{1}{2}$	$\frac{15}{2}$
	0	0	3	0	2	0	0	230

The system of constraints, as given, is in canonical form with basic variables x_7, x_2, and x_5, and the associated basic solution, (0,10,0,0,60,0,20), is feasible. However, the expression for the objective function contains the basic variables x_2 and x_5. By subtracting the second equation and adding twice the third equation to the equation

$$-4x_1 + x_2 + 30x_3 - 11x_4 - 2x_5 + 3x_6 = z$$

we have

$$13x_3 - 6x_4 + 2x_6 = 110 + z$$

Using this expression to define the objective function, the problem is in canonical form with basic variables x_7, x_2, and x_5, and the simplex method can be initiated. The corresponding tableaux are given in Table 3.6. As can be seen, the minimal value of the objective function is -230 and is attained at the point $(\frac{5}{4},0,0,\frac{45}{2},0,\frac{15}{2},0)$. Note the presence of degeneracy in the second and third steps.

PROBLEM SET 3.5

1. Solve the following.

(a) Maximize $10x_1 + 6x_2 - 8x_3$

subject to

$$\begin{aligned} 5x_1 - 2x_2 + 6x_3 &\le 20 \\ 10x_1 + 4x_2 - 6x_3 &\le 30 \\ x_1,x_2,x_3 &\ge 0 \end{aligned}$$

(b) Maximize $9x_2 + 2x_3 - x_5$

subject to

$$\begin{aligned} x_1 - 3x_2 \qquad - 4x_4 \qquad + 2x_6 &= 60 \\ 2x_2 \qquad - x_4 - x_5 + 4x_6 &= -20 \\ x_2 + x_3 \qquad\qquad + 3x_6 &= 10 \\ x_j \ge 0,\ 1 \le j \le 6 & \end{aligned}$$

(c) Minimize $-5x_1 - 14x_2 + 7x_3 + 9x_4$

subject to

$$\begin{aligned} x_1 - 2x_2 + x_3 + x_4 &= 5 \\ -x_1 - 3x_2 \qquad + 5x_4 &\le 10 \\ 2x_1 + 2x_2 \qquad - 4x_4 &\le 10 \\ x_1,x_2,x_3,x_4 &\ge 0 \end{aligned}$$

(d) Minimize $x_3 - x_4$

subject to

$$\begin{aligned} x_1 \qquad\qquad - 3x_4 + x_5 &= 1 \\ x_2 \qquad + 6x_4 - 5x_5 &= 6 \\ x_3 - 3x_4 + 2x_5 &= 5 \end{aligned}$$

$x_j \geq 0,\ 1 \leq j \leq 5$

2. Solve. Maximize $x_4 - x_5$

subject to

$$\begin{aligned} x_1 \qquad\qquad + x_4 - 2x_5 &= 1 \\ x_2 \qquad + x_4 \qquad &= 6 \\ x_3 + 2x_4 - 3x_5 &= 4 \end{aligned}$$

$x_j \geq 0,\ j = 1,\ldots,5$

Note that in this example a variable removed from the basis in one step of the pivot operation eventually returns to the basis. Compare with Problem 1 of Section 3.4.

3. Solve. Maximize $10x_3 + 3x_4$

subject to

$$\begin{aligned} x_1 \qquad + 10x_3 + 2x_4 &= 20 \\ x_2 - \quad x_3 + \quad x_4 &= 12 \end{aligned}$$

$x_1,x_2,x_3,x_4 \geq 0$

(If in your first iteration you put x_3 into the basis, you will have an example of a variable inserted into the basis in one step of the simplex algorithm being removed from the basis in the very next step.)

4. Consider the problem of Example 3. The minimum value of the objective function is -230 and is attained at $(\frac{5}{4},0,0,\frac{45}{2},0,\frac{15}{2},0)$. However, this optimal value is attained at other solution points to the system of constraints.

(a) The last tableau for the solution to this problem suggests that optimal basic feasible solutions exist with either x_2 or x_7 in the basis. Why?

(b) Use this last tableau to determine an optimal basic feasible solution with x_7 in the basis.

(c) Find an optimal solution with x_2 in the basis.

5. Consider the problem of Example 2 of this section.

(a) Show that any point of the form $(t,0,t)$, for $t \geq 0$, is a feasible solution.

(b) Show using this that the objective function is unbounded.

6. Compute the solution to Problem 6 of Section 2.3.

7. Compute the solution to Problem 5 of Section 2.6.

SECTION 3.6 ARTIFICIAL VARIABLES

As we have seen, many linear programming problems can be put into canonical form with little or no effort. For example, the addition of slack variables with positive coefficients can provide the basic variables necessary for the initial basic feasible solution. On the other hand, the system of constraints for many other problems contain no obvious basic feasible solutions. Problems of this type occur, for example, in production models involving output requirements and therefore ($\geq$) inequalities in the constraint set, such as Example 2 of Section 2.3, or in transportation problems involving fixed demands and therefore equalities in the constraint set, such as the example of Section 2.4. In fact, in any application of linear programming to a real-world problem, it would be rare to find the original formulation of the problem in canonical form.

What must be developed is a technique for determining an initial basic feasible solution for an arbitrary system of equations. This technique must also be capable of handling a problem that has no feasible solution. Such a problem could arise, for example, in a model containing a simple error in formulation or in a complicated production model where it is not obvious that the various output requirements can be met with the limited resources available. In this section we will introduce such a technique; in the next section we will discuss some of the complications that can occur.

The basic idea behind the method used to find an initial basic feasible solution is simple. We introduce into the problem a sufficient number of variables, called *artificial variables*, to put the system of constraints into canonical form with these variables as the basic variables. Then we apply the simplex method, not to the objective function of the original problem, but to a new function defined in such a way that its minimal value is attained at a feasible solution to the original problem. Thus the method of the last three sections applied to this new function drives the original problem to a basic feasible solution.

Consider the standard linear programming problem of (1) of Section 3.1. Find a nonnegative solution to the system

$$\begin{aligned} a_{11}x_1 + a_{12}x_2 + \cdots + a_{1n}x_n &= b_1 \\ a_{21}x_1 + a_{22}x_2 + \cdots + a_{2n}x_n &= b_2 \\ &\vdots \\ a_{m1}x_1 + a_{m2}x_2 + \cdots + a_{mn}x_n &= b_m \end{aligned} \tag{1}$$

that minimizes the function $z = c_1x_1 + c_2x_2 + \cdots + c_nx_n$. By multiplication of an equation by -1 if necessary, we may assume that all the constant terms b_i, $i = 1, \ldots, m$, are nonnegative. Now introduce into the system of constraints m

new variables, $x_{n+1}, x_{n+2}, \ldots, x_{n+m}$, called artificial variables, one to each equation. The resulting system is

$$\begin{aligned}
a_{11}x_1 + a_{12}x_2 + \cdots + a_{1n}x_n + x_{n+1} \qquad\qquad\qquad &= b_1 \\
a_{21}x_1 + a_{22}x_2 + \cdots + a_{2n}x_n \qquad + x_{n+2} \qquad\qquad &= b_2 \\
\vdots \qquad\qquad\qquad\qquad\qquad\qquad & \\
a_{m1}x_1 + a_{m2}x_2 + \cdots + a_{mn}x_n \qquad\qquad\qquad + x_{m+n} &= b_m
\end{aligned} \tag{2}$$

Notice that this system is in canonical form with basic variables $x_{n+1}, \ldots, x_{n+m}$, and that the associated basic solution is feasible, since we have assumed that the b_i's are nonnegative.

Now consider the problem of determining the minimal value of the function $w = x_{n+1} + x_{n+2} + \cdots + x_{n+m}$ on the set of all nonnegative solutions to the system of equations in (2). Since all variables are nonnegative, w can never be negative. The function w would assume the value zero at any feasible solution to (2) in which all the artificial variables were at zero level. Thus, the simplex method applied to this function should replace the artificial variables as basic variables with the variables from the original problem and will hopefully drive the system in (2) into canonical form with basic variables from the original set x_j, $j = 1, \ldots, n$. The value of w at the associated basic feasible solution would be 0, its minimal value, and the simplex method could then be initiated on the original problem as stated in (1). Furthermore, if the system of constraints in (1) does have at least one feasible solution, the system in (2) must have feasible solutions in which all the artificial variables equal zero. In this case the minimal value of w would be, in fact, zero. Thus, when applying the simplex method to the function w, if we reach a step at which we can pivot no more but the associated value of w is greater than zero, we can conclude that the original problem has no feasible solutions.

Before doing some examples, some remarks of a technical nature are in order. First, before the simplex method can be applied to the function $w = x_{n+1} + \cdots + x_{n+m}$, the problem must be in canonical form. The system of constraints in (2) is in canonical form with the artificial variables as basic variables and the associated basic solution is feasible, but the function w is not expressed in terms of only the nonbasic variables. To rectify this, each equation in the system of constraints must be subtracted from the equation

$$x_{n+1} + x_{n+2} + \cdots + x_{n+m} = w$$

Second, if the pivot operations dictated by the problem of minimizing w are also simultaneously performed on the equation $c_1x_1 + c_2x_2 + \cdots + c_nx_n = z$, defining the original objective function, this function will be expressed in terms of nonbasic variables at each step. Thus, if an initial basic feasible solution is found for the original problem, the simplex method can be initiated immediately on z. Therefore we incorporate this z equation into the notation and operations of the problem of minimizing w.

In sum, the first step in solving the general problem of (1) is to consider the problem of minimizing w with

$$\begin{aligned}
a_{11}x_1 + a_{12}x_2 + \cdots + a_{1n}x_n + x_{n+1} \qquad\qquad\qquad\qquad &= b_1 \\
a_{21}x_1 + a_{22}x_2 + \cdots + a_{2n}x_n \qquad\quad + x_{n+2} \qquad\qquad &= b_2 \\
\vdots \qquad\qquad\qquad\qquad\qquad\qquad\qquad\qquad & \\
a_{m1}x_1 + a_{m2}x_2 + \cdots + a_{mn}x_n \qquad\qquad\qquad + x_{n+m} &= b_m \\
c_1x_1 + c_2x_2 + \cdots + c_nx_n \qquad\qquad\qquad\qquad\qquad &= z \\
d_1x_1 + d_2x_2 + \cdots + d_nx_n \qquad\qquad\qquad\qquad\qquad &= w_0 + w \\
x_j \ge 0, j = 1, \ldots, n + m \qquad\qquad\qquad\qquad\qquad &
\end{aligned} \tag{3}$$

where $d_j = -(a_{1j} + a_{2j} + \cdots + a_{mj})$ and $w_0 = -(b_1 + b_2 + \cdots + b_m)$.

Example 1

$$\text{Minimize} \quad 2x_1 - 3x_2 + x_3 + x_4$$

subject to

$$\begin{aligned}
x_1 - 2x_2 - 3x_3 - 2x_4 &= 3 \\
x_1 - x_2 + 2x_3 + x_4 &= 11 \\
x_1, x_2, x_3, x_4 &\ge 0
\end{aligned} \tag{4}$$

Introducing artificial variables x_5 and x_6, we consider the problem of minimizing w where

$$\begin{aligned}
x_1 - 2x_2 - 3x_3 - 2x_4 + x_5 \qquad &= 3 \\
x_1 - x_2 + 2x_3 + x_4 + \quad + x_6 &= 11 \\
2x_1 - 3x_2 + x_3 + x_4 \qquad\qquad &= z \\
x_5 + x_6 &= w \\
x_j \ge 0, j = 1, \ldots, 6 \qquad\qquad &
\end{aligned} \tag{5}$$

Subtracting the first two equations from the w equation gives the system

$$\begin{aligned}
x_1 - 2x_2 - 3x_3 - 2x_4 + x_5 \qquad &= 3 \\
x_1 - x_2 + 2x_3 + x_4 \qquad + x_6 &= 11 \\
2x_1 - 3x_2 + x_3 + x_4 \qquad\qquad &= z \\
-2x_1 + 3x_2 + x_3 + x_4 \qquad\qquad &= -14 + w
\end{aligned}$$

This information can be recorded in tableau form by simply augmenting the notation of the previous section. See Table 3.7. The last row corresponds to the w equation, with the w suppressed from the notation. Now the simplex method is initiated, with the entires in the last row determining the pivoting column at each

Table 3.7

	x_1	x_2	x_3	x_4	x_5	x_6	
x_5	1	−2	−3	−2	1	0	3
x_6	1	−1	2	1	0	1	11
	2	−3	1	1	0	0	0
	−2	3	1	1	0	0	−14

step. The second last row, the z row, is operated on at each pivot operation, but is otherwise ignored for the time being. Table 3.8 gives the resulting tableaux.

Thus the minimal value of w is 0, and one point at which this value is attained is $(\frac{39}{5},0,\frac{8}{5},0,0,0)$. Since this point is a solution to the system of constraints in (5) and has as its last two coordinates zero, $(\frac{39}{5},0,\frac{8}{5},0)$ is a basic feasible solution to the system in (4), and the data for the tableau corresponding to the original problem expressed in canonical form with basic variables x_1 and x_3 are contained in the last

Table 3.8

	x_1	x_2	x_3	x_4	x_5	x_6	
x_5	(1)	−2	−3	−2	1	0	3
x_6	1	−1	2	1	0	1	11
	2	−3	1	1	0	0	0
	−2	3	1	1	0	0	−14
x_1	1	−2	−3	−2	1	0	3
x_6	0	1	(5)	3	−1	1	8
	0	1	7	5	−2	0	−6
	0	−1	−5	−3	2	0	−8
x_1	1	$-\frac{7}{5}$	0	$-\frac{1}{5}$	$\frac{2}{5}$	$\frac{3}{5}$	$\frac{39}{5}$
x_3	0	$\frac{1}{5}$	1	$\frac{3}{5}$	$-\frac{1}{5}$	$\frac{1}{5}$	$\frac{8}{5}$
	0	$-\frac{2}{5}$	0	$\frac{4}{5}$	$-\frac{3}{5}$	$-\frac{7}{5}$	$-\frac{86}{5}$
	0	0	0	0	1	1	0

tableau. In fact, translating these data back into equation form gives the following system, equivalent to (4).

$$\begin{aligned} x_1 - \tfrac{7}{5}x_2 \qquad\quad - \tfrac{1}{5}x_4 &= \tfrac{39}{5} \\ \tfrac{1}{5}x_2 + x_3 + \tfrac{3}{5}x_4 &= \tfrac{8}{5} \\ -\tfrac{2}{5}x_2 \qquad\quad + \tfrac{4}{5}x_4 &= -\tfrac{86}{5} + z \end{aligned}$$

The second stage of the problem, the application of the simplex process to the problem of minimizing z, can be initiated immediately. See Table 3.9.

Table 3.9

	x_1	x_2	x_3	x_4	
x_1	1	$-\frac{7}{5}$	0	$-\frac{1}{5}$	$\frac{39}{5}$
x_3	0	$\textcircled{\frac{1}{5}}$	1	$\frac{3}{5}$	$\frac{8}{5}$
	0	$-\frac{2}{5}$	0	$\frac{4}{5}$	$-\frac{86}{5}$
x_1	1	0	7	4	19
x_2	0	1	5	3	8
	0	0	2	2	−14

The minimal value of z is 14 and is attained at the point (19,8,0,0).

The above computational procedure can be streamlined somewhat. First, there is no need to make a formal break in the tableau notation when passing from the first stage of a linear programming problem, the minimization of the w function, to the second stage, the minimization of the z function. Once a basic feasible solution to the original problem has been found, the w row of the augmented tableau notation can be dropped and the problem continued directly using the z row.

Second, once an artificial variable is extracted from the basis, there is no need to reenter it in any future step. To prove this, consider the original problem augmented with only a subset of the full complement of artificial variables. This is the situation that one would have, for example if, after several steps of the pivot operation applied to the w function, those artificial variables not in the basis were set equal to zero (and thus disregarded). If the original problem has a feasible solution, the minimal value of the modified w function defined as the sum of the remaining artificial variables would still be zero, and the simplex method applied to this modified problem would drive the modified w function to zero using only the variables at hand. Thus, once taken from the basis, artificial variables need never return to the basis. As a result, it is never necessary to use the information in the artificial variables columns of the tableau, so these data need not be calculated at each pivot step.

Example 2

$$\text{Minimize } x_1 + x_2 + x_3 = z$$

subject to

$$\begin{aligned} -x_1 + 2x_2 + x_3 &\leq 1 \\ -x_1 + 2x_3 &\geq 4 \\ x_1 - x_2 + 2x_3 &= 4 \\ x_1, x_2, x_3 &\geq 0 \end{aligned}$$

Adding two slack variables, the problem in standard form becomes:

$$\text{Minimize } x_1 + x_2 + x_3 = z$$

subject to

$$\begin{aligned} -x_1 + 2x_2 + x_3 + x_4 &= 1 \\ -x_1 + 2x_3 - x_5 &= 4 \\ x_1 - x_2 + 2x_3 &= 4 \\ x_1, x_2, x_3, x_4, x_5 &\geq 0 \end{aligned}$$

Note that the x_4 variable can serve as a basic variable. Thus it is sufficient to add only two artificial variables, say x_6 and x_7, to the problem, and at the first stage minimize the function $w = x_6 + x_7$. The problem is then:

$$\begin{aligned} -x_1 + 2x_2 + x_3 + x_4 &= 1 \\ -x_1 + 2x_3 - x_5 + x_6 &= 4 \\ x_1 - x_2 + 2x_3 + x_7 &= 4 \\ x_1 + x_2 + x_3 &= z \\ x_6 + x_7 &= w \end{aligned}$$

Subtracting the second and third equations from the w equation gives the equation $x_2 - 4x_3 + x_5 = -8 + w$. Now the expression for w does not contain the initial basic variables x_4, x_6, and x_7, and the simplex method can be initiated. The resulting tableaux are given in Table 3.10.

The minimal value for the function $w = x_6 + x_7$ is $\frac{4}{3}$, and this value is attained at the point $(\frac{2}{3},0,\frac{5}{3},0,0,\frac{4}{3},0)$. Therefore we can conclude that the original problem has no feasible solution.

Table 3.10

	x_1	x_2	x_3	x_4	x_5	x_6	x_7	
x_4	-1	2	①	1	0	0	0	1
x_6	-1	0	2	0	-1	1	0	4
x_7	1	-1	2	0	0	0	1	4
	1	1	1	0	0			0
	0	1	-4	0	1			-8
x_3	-1	2	1	1	0			1
x_6	1	-4	0	-2	-1			2
x_7	③	-5	0	-2	0			2
	2	-1	0	-1	0			-1
	-4	9	0	4	1			-4
x_3	0	$\frac{1}{3}$	1	$\frac{1}{3}$	0			$\frac{5}{3}$
x_6	0	$-\frac{7}{3}$	0	$-\frac{4}{3}$	-1			$\frac{4}{3}$
x_1	1	$-\frac{5}{3}$	0	$-\frac{2}{3}$	0			$\frac{2}{3}$
	0	$\frac{7}{3}$	0	$\frac{1}{3}$	0			$-\frac{7}{3}$
	0	$\frac{7}{3}$	0	$\frac{4}{3}$	1			$-\frac{4}{3}$

PROBLEM SET 3.6

1. Using the technique described in this section, find solutions with nonnegative coordinates to the following systems of equations.

(a)
$$\begin{aligned} x_1 - x_2 \quad\quad &= 1 \\ 2x_1 + x_2 - x_3 &= 3 \end{aligned}$$

(b)
$$\begin{aligned} x_1 + x_2 \quad\quad &= 1 \\ 2x_1 + x_2 - x_3 &= 3 \end{aligned}$$

2. Solve the following.

(a) Minimize $2x_1 + 2x_2 - 5x_3$

subject to

$$3x_1 + 2x_2 - 4x_3 = 7$$
$$x_1 - x_2 + 3x_3 = 2$$
$$x_1, x_2, x_3 \geq 0$$

(b) Minimize $x_1 - 3x_3$

subject to

$$x_1 + 2x_2 - x_3 \leq 6$$
$$x_1 - x_2 + 3x_3 = 3$$
$$x_1, x_2, x_3 \geq 0$$

(c) Maximize $3x_1 - x_2$

subject to

$$x_1 - x_2 \leq 3$$
$$2x_1 \leq x_2$$

subtract x_2 from both sides

$$x_1 + x_2 \geq 12$$
$$x_1, x_2 \geq 0$$

(d) Maximize $x_1 + 2x_2 + 3x_3 + 4x_4$

subject to

$$x_1 + x_3 - 4x_4 = 2$$
$$x_2 - x_3 + 3x_4 = 9$$
$$x_1 + x_2 - 2x_3 - 3x_4 = 21$$
$$x_1, x_2, x_3, x_4 \geq 0$$

3. Consider the tableaux for the first stage of the problem discussed in Example 1 of this section. The very last row, the w row after the third pivot step, contains 4 zeros, 2 ones, and 1 zero. This row corresponds to what function? Why was this result expected?

4. Suppose that the objective function z for a linear programming problem is unbounded. Show that this can be learned from the simplex method only after the first stage of the method is completed.

5. Show that at each step of the first stage of the simplex method, the coefficients d_j^* of the w function are equal to $-\sum a_{ij}^*$, where the sum is over those rows i that isolate the remaining artificial variables.

6. Prove that Theorem 2 of Section 3.4 can never apply to the w function (i.e., if a $d_s^* < 0$, there must exist an $a_{is}^* > 0$).

7. Compute the solution to Problem 7 of Section 2.3.

8. In the first tableau for the problem of Example 1 of this section, the second column contains three negative entries, the a_{12} and a_{22} entries and the c_2 entry. Evidently the objective function of the problem is bounded below and Theorem 2 of Section 3.4 does not apply. Why not?

SECTION 3.7 REDUNDANT SYSTEMS

In the last section it was seen that by introducing artificial variables, a linear programming problem could be put into canonical form by means of the simplex method applied to the function w, defined to be the sum of the artificial variables. If the original problem has no feasible solutions, this method would also make evident that fact. In this section we will discuss one minor complication, the problem of redundancy, that could occur with the original system of constraints.

It could very well be that an equation or some equations in the original system of constraints are linear combinations of the remaining equations in the system. This often occurs when, for ease of formulation, more than the minimal number of necessary variables are introduced into a problem. For example, the five constraints of the transportation problem formulated in Section 2.4 contain one redundant equation. Now the simplex method described in Sections 3.3–3.5 could begin only with the original problem in canonical form. Clearly, however, if a system of equations is in canonical form, there can be no redundant equations because the isolated nature of the basic variables. It would seem at first glance that it would be necessary to ferret out redundant equations from the system of constraints before the machinery of the simplex method could be applied to a linear programming problem. Fortunately, this is not the case. In this section we will show that from the first stage of the simplex method using artificial variables, redundancies in the original system of constraints can be discovered and deficiencies caused by the lack of a complete set of basic variables from the original set of variables can be compensated for by the presence of artificial variables.

Suppose now we apply the two stage simplex method to an arbitrary linear programming problem. We have already shown that if the simplex process applied to the w function stops without driving w to zero (e.g., we reach a step in which Theorem 1 of Section 3.4 applies to the w function but $w_0 > 0$), the original problem has no feasible solution. Thus suppose that after several pivot operations, the minimal value of w is determined to be zero. If at this time no artificial variables remain in the basis, the original system of constraints must be in canonical form and so contains no redundant equations. Stage two of the simplex method can now begin.

Consider the remaining case; the value of w is driven to zero, but artificial variables remain as basic variables. Note that since w is the sum of the artificial variables, and that we have reached a point in the simplex process in which the value of w at the associated basic feasible solution is zero, those remaining artificial variables must be at zero level; that is, the constant terms b_i^* in those constraining equations containing the artificial variables must be zero. We now attempt to replace these artificial variables in the basis with variables from the original set.

Suppose the ith equation in the set of constraints defined by this tableau we have reached contains one of the remaining artificial variables. Consider the coefficients, say a_{ij}^*, $1 \leq j \leq n$, in this ith row. If any $a_{ij}^* \neq 0$, pivot at this term, replacing the artificial variable associated with that row in the basis with a variable from the original set. Since the artificial variable was at zero level, this pivot operation leaves the constant term column, the right-hand column, unchanged (see Problem 4 of Section 3.4). Continue this process wherever possible. If, however, a point is reached in which the ith row contains a remaining artificial variable but $a_{ij}^* = 0$ for all j, $1 \leq j \leq n$, we can conclude that because of redundancies, it is impossible to find a complete set of m basic variables from the original set. In fact, the number of redundant equations would equal the number of artificial variables remaining with coefficient row zero. However, stage two of the simplex method can still be initiated on the tableau at hand, and the rows of zeros corresponding to the remaining artificial variables ignored. In essence, this tableau, ignoring the zero rows, corresponds to an independent system of equations in canonical form and equivalent to the original system of constraints.

Example 1

$$\text{Minimize} \quad z = 2x_1 - x_2 + x_3$$

subject to

$$\begin{aligned} x_1 - 2x_2 + 3x_3 + x_4 &= 6 \\ -x_1 + x_2 + 2x_3 + \tfrac{2}{3}x_4 &= 4 \\ x_1, x_2, x_3, x_4 &\geq 0 \end{aligned}$$

Adding two artificial variables x_5 and x_6 for initial basic variables, and expressing $w = x_5 + x_6$ in terms of the original variables, we have the system

$$\begin{aligned} x_1 - 2x_2 + 3x_3 + x_4 + x_5 \qquad &= 6 \\ -x_1 + x_2 + 2x_3 + \tfrac{2}{3}x_4 \qquad + x_6 &= 4 \\ 2x_1 - x_2 + x_3 \qquad &= z \\ x_2 - 5x_3 - \tfrac{5}{3}x_4 \qquad &= -10 + w \end{aligned}$$

From the tableaux of Table 3.11, we see that the minimal value of the objective function z is 0, and is attained at the point (0,0,0,6). Notice that the first pivot term could have been either the 3 or 2 of the x_3 column (or either term in the x_4 column, for that matter). The purpose of the second pivot step is to eliminate the artificial variable x_6 from the basis, and this pivot could have been made at either nonzero entry in the x_6 row of the second tableau. Since both artificial variables were

Table 3.11

	x_1	x_2	x_3	x_4	x_5	x_6	
x_5	1	-2	$\textcircled{3}$	1	1	0	6
x_6	-1	1	2	$\frac{2}{3}$	0	1	4
	2	-1	1	0			0
	0	1	-5	$-\frac{5}{3}$			-10
x_3	$\frac{1}{3}$	$-\frac{2}{3}$	1	$\frac{1}{3}$			2
x_6	$-\frac{5}{3}$	$\textcircled{\frac{7}{3}}$	0	0			0
	$\frac{5}{3}$	$-\frac{1}{3}$	0	$-\frac{1}{3}$			-2
	$\frac{5}{3}$	$-\frac{7}{3}$	0	0			0
x_3	$-\frac{1}{7}$	0	1	$\textcircled{\frac{1}{3}}$			2
x_2	$-\frac{5}{7}$	1	0	0			0
	$\frac{10}{7}$	0	0	$-\frac{1}{3}$			-2
	0	0	0	0			0
x_4	$-\frac{3}{7}$	0	3	1			6
x_2	$-\frac{5}{7}$	1	0	0			0
	$\frac{9}{7}$	0	1	0			0

extracted from the basis, the original system of constraints contained no redundancies.

Example 2

Minimize $z = x_1 + 4x_2 + 3x_3 + 2x_4$

subject to

$$
\begin{aligned}
x_1 + 2x_2 \qquad\quad + \ x_4 &= 20 \\
2x_1 + \ x_2 + \ x_3 \qquad\quad &= 10 \\
-x_1 + 4x_2 - 2x_3 + 3x_4 &= 40 \\
x_1, x_2, x_3, x_4 &\geq 0
\end{aligned}
$$

Adding artificial variables x_5, x_6, and x_7 and expressing $w = x_5 + x_6 + x_7$ in terms of x_1, x_2, x_3, and x_4, the system becomes

$$\begin{aligned}
x_1 + 2x_2 \qquad + x_4 + x_5 \qquad\qquad &= 20\\
2x_1 + x_2 + x_3 \qquad\qquad + x_6 \qquad &= 10\\
-x_1 + 4x_2 - 2x_3 + 3x_4 \qquad\qquad + x_7 &= 40\\
x_1 + 4x_2 + 3x_3 + 2x_4 \qquad\qquad &= z\\
-2x_1 - 7x_2 + x_3 - 4x_4 \qquad\qquad &= -70 + w
\end{aligned}$$

Table 3.12

	x_1	x_2	x_3	x_4	x_5	x_6	x_7	
x_5	1	2	0	1	1	0	0	20
x_6	2	(1)	1	0	0	1	0	10
x_7	−1	4	−2	3	0	0	1	40
	1	4	3	2				0
	−2	−7	1	−4				−70
x_5	−3	0	−2	(1)				0
x_2	2	1	1	0				10
x_7	−9	0	−6	3				0
	−7	0	−1	2				−40
	12	0	8	−4				0
x_4	−3	0	−2	1				0
x_2	(2)	1	1	0				10
x_7	0	0	0	0				0
	−1	0	3	0				−40
	0	0	0	0				0
x_4	0	$\frac{3}{2}$	$-\frac{1}{2}$	1				15
x_1	1	$\frac{1}{2}$	$\frac{1}{2}$	0				5
x_7	0	0	0	0				0
	0	$\frac{1}{2}$	$\frac{7}{2}$	0				−35

The steps of the simplex method are displayed in the tableaux of Table 3.12. The minimal value of the objective function z is 35, and this value is attained at the point (5,0,0,15). Since an artificial variable cannot be removed from the basis, the original system of constraints contains one redundant equation. Notice that any non-zero term in the x_5 or x_7 row of the second tableau could have been used as the pivot term of the second step.

PROBLEM SET 3.7

1. Show that the third equation in the set of constraints for the problem of Example 2 is a linear combination of the other two equations.

2. Show that the system of constraints for the transportation problem formulated in Section 2.4 is redundant by:
   (a) Exhibiting a relationship between the equations.
   (b) Solving the problem using the simplex method.

3. Solve the following.

   (a) Minimize $x_1 + 2x_2 - x_4$

   subject to

$$\begin{aligned} x_1 - x_2 + 3x_3 \qquad\quad &= 1 \\ 2x_2 - 4x_3 + 2x_4 &= 2 \\ 3x_1 + x_2 + x_3 + 4x_4 &= 7 \\ x_1,x_2,x_3,x_4 &\geq 0 \end{aligned}$$

   (b) Maximize $3x_1 - 2x_2 - 2x_3 + 2x_4$

   subject to

$$\begin{aligned} -x_1 + 3x_2 - x_3 + 2x_4 &= 1 \\ -2x_1 \qquad\quad + 4x_3 - x_4 &= 8 \\ 2x_1 - 2x_2 + 2x_3 + x_4 &= 2 \\ x_1,x_2,x_3,x_4 &\geq 0 \end{aligned}$$

4. True or false: Suppose that the system of constraints for a linear programming problem in standard form is not redundant and has no degenerate solutions. Then at that step in the first stage of the simplex method when the w function first attains value zero, no artificial variables can remain in the basis?

5. (Requires linear algebra.) Show that if, when applying the simplex method to a linear programming problem with m constraining equations and n unknowns, it is impossible to drive r artificial variables from the basis, then the rank of the coefficient matrix of the original system of constraints is $m - r$.

SECTION 3.8 A CONVERGENCE PROOF

Consider a linear programming problem presented in canonical form, as in (1) of Section 3.4. We proved in that section that, under the assumption that no basic feasible solution to the problem was degenerate, the simplex process must terminate after a finite number of steps, with the process driving to either the minimal value of the objective function (Theorem 1 of Section 3.4 applying) or a set of feasible solutions on which the objective function is unbounded below (Theorem 2 of Section 3.4 applying). However, the proof breaks down if degeneracy is present because there could exist a sequence of pivot steps for which the associated value of the objective function remains fixed and the basic feasible solutions repeat, that is, the process cycles.

However, as already shown in our examples, at any step of the simplex process there may be more than one term qualified to serve as pivot term. For example, pivoting can occur in any column with a negative c_j (and a positive a_{ij}), and if c_s is negative and the minimum of $\{b_i/a_{is} \mid a_{is} > 0\}$ is not attained in a unique row, pivoting can occur in the s column at any row attaining this minimum. We will show in this section that for any problem, degenerate or not, it is always possible to select a finite sequence of pivot steps that leads to either the minimal value of the objective function or to a set of feasible solutions on which the objective function is unbounded. Thus we will prove that although cycling is possible, by a proper choice of pivot terms it can be avoided and a step reached in the simplex process where either Theorem 1 or 2 of Section 3.4 applies. The proof that we give is by induction on m, the number of equations in the system of constraints, and is due to Dantzig ([5], or see [6]).

In the following we will continue our present use of the notation. Thus the constants a_{ij}, b_i, and c_j, $1 \le i \le m$, $1 \le j \le n$, refer to the coefficients of the constraining equations, the constant terms for these equations, and the coefficients for the objective function, as in (1) of Section 3.4. After the system has been modified by the application of the simplex method, the corresponding constants will be denoted by a_{ij}^*, b_i^*, c_j^*.

Theorem *For any linear programming problem presented in canonical form, there exists a finite sequence of pivot steps that leads to either of the following.*

(a) *The minimal value of the objective function with Theorem 1 of Section 3.4 applying.*

(b) *A set of feasible solutions on which the objective function is unbounded below, with Theorem 2 of Section 3.4 applying.*

That is, the sequence leads to a presentation of the problem in canonical form with either of the following.

(a) *All the coefficients c_j^* of the nonbasic variables in the expression for the objective function ≥ 0.*

(b) *A column index s such that $c_s^* < 0$ and $a_{is}^* \leq 0$ for all i.*

Before proving the theorem, we state two lemmas. The proofs are left to the reader.

Lemma 1 *If all the constant terms b_i, $1 \leq i \leq m$, of the system of constraints for a linear programming problem equal zero, then after a pivot operation, all the $b_i^* = 0$. If initially at least one $b_i \neq 0$, then after a pivot operation at least one $b_i^* \neq 0$.*

Lemma 2 *Given a linear programming problem with at least one non-zero term b_i, if there exists a sequence of pivot steps leading to the completion of the problem [reaching either (a) or (b) of the above theorem], this same sequence of pivot steps leads to the completion of that problem derived from the given problem by replacing all the non-zero b_i's with zero.*

Proof of Theorem By induction on m, the number of equations in the system of constraints. For the case $m = 1$, we have a problem with only one constraining equation. If the constant term b_1 is initially non-zero, Lemma 1 guarantees that this constant term will remain non-zero after a pivot step. Thus the former argument valid in the nondegenerate case applies here and shows that the simplex process must terminate after a finite number of steps. Now if $b_1 = 0$, we can replace it with any positive constant and then apply Lemma 2.

Now we prove the induction step. Thus assume that the theorem is valid for any problem with a system of constraints containing $m - 1$ or fewer equations. With this assumption the theorem will be proven for a system with m constraining equations.

Consider now any linear programming problem with m constraints, and assume that at least one $b_i \neq 0$. We apply the simplex process until, due to degeneracy, it is not possible to find a pivot operation that reduces the value of the objective function at that step. Rearrange the constraints so that the constant terms for exactly the first r equations are zero, for example, $b_i = 0$, $1 \leq i \leq r$, and $b_i > 0$, $r + 1 \leq i \leq m$. Notice that from Lemma 1, $r < m$. Let us call this canonical form of the problem Form I.

Consider that linear programming problem derived from the problem of Form I by deleting the last $m - r$ equations in the system of constraints. Notice that the last $m - r$ basic variables from Form I do not appear anywhere in this problem, and the problem is in canonical form with basic variables the first r basic variables of Form I. To this problem we can apply the induction assumption.

Using it, we find a sequence of pivot steps that leads to a canonical form of the problem with either of the following.

(a) All $c_j^* \geq 0, 1 \leq j \leq n.$

(b) A $c_s^* < 0$ and all $a_{is}^* \leq 0, 1 \leq i \leq r.$

Now apply these same pivot steps to the full problem in Form I. The resulting problem will be in canonical form as, first, the last $m - r$ basic variables from Form I combine with the r basic variables from the first r equations to give m distinct basic variables and, second, since each pivot term is in a row with a zero constant term, the constant term column remains unchanged and the associated basic solution feasible. Notice also that the effect of these pivot operations on the c_j^* row is completely independent of the addition of the $m - r$ constraints. There are now three possibilities.

1. If the sequence of pivot steps on the r equations reached condition (a) above, then all $c_j^* \geq 0$ implies that the minimal value of the objective function on the full constraint set has been attained.
2. If condition (b) was reached and $a_{is}^* \leq 0$ also for $r + 1 \leq i \leq m$, the objective function is unbounded below on the full system of constraints.
3. If condition (b) was reached but $a_{is}^* > 0$ for some i, $r + 1 \leq i \leq m$, then a new pivot term can be found in the s column at that row where

$$\operatorname{Min}\left\{\frac{b_i^*}{a_{is}^*} \,\middle|\, a_{is}^* > 0\right\}$$

is attained. Since $b_i^* > 0$, pivoting here reduces the value of the objective function at the associated basic feasible solution.

In sum we have shown that, for any problem with at least one $b_i \neq 0$, at a given step of the simplex process, either:

1. The minimum value of the objective function or a set on which the objective function is unbounded is attained.
2. A single pivot term or a sequence of pivot steps can be found that leads to a reduced value for the objective function.

Since there are only a finite number of basic feasible solutions and the reduced value for the objective function guarantees that they cannot repeat, the simplex process must eventually terminate.

Finally, to complete the proof, we must prove the theorem for a problem with m constraints and all constant terms equal to zero. But in this case we can simply replace a constant term with any positive constant and use what we have already proven and Lemma 2. # # #

Corollary

Given a linear programming problem with a system of constraints that has feasible solutions and an objective function to be minimized that is bounded below, then there exists at least one feasible solution (in fact, a basic feasible solution) at which the objective function attains its minimal value.

Thus we have shown that a bounded objective function of a linear programming problem with feasible solutions must reach its optimal value, a property shared by continuous functions on closed and bounded sets (compact sets) and in contrast to the problem of optimizing $f(x) = x$ on the set $0 < x < 1$. Moreover, we know that this optimal value can be attained at a point with, at most, m non-zero coordinates, where m is the number of equations in the constraints.

One final note in passing. The proof of convergence of the simplex algorithm given in this section is an existence proof, but not a constructive proof. This is because the induction proof only demonstrates the existence of a sequence of pivot steps leading to the termination of the simplex algorithm applied to any linear programming problem, and does not prescribe a constructive method to use to actually determine this sequence. However, constructive proofs certainly do exist. In fact, a constructive procedure that is related to the inductive proof of this section and involves the modification (perturbation) of the constant terms of the problem has been developed by Philip Wolfe [34]. Other constructive techniques to prevent cycling have also been developed. For a discussion of some of these, see, for example, the texts of Dantzig [6] or Hadley [15]. Or, see the recent article by R. Bland [2], in which is presented a simple rule, in terms of the indices of the involved variables, for selecting the pivot term in the simplex algorithm so as to avoid cycling.

PROBLEM SET 3.8

1. Prove Lemma 1.

2. Prove Lemma 2. *Hint*. Consider the effect or noneffect of these pivot operations on the b_i column and the c_j row.

3. Prove the corollary. Note that the theorem of this section applies only to a problem presented in canonical form.

4. True or false: Suppose the simplex method is applied to a linear programming problem presented in canonical form, and such that at each step, there is at most one term that could serve as pivot term. Then for this problem, cycling is impossible?

5. True or false: Given a linear programming problem with $n = m + 1$ and presented in canonical form, at most one step in the simplex method is necessary to drive the process to termination?

SECTION 3.9 LINEAR PROGRAMMING AND CONVEXITY

In Section 2.2 we consider a simplex linear programming problem involving only two variables. We were able to graph the set of feasible solutions to the set of constraints (Figure 2.3 of Section 2.2) and, by a geometric argument, show that the optimal value of the linear objective function must be attained at a corner or vertex to this solution set. This result generalizes, as suggested in Section 3.2. In this section we will first define the concept of convexity and show that the solution set to a system of equations and inequalities is convex. Then we will define the concept of a vertex of a convex set and relate the basic feasible solutions of a system of constraints to the vertices of the solution set to this system. The corollary of the last section will then give directly the generalization of the above result.

Only the concept of convexity will be used later in the book, and that not until Chapter 9. We present these ideas here primarily to initiate an appreciation of some of the geometry underlying the linear programming problem.

For two points P and Q in R^n, the *line segment* between P and Q is that set of points in R^n of the form $tP + (1 - t)Q$ for $0 \leq t \leq 1$ (see Problem 1). A subset S of R^n is said to be *convex* if, for any two points of S, the line segment between these two points is also in S.

Example 1

The subsets of R^2 depicted in Figure 3.4 are convex, the subsets in Figure 3.5 are not.

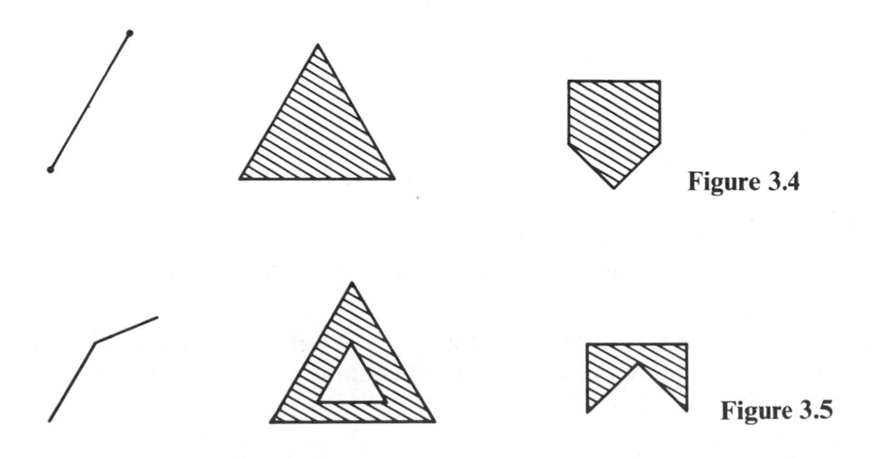

Figure 3.4

Figure 3.5

Example 2

Let $S = \{(x_1,x_2) \in R^2 | x_1 + x_2 \geq 2\}$. Then S is convex, a fact obvious from a graph of S. To prove this algebraically using only our definitions, take any two points $P = (p_1,p_2)$ and $Q = (q_1,q_2)$ in S. Then $p_1 + p_2 \geq 2$ and $q_1 + q_2 \geq 2$. Take any point

$$tP + (1 - t)Q = [tp_1 + (1 - t)q_1, tp_2 + (1 - t)q_2], \qquad 0 \leq t \leq 1$$

on the line segment between P and Q. We have

$$\begin{aligned} &tp_1 + (1 - t)q_1 + tp_2 + (1 - t)q_2 \\ &= t(p_1 + p_2) + (1 - t)(q_1 + q_2) \\ &\geq 2t + 2(1 - t) = 2 \end{aligned}$$

using the fact that t and $1 - t$ are nonnegative. Thus $tP + (1 - t)\,Q$ is in S, and we have an algebraic proof that S is convex.

The set of feasible solutions to a linear programming problem is convex, since it is the intersection of a collection of hyperplanes and half-spaces. We state these results in the following, leaving the proofs of the theorems for the reader.

Definition A subset of R^n of the form

$$X = \{(x_1,\ldots,x_n) | a_1x_1 + a_2x_2 + \cdots + a_nx_n = b\}$$

for constants $a_1,a_2,\ldots,a_n$, and b is called a *hyperplane*.

A subset of the form

$$X = \{(x_1,\ldots,x_n) | a_1x_1 + a_2x_2 + \cdots + a_nx_n \leq b\}$$

for constants $a_1,a_2,\ldots,a_n$, and b is called a *half-space*.

Theorem 1 *A half-space is convex.*

Theorem 2 *The intersection of two convex sets is convex.*

Corollary

The set of feasible solutions to a linear programming problem is convex.

Intuitively, the corners or vertices of a convex set are those points of the set that do not lie on the interior of a line segment contained in the set. This suggests the following.

Definition A point P of a convex set S is a *vertex* of S if P is not the midpoint of a line segment connecting two other points of S.

Example 1 (continued)

For the three convex figures of Example 1, the line segment has two vertices (the two end points), the triangle has three (the three corners), and the home plate has five.

Theorem 3 *Let S be the set of feasible solutions to the system of constraints of a linear programming problem in a standard form. Then any basic feasible solution to the problem is a vertex of S.*

Proof Let X be a basic feasible solution, and suppose the first m variables are the basic variables, with n the total number of variables. Assume $X = (P + Q)/2$, where $P = (p_1, \ldots, p_n)$ and $Q = (q_1, \ldots, q_n)$ are in S. Then

$$X = (x_1, \ldots, x_m, 0, \ldots, 0) = \tfrac{1}{2}(p_1 + q_1, \ldots, p_m + q_m, p_{m+1} + q_{m+1}, \ldots, p_n + q_n).$$

Since all the coordinates of P and Q are nonnegative,

$$p_j = q_j = 0 \qquad \text{for} \qquad j = m + 1, \ldots, n$$

But there is only one basic feasible solution, X, with all these coordinates equal to zero (see Problem 10 of Section 3.2). Thus $P = Q = X$. Hence X is a vertex of S.

###

Corollary

If the objective function of a linear programming problem has a finite optimal value, this value is assumed by at least one vertex of the set of feasible solutions to the system of constraints.

Proof This follows directly from the above theorem and the corollary of the last section.

###

In the simplex algorithm we move from basic feasible solution to basic feasible solution by replacing at each step one variable in the basis. From Theorem 3, we see that we are simply moving from vertex to vertex in the convex set of feasible solutions to the system of constraints. In fact, since at each step exactly one basic variable is replaced, we are actually moving between adajcent vertices. See Problem 10 for a development of these ideas.

By using the corollary of the last section in the proof of the above corollary we have made use of the central theorems of this chapter, theorems that have been proved algebraically. In fact, the above result can also be proved independently using only the theory of convex sets. (See, for example, Problem 11). This suggests an alternate, theoretically sound approach to the linear optimization problem. First, compute all the basic feasible solutions to the problem; second, compare the value of the objective function at each of these points. As long as we know that the function has an optimal value, it must be the optimal value in this set. However, this technique is far from practical; if the constraint system has m equations and n unknowns, there could be up to $\binom{n}{m}$ basic feasible solutions, where

$$\binom{n}{m} = \frac{n!}{m!(n-m)!}$$

is the binomial coefficient. For example,

$$\binom{15}{5} = 3003 \quad \text{and} \quad \binom{20}{10} = 184{,}756$$

PROBLEM SET 3.9

1. Suppose P and Q are points in R^n. Show geometrically that the set $tP + (1-t)Q = Q + t(P-Q)$, $0 \le t \le 1$, is the line segment connecting P and Q.

2. Prove Theorem 1. (*Hint.* Use Example 2 as a model.)

3. Prove Theorem 2.

4. Prove the Corollary to Theorem 2.

5. Theorems 1 and 2 imply immediately that a hyperplane is convex. Why?

6. True or false:
(a) The union of two convex sets is convex?
(b) The complement of a convex set is convex?

7. True or false: A point P is a vertex of a convex set S if and only if P is not the interior point of any line segment in S? (An interior point of a line segment L is any point of L other than the two end points.)

8. Prove that if P and Q are vertices of a convex set S and $X = P + t(Q-P)$ is a point of S, then $0 \le t \le 1$.

9. Consider the general linear programming problem (1) of Section 3.4. Suppose $P = (b_1, \ldots, b_m, 0, \ldots, 0)$ and $Q = (0, b_2^*, \ldots, b_m^*, b_{m+1}^*, \ldots, 0)$ are distinct basic feasible solutions, and $X = (x_1, x_2, \ldots, x_m, x_{m+1}, 0, \ldots, 0)$ is a feasible solution. Show that $X = P + t(Q-P)$

for some t, $0 \leq t \leq 1$. (*Hints.* For each solution Q and X, use the equations of (1) of Section 3.4 to express the first m coordinates of the solution in terms of the $(m + 1)$ coordinate. Problem 8 is also of use.)

10. Let S be the set of feasible solutions to a linear programming problem. A line segment L joining two vertices of S is an *edge* of S if no point of L is the midpoint of a line segment between two points of $S - L$. Two such vertices of S joined by an edge are said to be *adjacent* vertices. Show that in each step of the simplex algorithm, we move from a vertex of S to an adjacent vertex. (*Hint.* Use Problem 9.)

11. Define a function $f(X) = c \cdot X = c_1x_1 + c_2x_2 + \cdots + c_nx_n$ for $X = (x_1, \ldots, x_n) \in R^n$, where $c = (c_1, \ldots, c_n) \in R^n$ is a constant.

(a) Show that for any P and Q in R^n and $t \in R$, $0 \leq t \leq 1$, $f[tP + (1 - t)Q] = tf(P) + (1 - t)f(Q)$. Note that $tf(P) + (1 - t)f(Q)$ is a real number between $f(P)$ and $f(Q)$.

(b) What does this suggest about the optimal value of f on a convex set S in R^n?

12. Suppose T is the set of those feasible solutions to a linear programming problem at which the objective function of the problem attains its optimal value. What does Problem 11a say about T?

4 Duality

SECTION 4.1 DEFINITION OF THE DUAL PROBLEM

Frequently in mathematics there exist relationships between concepts, systems, or problems that are not immediately apparent but, once understood, reap many dividends. For example, consider in calculus the relationship between the integral and the derivative expressed in the Fundamental Theorem of Calculus, or in linear algebra, the relationship between linear transformations and matrices. Relationships such as these not only can be used for practical or computational purposes, but also can provide a unified and coherent theory, so that insights and techniques from one system can contribute to the understanding and usefulness of another.

In this chapter we will develop one such unifying notion, the concept of duality. In this first section, for any linear programming problem the dual problem will be defined, and it will not be immediately obvious why this dual problem should be of interest or usefulness. However, in the second section, it will be shown that in certain economic situations, the dual problem arises quite naturally. In the third section, the fundamental Duality Theorem, which states roughly that a linear programming problem and its dual have the same optimal solutions, will be proven. Later, when we consider two-person, zero-sum games, it will be seen that the problem of solving such a game is equivalent to a linear programming problem and, moreover, the existence of a solution to a game will follow from this Duality Theorem. Thus the concept of duality relates and unites much of the material of this text.

In order to define the dual of a linear programming problem, we will require first that the problem be expressed in a special form, called the *primal form* of the problem.

Definition A linear programming problem stated in the following form is said to be in *primal form.*

$$\text{Maximize} \quad z = c_1x_1 + c_2x_2 + \cdots + c_nx_n$$

subject to

$$\begin{aligned} a_{11}x_1 + a_{12}x_2 + \cdots + a_{1n}x_n &\leq b_1 \\ a_{21}x_1 + a_{22}x_2 + \cdots + a_{2n}x_n &\leq b_2 \\ &\vdots \\ a_{m1}x_1 + a_{m2}x_2 + \cdots + a_{mn}x_n &\leq b_m \\ x_1, x_2, \ldots, x_n &\geq 0 \end{aligned} \tag{1}$$

Thus the primal form of a linear programming problem, called simply the *primal problem*, is a maximization problem with a system of constraints consisting of only $(\leq)$ inequalities. Note that here there are no restrictions on the signs of the coefficients a_{ij}, constant terms b_i, and coefficients c_j. Now to this primal problem we associate its *dual problem.*

Definition The *dual* of the primal problem is the following linear programming problem.

$$\text{Minimize} \quad v = b_1y_1 + b_2y_2 + \cdots + b_my_m$$

subject to

$$\begin{aligned} a_{11}y_1 + a_{21}y_2 + \cdots + a_{m1}y_m &\geq c_1 \\ a_{12}y_1 + a_{22}y_2 + \cdots + a_{m2}y_m &\geq c_2 \\ &\vdots \\ a_{1n}y_1 + a_{2n}y_2 + \cdots + a_{mn}y_m &\geq c_n \\ y_1, y_2, \ldots, y_m &\geq 0 \end{aligned} \tag{2}$$

Thus the components of the dual problem relate to the primal problem as follows.

1. Variables. The number of nonnegative variables of the dual equals the number of $(\leq)$ inequalities of the primal.
2. Constraints.
   (a) The number of $(\geq)$ inequalities of the dual equals the number of nonnegative variables of the primal.

(b) The coefficients for the inequalities of the dual are the coefficients of the inequalities of the primal, with the rows and columns interchanged (transposed).
(c) the constant terms for the inequalities of the dual are the coefficients of the objective function of the primal.

3. Objective function.
(a) The coefficients for the dual objective function are the constant terms of the constraints of the primal.
(b) The objective function of the dual is to be minimized, whereas the objective function of the primal is to be maximized.

In fact, the dual variable y_i corresponds to the ith inequality of the constraints of the primal. And, reciprocally, the coefficients of the jth inequality of the dual correspond to the coefficients of the x_j variable in the primal problem. We summarize these relationships in Table 4.1.

Table 4.1

Primal	*Dual*
Maximization problem	Minimization problem
ith ($\leq$) inequality	ith nonnegative variable
jth nonnegative variable	jth ($\geq$) inequality
Objective function coefficients	Constant terms of constraints
Constant terms of constraints	Objective function coefficients

Example

The linear programming of

$$\text{Minimizing} \quad 6x_1 + x_2 + 4x_3$$

subject to

$$3x_1 + 7x_2 + x_3 \leq 15$$
$$x_1 - 2x_2 + 3x_3 \leq 20$$
$$x_1, x_2, x_3 \geq 0$$

has as its dual the problem of

$$\text{Minimizing} \quad 15y_1 + 20y_2$$

subject to

$$3y_1 + y_2 \geq 6$$
$$7y_1 - 2y_2 \geq 1$$
$$y_1 + 3y_2 \geq 4$$
$$y_1, y_2 \geq 0$$

Matrix notation can be used to express any linear programming problem and, in particular, the primal and dual problems, succinctly. Define, from (1), the coefficient matrix A and column vectors b, c, and X as follows.

$$A = \begin{bmatrix} a_{11} & a_{12} & \cdots & a_{1n} \\ a_{21} & a_{22} & \cdots & a_{2n} \\ \vdots & & & \\ a_{m1} & a_{m2} & \cdots & a_{mn} \end{bmatrix}, \quad b = \begin{bmatrix} b_1 \\ b_2 \\ \vdots \\ b_m \end{bmatrix}, \quad c = \begin{bmatrix} c_1 \\ c_2 \\ \vdots \\ c_n \end{bmatrix}, \quad X = \begin{bmatrix} x_1 \\ x_2 \\ \vdots \\ x_n \end{bmatrix}$$

Let A^t denote the transpose of matrix A, and $c \cdot X$ denote the dot or scalar product of vectors c and X. Then

$$A^t = \begin{bmatrix} a_{11} & a_{21} & \cdots & a_{m1} \\ a_{12} & a_{22} & \cdots & a_{m2} \\ \vdots & & & \\ a_{1n} & a_{2n} & \cdots & a_{mn} \end{bmatrix}$$

and

$$c \cdot X = c_1x_1 + c_2x_2 + \cdots + c_nx_n = c^tX = X^tc = X \cdot c$$

The primal problem of (1) is simply to maximize $z = c \cdot X$ subject to $AX \leq b$, $X \geq 0$, where $AX \leq b$ means that each component of the column vector AX is less than or equal to the corresponding component of the vector b, and $X \geq 0$ is defined similarly, with 0 in this case being the n-dimensional zero vector. Let Y be the m-dimensional column vector $(y_1, y_2, \ldots, y_m)^t$. Then the dual problem of (2) is to minimize $v = b \cdot Y$ subject to $A^tY \geq c$, $Y \geq 0$.

In summary, we have the following.

$$\begin{array}{ll} \text{Primal problem:} & \text{Maximize } z = c \cdot X \text{ subject to } AX \leq b,\ X \geq 0 \\ \text{Dual problem:} & \text{Minimize } v = b \cdot Y \text{ subject to } A^tY \geq c,\ Y \geq 0 \end{array} \tag{3}$$

Any linear programming problem is equivalent to a problem stated in primal form. For example, we have already seen how a minimization problem can be transformed into an equivalent maximization problem and unrestricted variables replaced by variables restricted in sign. A constraint involving an equation can be replaced by two inequalities in opposite directions. For example, the set of points $(x_1, x_2) \in R^2$ such that $3x_1 + 2x_2 = 5$ equals the set of (x_1, x_2) such that $3x_1 + 2x_2 \geq 5$ and $3x_1 + 2x_2 \leq 5$. Finally, the direction of an inequality can always be changed by multiplication by -1.

As a result, it follows that every linear programming problem has a dual. As an application, let us determine the dual to the general linear programming problem of (2). The problem as stated is to minimize $b \cdot Y$ subject to $A^tY \geq c$, $Y \geq 0$. Letting $-M$ denote that matrix found by multiplying all the entries of a matrix M by -1, the problem of (2) is equivalent to the problem of

$$\text{Maximizing } (-b) \cdot Y \text{ subject to } (-A^t)Y \leq -c,\ Y \geq 0$$

But this problem is in primal form, and its dual is, using (3), to

$$\text{Minimize } (-c) \cdot X \text{ subject to } (-A^t)^t X \geq -b,\ X \geq 0$$

Now, using the fact that for any matrix M, $(M^t)^t = M$, this problem is equivalent to the problem of

$$\text{Maximizing } c \cdot X \text{ subject to } AX \leq b,\ X \geq 0$$

Note that this is precisely the problem of (1). Thus we have proven that for any linear programming problem, the dual of the dual is the original problem. Hence, repeated application of this operation of constructing the dual problem to a given problem does not lead to a chain of distinct problems but, instead, cycles after two steps, resulting in exactly two problems, each the dual of the other.

As a final remark to this section, note that the dual of any linear programming problem can also be found directly without the intermediate step of determining the equivalent problem in primal form. All we need do is record the results of the dual operation as defined here on the various components of a problem, and then use these rules. (See Problem 4.) The dual operation as we have defined it is sometimes called the symmetric form of the primal-dual problems, and it is this form that is especially useful for our applications in game theory.

PROBLEM SET 4.1

1. Determine the dual problems for the following linear programming problems.

   (a) Maximize $20x_1 + 30x_2$

   subject to

   $$\begin{aligned} 5x_1 - 4x_2 &\leq 100 \\ -x_1 + 12x_2 &\leq 90 \\ x_2 &\leq 500 \end{aligned}$$

   $x_1, x_2 \geq 0$

   (b) Maximize $-x_1 + 2x_2$

   subject to

   $5x_1 + x_2 \leq 60$

   $3x_1 - 8x_2 \geq 10$

   $x_1, x_2 \geq 0$

   (c) Minimize $8x_1 + 11x_2 - 14x_3$

   subject to

   $$\begin{aligned} -3x_1 \quad + x_3 &\geq -63 \\ x_2 - 2x_3 &\geq 52 \end{aligned}$$

   $x_1, x_2, x_3 \geq 0$

(d) Minimize $100x_1 + 90x_2 + 500x_3$

subject to

$$\begin{aligned} 5x_1 - x_2 &\geq 20 \\ -4x_1 + 12x_2 + x_3 &\geq 30 \\ x_1, x_2, x_3 &\geq 0 \end{aligned}$$

2. Consider the linear programming problem of the example in this section.
(a) Show that the objective function of the dual problem is bounded below.
(b) Solve the dual problem graphically.
(c) Solve the primal problem using the simplex method. Note that the optimal values of the objective functions are equal.
(d) Compare the bottom two entries in the slack variable columns of the last simplex tableau of part c with the point in part b that yielded the minimal value.

3. Prove the following lemma, which will be used in Section 4.3. If X, Y, and W are vectors all of the same dimension such that $X \geq Y$ and $W \geq 0$, then $X \cdot W \geq Y \cdot W$.

4. (a) Show that the linear programming problem of

Maximizing $5x_1 + 3x_2 + 7x_3$

subject to

$$\begin{aligned} x_1 - 2x_2 + 3x_3 &\leq 12 \\ 3x_1 + 2x_2 + x_3 &= 30 \\ x_1, x_2 \geq 0,\ x_3 &\text{ unrestricted} \end{aligned} \tag{4}$$

is equivalent to the problem of

Maximizing $5x_1 + 3x_2 + 7x_3' - 7x_3''$

subject to

$$\begin{aligned} x_1 - 2x_2 + 3x_3' - 3x_3'' &\leq 12 \\ 3x_1 + 2x_2 + x_3' - x_3'' &\leq 30 \\ -3x_1 - 2x_2 - x_3' + x_3'' &\leq -30 \\ x_1, x_2, x_3', x_3'' &\geq 0 \end{aligned} \tag{5}$$

(b) Show that the dual problem to the problem of (5) is to

Minimize $12y_1 + 30y_2' - 30y_2''$

subject to

$$\begin{aligned} y_1 + 3y_2' - 3y_2'' &\geq 5 \\ -2y_1 + 2y_2' - 2y_2'' &\geq 3 \\ 3y_1 + y_2' - y_2'' &\geq 7 \\ -3y_1 - y_2' + y_2'' &\geq -7 \\ y_1, y_2', y_2'' &\geq 0 \end{aligned} \tag{6}$$

(c) Show that (6) is equivalent to the problem of

$$\begin{aligned} &\text{Minimizing} \quad 12y_1 + 30y_2 \\ &\text{subject to} \\ &\quad y_1 + 3y_2 \geq 5 \\ &-2y_1 + 2y_2 \geq 3 \\ &\quad 3y_1 + y_2 = 7 \\ &y_1 \geq 0,\ y_2 \text{ unrestricted} \end{aligned} \tag{7}$$

(d) Compare the problem of (4) with its dual, the problem of (7).

(e) *Conclusion.* Table 4.1 can be extended as follows.

Table 4.1 (continued)

Primal	*Dual*
*i*th equality of constraints	*i*th unrestricted variable
*j*th unrestricted variable	*j*th equality of constraints

5. Determine the dual problems for the following.

(a) Minimize $-50x_1 + 100x_2$

subject to

$$\begin{aligned} 4x_1 - x_2 &= 3 \\ x_1 + 5x_2 &\geq 2 \\ x_1 + x_2 &\leq 4 \\ x_1 \geq 0,\ x_2 &\text{ unrestricted} \end{aligned}$$

(b) Minimize $3y_1 - 2y_2 + 4y_3$

subject to

$$\begin{aligned} 4y_1 - y_2 + y_3 &\geq 50 \\ y_1 + 5y_2 - y_3 &= 100 \\ y_2, y_3 \geq 0,\ y_1 &\text{ unrestricted} \end{aligned}$$

How are these two problems related?

SECTION 4.2 EXAMPLES AND INTERPRETATIONS

In this section the dual problems to some specific linear programming examples will be defined and discussed. The examples, using the categories of Chapter 2, are from the classes of blending problems, production problems (minimization of costs while meeting given demands), and transportation problems. Additional examples, and in particular a production problem involving the maximization of profits using limited resources, are contained in the problems at the end of this section.

Example 1 ***A Blending Problem***

The diet problems that we have already seen lead to dual problems that have a standard but still extremely interesting interpretation. Consider, for example, the problem in Section 2.2 of the farmer wishing to feed his stock. The farmer's problem was to determine a diet using two feeds that minimized cost and satisfied three nutritional requirements. Here, letting x_1 and x_2 denote the amounts in pounds of Feeds 1 and 2 to use, respectively, the mathematical problem was to

$$\begin{aligned} &\text{Minimize} \quad 10x_1 + 4x_2 \\ &\text{subject to} \\ &3x_1 + 2x_2 \geq 60 \\ &7x_1 + 2x_2 \geq 84 \\ &3x_1 + 6x_2 \geq 72 \\ &x_1, x_2 \geq 0 \end{aligned} \tag{1}$$

The three inequalities in the system of constraints result from the requirement that the diet provide specified amounts of the nutritional elements A, B, and C. Now this problem is in the form of the general dual problem of (2) of Section 4.1, and it was shown in that section that the dual of (2) is the problem of (1) of Section 4.1, the primal problem. Using this, it follows immediately that the dual to the above problem is to

$$\begin{aligned} &\text{Maximize} \quad 60y_1 + 84y_2 + 72y_3 \\ &\text{subject to} \\ &3y_1 + 7y_2 + 3y_3 \leq 10 \\ &2y_1 + 2y_2 + 6y_3 \leq 4 \\ &y_1, y_2, y_3 \geq 0 \end{aligned} \tag{2}$$

In order to provide an interpretation of the dual, consider the problem of a traveling salesman dealing in nutrition tablets for cattle. Suppose the salesman has to offer the farmer three types of pure tablet, one type containing exactly 1 unit of nutritional element A and nothing else, one containing exactly 1 unit of B and nothing else, and the last containing exactly 1 unit of C and nothing else. Now the salesman hopes to convince the farmer that it is to his advantage to nourish his cattle by using these tablets instead of any combination of the two feeds 1 and 2. Although the farmer is probably somewhat set in his ways, the salesman believes that due to the problems of maintaining a small farm today, he can still appeal to the farmer's frugality. Thus the salesman attempts to set the prices for each of three types of tablets in such a way that the tablets can compete favorably with the two feeds, and he can realize the greatest income. To do this, he lets y_1, y_2, and y_3 denote the cost in cents to the farmer of one tablet of nutritional elements A, B, and C, respectively.

Now 1 lb of Feed 1 provides 3, 7, and 3 units of A, B, and C, and costs 10 cents. In order to replace 1 lb of this feed with tablets, the farmer would need three tablets each of the first and third types, and seven of the second type. This would cost $3y_1 + 7y_2 + 3y_3$, and so, in order to be competitive, the salesman must have

$$3y_1 + 7y_2 + 3y_3 \leq 10$$

Similarly, 1 lb of Feed 2 provides 2, 2, and 6 units of A, B, and C, and cost 4 cents. Thus the inequality

$$2y_1 + 2y_2 + 6y_3 \leq 4$$

Since the farmer has determined that the daily requirements of elements A, B, and C are 60, 84, and 72 units, respectively, the cost of meeting these requirements by using the tablets would be $60y_1 + 84y_2 + 72y_3$. Thus the salesman wishes to maximize this function subject to the above two inequalities. Notice that this is precisely the dual of the original problem.

The salesman, upon further reflection, realizes that there is one more restriction to be placed on the prices he sets, that the cost to the farmer of providing an adequate diet by use of his tablets cannot exceed the cost of a diet consisting of any combination of the two feeds that meets the nutritional requirements. In other words, for any pair x_1 and x_2 that satisfies the constraints of (1), y_1, y_2, and y_3 must be chosen so that

$$60y_1 + 84y_2 + 72y_3 \leq 10x_1 + 4x_2$$

In fact, since the solution of the original problem, as worked out in Section 2.2, provided that the minimal value of the objective function $10x_1 + 4x_2$ was 144 and was attained at the point $x_1 = 6$, $x_2 = 21$, this last restriction simplifies to

$$60y_1 + 84y_2 + 72y_3 \leq 144$$

as long as the farmer is aware of this minimal cost diet.

The salesman, with the hope that the farmer will be satisfied if his prices compete favorably with each of the two feeds individually, neglects for the time being this last restriction and attempts to solve the linear programming problem of (2). Adding two slack variables and using the simplex method, he gets the tableaux of Table 4.2.

It should be remarked that the salesman chose the first pivot term primarily to avoid fractions he believed difficult to work with. Now, from the final tableau,

Table 4.2

	y_1	y_2	y_3	y_4	y_5	
y_4	3	7	3	1	0	10
y_5	(2)	2	6	0	1	4
	-60	-84	-72	0	0	0
y_4	0	(4)	-6	1	$-\frac{3}{2}$	4
y_1	1	1	3	0	$\frac{1}{2}$	2
	0	-24	108	0	30	120
y_2	0	1	$-\frac{3}{2}$	$\frac{1}{4}$	$-\frac{3}{8}$	1
y_1	1	0	$\frac{9}{2}$	$-\frac{1}{4}$	$\frac{7}{8}$	1
	0	0	72	6	21	144

the salesman sees that he should charge the farmer 1 cent each for the tablets of A and B, and nothing for the tablets of C, and in doing this he will realize his maximum income of \$1.44. He notes that this maximum income of \$1.44 equals the minimum cost to the farmer of an adequate diet using Feeds 1 and 2 and, since this is the maximum possible income for any y_1, y_2, and y_3 satisfying (2), his last restriction, that $60y_1 + 84y_2 + 72y_3 \leq 144$, will always be satisfied.

In the next section we will prove that the above results are not just coincidental. We will show first that, for primal-dual problems, any possible value of the objective function to be maximized [the function of (1) of Section 4.1] is less than or equal to any possible value of the objective function to be minimized [the function of (2) of Section 4.1]. Here that means that $60y_1 + 84y_2 + 72y_3 \leq 10x_1 + 4x_2$ for any points (y_1,y_2,y_3) and (x_1,x_2) that satisfy their respective constraints. Second, we will prove the Duality Theorem, which states that the optimal values of the two objective functions must be equal. Here the common value was \$1.44. The alert reader may have already noticed one other curious fact about the example we have worked out. The minimal value of the first objective function, determined graphically in Section 2.2, was attained at the point $x_1 = 6$, $x_2 = 21$. Note that these are the last entries in the slack variable columns of the final tableau of the dual problem. This is not coincidental and, in fact, it is this relationship on which our proof of the Duality Theorem will rest.

Example 2

Consider the situation described in Problem 7 of Section 2.3 (see also Problem 7 of Section 3.6). An oil refinery, with three available processes, produces three grades of gasoline. The problem is to determine the operation that minimizes cost and satisfies specified demands. Using the data from the problem of Section 2.3, and letting x_j denote the number of hours of operation of Process j, $j = 1, 2,$ and 3, the resulting linear programming problem is to

$$\begin{aligned}
&\text{Minimize} \quad 160x_1 + 400x_2 + 300x_3 \\
&\text{subject to} \\
&3x_1 + 6x_2 + 6x_3 \geq 36 \\
&4x_1 + 6x_2 + 3x_3 \geq 20 \\
&2x_1 + 8x_2 + 4x_3 \geq 30 \\
&x_1, x_2, x_3 \geq 0
\end{aligned}$$

(As determined in Problem 7 of Section 3.6, the minimal cost is \$1950 and is attained by using Process 2 for $1\frac{1}{2}$ hr and Process 3 for $4\frac{1}{2}$ hr.)

This problem is in the form of (2) of Section 4.1, and thus its dual is to

$$\begin{aligned}
&\text{Maximize} \quad 36y_1 + 20y_2 + 30y_3 \\
&\text{subject to} \\
&3y_1 + 4y_2 + 2y_3 \leq 160 \\
&6y_1 + 6y_2 + 8y_3 \leq 400 \\
&6y_1 + 3y_2 + 4y_3 \leq 300 \\
&y_1, y_2, y_3 \geq 0
\end{aligned}$$

To provide an interpretation of this problem, suppose that the management of the refinery wishes to determine a market price for each grade of gasoline. Although the market prices certainly should reflect the level of the grade, another factor to be considered is the actual cost of production. Thus, with the refinery operating to meet these specified demands, some estimate of the costs of each grade of gasoline must be made. Let y_1, y_2, and y_3 denote the cost in dollars of 100 gal of regular, low-lead, and premium gasoline, respectively. Now 1 hr of operation of Process 1 produces 300, 400, and 200 gal of regular, low-lead, and premium gasoline, and costs \$160. Thus it is reasonable to demand that these estimated production costs satisfy the inequality

$$3y_1 + 4y_2 + 2y_3 \leq 160$$

Similarly, considering the output and costs of Processes 2 and 3, we see that y_1, y_2, and y_3 must satisfy

$$\begin{aligned}
6y_1 + 6y_2 + 8y_3 &\leq 400 \\
6y_1 + 3y_2 + 4y_3 &\leq 300
\end{aligned}$$

In order to justify a price increase to the government, say, the management would desire that these operational costs are as high as possible. More precisely, since the refinery is delivering weekly 3600, 2000, and 3000 gal of regular, low-lead, and premium gasolines, it should choose y_1, y_2, and y_3 to maximize the function $36y_1 + 20y_2 + 30y_3$. Thus we have the dual problem.

Example 3

Consider the transportation problem in Problem 2 of Section 2.4. Letting x_{1j} denote the number of cases shipped from the Eastern Warehouse to Outlet j, $1 \leq j \leq 4$, and x_{2j} the number from the Western Warehouse, the resulting mathematical problem is to

Minimize $\quad 20x_{11} + 16x_{12} + 30x_{13} + 20x_{14} + 45x_{21} + 40x_{22} + 50x_{23} + 44x_{24}$

subject to

$$\begin{array}{llllllllll} x_{11} + x_{12} + x_{13} + x_{14} & & & & & & & & \leq & 600 \\ & & & & x_{21} + x_{22} + x_{23} + x_{24} & & & & \leq & 1000 \\ x_{11} & & & + x_{21} & & & & & \geq & 300 \\ & x_{12} & & & + x_{22} & & & & \geq & 350 \\ & & x_{13} & & & + x_{23} & & & \geq & 400 \\ & & & + x_{14} & & & & + x_{24} & \geq & 450 \end{array}$$

$x_{ij} \geq 0, 1 \leq i \leq 2, 1 \leq j \leq 4$

(You may feel that the last four inequalities in the system of constraints should be equalities, but the nature of the objective function guarantees that its minimal value will be attained at a point at which these inequalities are, in fact, equalities. Thus we may formulate the problem as above, with the advantage being that the dual to the above problem is readily determined.)

Multiplying the first two inequalities by -1 puts this problem in the form of (2) of Section 1, and so its dual is to

Minimize $\quad -600y_1 - 1000y_2 + 300y_3 + 350y_4 + 400y_5 + 450y_6$

subject to

$$\begin{array}{lllllll} -y_1 & & + y_3 & & & & \leq 20 \\ -y_1 & & & + y_4 & & & \leq 16 \\ -y_1 & & & & + y_5 & & \leq 30 \\ -y_1 & & & & & + y_6 & \leq 20 \\ & - y_2 & + y_3 & & & & \leq 45 \\ & - y_2 & & + y_4 & & & \leq 40 \\ & - y_2 & & & + y_5 & & \leq 50 \\ & - y_2 & & & & + y_6 & \leq 44 \end{array}$$

$y_1, y_2 y_3, y_4, y_5, y_6 \geq 0$

Suppose now that a national shipping company, wanting to expand, offers to deliver the canned goods for the supplier. Instead of charging normal transportation costs, however, the shipper proposes to buy from the supplier all the available cases of canned goods, paying y_1 cents/case for those cases at the Eastern Warehouse and y_2 cents for the cases at the Western Warehouse. He guarantees delivery of the required number of cases at each of the four outlets, selling the cases back to the supplier at a cost of y_{i+2} cents/case at Outlet i, $1 \leq i \leq 4$.

Now the shipper must determine these six prices in such a way that they are competitive and realize the maximum income. To be competitive, for example, since it costs the supplier 20 cents to ship a case from the Eastern Warehouse to Outlet 1, y_1 and y_3 must be chosen so that $y_3 - y_1 \leq 20$. Consideration of the other seven shipping costs of the supplier leads to the other seven inequalities in the above dual problem. And as long as the y_i satisfy these inequalities, the shipper can assure the supplier that his offer certainly can cost the supplier no more than he is already paying for transportation and may save him money.

The income the shipper is going to realize from this venture is simply the difference between the total amount he pays at the two warehouses and the total amount he receives at the four outlets. But this difference is precisely the quantity measured by the objective function in the above problem. Thus, in determining the y_i, the shipper encounters the dual of the original transportation problem.

PROBLEM SET 4.2

1. Using the simplex method, show that the dual problem of Example 2 has as solution a maximum value of 1950 attained at the point $y_1 = 33\frac{1}{3}$, $y_2 = 0$, $y_3 = 25$. Interpret the fact that the production cost of low-lead gasoline is zero. How would the vice president in charge of sales react?

2. Consider the problem of Example 1 of Section 2.3, a production problem of maximizing profits using limited resources.

(a) Show that the dual problem is to

$$\text{Minimize} \quad 2000y_1 + 300y_2 + 200y_3$$

subject to

$$\begin{aligned} 50y_1 + 6y_2 + 3y_3 &\geq 50 \\ 30y_1 + 5y_2 + 5y_3 &\geq 60 \\ y_1, y_2, y_3 &\geq 0 \end{aligned}$$

Now suppose a competitor approaches the boat manufacturer, offering to operate the small boat division under a lease agreement. The competitor claims that he will pay the manufacturer y_1 dollars for each pound of aluminum available, y_2 dollars for each minute of machine time, and y_3 dollars/hour for the finishing labor, and that these prices will guarantee that the manufacturer realize at least as much income as he could by operating the plant himself. Thus the competitor offers the manufacturer an income comparable to any profit the manufacturer could realize himself, while freeing him from determining optimal schedules, the actual running of the plant, and the selling of the boats.

(b) Considering separately the profit the manufacturer realizes from the sale of a rowboat and a canoe, construct two inequalities that the prices y_1, y_2, and y_3 should satisfy. Considering the total amount the competitor pays the manufacturer, compare the problem of determining y_1, y_2, and y_3 with the problem of part a above.

(c) Solve part a by the simplex method. Comparing your solution to the original problem (see Problem 4 of Section 2.3) with the solution of its dual (part a), show that the competitor's offer of a comparable if not favorable income will be realized.

(d) Compare the entries in the slack variable columns in the bottom row of the last tableau computed in part c with the values for R and C found in Problem 4 of Section 2.3.

3. Consider Problem 8 of Section 2.3. Letting x_1 and x_2 denote the number of hours of operation of Systems 1 and 2, the resulting mathematical problem is to

$$\text{Minimize} \quad 2x_1 + 11x_2$$

subject to

$$\begin{aligned} x_1 + 4x_2 &\le 100 \\ 4x_1 + 20x_2 &\ge 480 \\ 2x_1 + 40x_2 &\ge 800 \\ x_1, x_2 &\ge 0 \end{aligned}$$

(a) Show that the dual problem is to

$$\text{Maximize} \quad -100y_1 + 480y_2 + 800y_3$$

subject to

$$\begin{aligned} -y_1 + 4y_2 + 2y_3 &\le 2 \\ -4y_1 + 20y_2 + 40y_3 &\le 11 \end{aligned}$$

(b) A college student, working for the fruit grower for the summer, believes she can have the fruit picked more efficiently than the grower by using her own system and equipment. Fearing that she has nothing to gain financially by simply revealing her plan to the grower, she suggests to the grower that she will supervise the picking of the crop, paying the grower a set amount for each available man-hour of labor and then selling back to the grower the fruit, using two prices, one for a bushel of choice produce, and the other for regular. Considering that the student must convince the grower that it is to his advantage to let the student supervise the harvest, how should she set these three costs?

4. In the third example of this section, if the supplier lets the shipper transport the canned goods as described, there will result 100 extra cases in the hands of the shipper. How might the two parties resolve this difficulty?

5. Consider Problem 6 of Section 2.3.
   (a) Formulate the associated linear programming problem.
   (b) Determine the dual problem.
   (c) Suppose the manager of the electronics firm wants to assess the value of a unit of material and a unit of labor in the production and sale of the circuits. To do this he lets $\$y_1$ and $\$y_2$ denote these two values. The circuit for a radio requires 2 units of material and 1 unit of labor and sells for \$8. The manager reasons therefore that 2 units of material plus 1 unit of labor must be worth at least \$8, but could be worth more if these units can be used in the production of other types of circuits that are more profitable. Thus hc sets $2y_1 + y_2 \geq 8$. The manager continues in this manner. Compare the resulting problem with the problem determined in part b above. (Note that the Duality Theorem guarantees that the optimal values for the problems of parts a and b are equal.)

SECTION 4.3 THE DUALITY THEOREM

In this section we prove the celebrated Duality Theorem. It is generally accepted that John von Newmann was the first mathematician to recognize the significance of the duality principle and endeavor to develop a proof of the Duality Theorem.

We start with the general primal problem of (1) of Section 4.1 to maximize $z = c \cdot X$ subject to $AX \leq b, X \geq 0$. The dual problem, as defined by (2) of Section 4.1, is to minimize $v = b \cdot Y$ subject to $A^tY \geq c$, $Y \geq 0$. We will show first that the set of possible values for the function z lies to the left of the set of possible values for the function v. Then, with this result, we will prove the Duality Theorem using the simplex method and, in particular, the theorem of Section 3.8.

Theorem 1 *Suppose X_0 is a feasible solution to the primal problem and Y_0 a feasible solution to the dual problem. Then*

$$c \cdot X_0 \leq b \cdot Y_0$$

Proof (Remember that in general, a vector is written as a column vector, and thus its transpose is the corresponding row vector.) Now X_0 a solution to the primal problem implies that $AX_0 \leq b$. Since $Y_0 \geq 0$, $Y_0{}^tAX_0 \leq Y_0{}^tb$ (see Problem 3 of Section 4.1). Similarly Y_0 a solution to the dual and $X_0 \geq 0$ implies that $A^tY_0 \geq c$ and $X_0{}^tA^tY_0 \geq X_0{}^tc$. Now $X_0{}^tA^tY_0$ is simply a real number, and so

$$(X_0{}^tA^tY_0) = (X_0{}^tA^tY_0)^t = Y_0{}^tAX_0$$

Therefore

$$c \cdot X_0 = X_0{}^tc \leq X_0{}^tA^tY_0 = Y_0{}^tAX_0 \leq Y_0{}^tb = b \cdot Y_0 \qquad \#\#\#$$

While the above proof is in terms of matrices and their transposes, the idea behind the proof is quite elementary, as the following example demonstrates.

Example 1

Problem P:

$$\text{Maximize} \quad f(x_1,x_2,x_3) = 13x_1 - 2x_2 + 5x_3$$

subject to

$$\begin{aligned} 2x_1 - 6x_2 + 7x_3 &\leq 100 \\ x_1 + 9x_2 - 8x_3 &\leq 150 \\ x_1,x_2,x_3 &\geq 0 \end{aligned}$$

Problem D:

$$\text{Minimize} \quad g(y_1,y_2) = 100y_1 + 150y_2$$

subject to

$$\begin{aligned} 2y_1 + y_2 &\geq 13 \\ -6y_1 + 9y_2 &\geq -2 \\ 7y_1 - 8y_2 &\geq 5 \\ y_1,y_2 &\geq 0 \end{aligned}$$

Problems P and D are dual. Suppose now that (x_1,x_2,x_3) is a feasible solution to the constraints of Problem P, and (y_1,y_2) is a feasible solution to the constraints of Problem D. Then, since the components of (x_1,x_2,x_3) are nonnegative and satisfy the inequalities of Problem P, and the components of (y_1,y_2) are nonnegative and satisfy the inequalities of Problem D, we have

$$\begin{aligned} f(x_1,x_2,x_3) &= 13x_1 - 2x_2 + 5x_3 \\ &= 13x_1 + (-2)x_2 + 5x_3 \\ &\leq (2y_1 + y_2)x_1 + (-6y_1 + 9y_2)x_2 + (7y_1 - 8y_2)x_3 \\ &= 2y_1x_1 + y_2x_1 - 6y_1x_2 + 9y_2x_2 + 7y_1x_3 - 8y_2x_3 \\ &= (2x_1 - 6x_2 + 7x_3)y_1 + (x_1 + 9x_2 - 8x_3)y_2 \\ &\leq 100y_1 + 150y_2 \\ &= g(y_1,y_2) \end{aligned}$$

The proof of Theorem 1 is simply a generalization of these steps.

Corollary 1

If X_0 and Y_0 are feasible solutions to the primal and dual problems, respectively, and if $c \cdot X_0 = b \cdot Y_0$, the optimal values of the objective functions z and v equal this common value; that is, maximum $z = c \cdot X_0 = b \cdot Y_0 =$ minimum v.

Proof Suppose X_1 is any feasible solution to the primal problem. Then, from the above theorem, $c \cdot X_1 \leq b \cdot Y_0$, and so $c \cdot X_1 \leq c \cdot X_0$. Thus the maximum value of the function $z = c \cdot X$ is $c \cdot X_0$. Similarly for the dual function v. # # #

Corollary 2

If the objective function z of the primal problem is not bounded above, the dual problem has no feasible solutions. Similarly, if the objective function v of the dual problem is not bounded below, the primal problem has no feasible solutions.

The proof of the second corollary is left to the reader (Problem 1). The converse to this corollary is false. There are examples of primal-dual problems such that neither problem has any feasible solutions (see Problem 2).

Duality Theorem If either the primal problem or the dual problem has a finite optimal solution, then so also does the other; moreover, the optimal values of the objective functions are equal; that is,

$$\text{Max } z = \text{Min } v$$

Proof Assume first that the primal problem of maximizing $c \cdot X$ subject to $AX \leq b$, $X \geq 0$ has a finite optimal solution. Thus we assume the existence an X_0 such that $AX_0 \leq b$, $X_0 \geq 0$ and, for any other X with $AX \leq b$, $X \geq 0$, we have $c \cdot X \leq c \cdot X_0$.

Now the solution to the dual problem is found by applying the simplex method to the primal problem. To do this we first write the primal problem in standard form by adding m slack variables x_j, $n + 1 \leq j \leq n + m$, and multiplying the objective function by -1. This gives the problem of

$$\text{Minimizing } -c_1x_1 - c_2x_2 - \cdots - c_nx_n = -z$$

subject to

$$\begin{aligned}
&a_{11}x_1 + a_{12}x_2 + \cdots + a_{1n}x_n + x_{n+1} &&= b_1 \\
&a_{21}x_1 + a_{22}x_2 + \cdots + a_{2n}x_n + \qquad x_{n+2} &&= b_2 \\
&\vdots \\
&a_{m1}x_1 + a_{m2}x_2 + \cdots + a_{mn}x_n + \qquad\qquad x_{n+m} &&= b_m \\
&x_j \geq 0,\ 1 \leq j \leq n + m
\end{aligned} \tag{1}$$

We now assume in our proof that the constants b_i, $1 \leq i \leq m$, are nonnegative. If this is the case, the above problem is in canonical form with basic variables $x_{n+1}, x_{n+2}, \ldots, x_{n+m}$, since the associated basic solution is feasible, and the simplex method can be initiated directly commencing with the second stage.

(Recall that in Section 4.1 when the primal and dual problems were defined, no restrictions were placed on the constants. Thus, with this assumption, our proof loses some generality. However, for those applications of the Duality Theorem in this text involving game theory, the corresponding constant terms b_i will be, in fact, nonnegative. And, in Problem 5 at the end of this section, a technique for generalizing this proof to the general case is discussed.)

From the theorem of Section 3.8, we know that there is a finite sequence of pivot operations driving the problem of (1) to the optimal value of the objective function. The initial and final tableaux for such a sequence would look like the following.

Initial Tableau

	x_1	x_2	$\cdots$	x_n	x_{n+1}	$\cdots$	x_{n+m}	
x_{n+1}	a_{11}	a_{12}	$\cdots$	a_{1n}	1	$\cdots$	0	b_1
x_{n+2}	a_{21}	a_{22}	$\cdots$	a_{2n}	0	$\cdots$	0	b_2
$\vdots$								
x_{n+m}	a_{m1}	a_{m2}	$\cdots$	a_{mn}	0	$\cdots$	1	b_m
	$-c_1$	$-c_2$	$\cdots$	$-c_n$	0	$\cdots$	0	0

Final Tableau

	x_1	x_2	$\cdots$	x_n	x_{n+1}	$\cdots$	x_{n+m}	
	r_1	r_2	$\cdots$	r_n	s_1	$\cdots$	s_m	$c \cdot X_0$

Since our concern will be with only the bottom row of this last tableau, we have used the symbols r_j, $1 \leq j \leq n$ and s_i, $1 \leq i \leq m$ to denote the numbers appearing in these positions and left the other positions of the tableau blank. Since this tableau represents the final step of the simplex process on the problem of (1), we have $r_j \geq 0$ and $s_i \geq 0$ for $1 \leq j \leq n$, $1 \leq i \leq m$, and the minimum of $-z$ is $-c \cdot X_0$.

Let Y_0 be the column vector $(s_1, s_2, \ldots, s_m)^t$. We will show that

(a) $Y_0 \geq 0$

(b) $A^t Y_0 \geq c$

(c) $b \cdot Y_0 = c \cdot X_0$

As has already been mentioned, $Y_0 \geq 0$. To show (b) and (c), consider the equation represented by the bottom row of the final tableau.

$$r_1 x_1 + r_2 x_2 + \cdots + r_n x_n + s_1 x_{n+1} + s_2 x_{n+2} + \cdots + s_m x_{n+m} = c \cdot X_0 + (-z)$$

This equation represents the result of all the pivot operations on the initial equation for the objective function

$$-c_1x_1 - c_2x_2 - \cdots - c_nx_n = 0 + (-z)$$

And, at each pivot step, some linear combination of the original constraining equations was added to this equation for the objective function. Thus there exist m constants, t_i, $1 \le i \le m$, such that when the $(m + 1)$ equations

$$\begin{aligned}
&t_1(a_{11}x_1 + a_{12}x_2 + \cdots + a_{1n}x_n + x_{n+1} && = b_1) \\
&t_2(a_{21}x_1 + a_{22}x_2 + \cdots + a_{2n}x_n + \qquad x_{n+2} && = b_2) \\
&\vdots \\
&t_m(a_{m1}x_1 + a_{m2}x_2 + \cdots + a_{mn}x_n + \qquad\qquad x_{n+m} && = b_m) \\
&(-c_1x_1 - c_2x_2 - \cdots - c_nx_n = -z)
\end{aligned}$$

are added together, the result is the equation

$$r_1x_1 + r_2x_2 + \cdots + r_nx_n + s_1x_{n+1} + \cdots + s_mx_{n+m} = -z + c \cdot X_0$$

Comparing coefficients of the slack variables, we see that $s_i = t_i$ for $1 \le i \le m$. Using this result and comparing the coefficients of x_1, we have

$$s_1a_{11} + s_2a_{21} + \cdots + s_ma_{m1} - c_1 = r_1 \ge 0$$

and so

$$s_1a_{11} + s_2s_{21} + \cdots + s_ma_{m1} \ge c_1$$

Similarly, comparing the coefficients of x_j for any j, $1 < j \le n$, we have

$$s_1a_{1j} + s_2a_{2j} + \cdots + s_ma_{mj} - c_j = r_j \ge 0$$

and so

$$s_1a_{1j} + s_2a_{2j} + \cdots + s_ma_{mj} \ge c_j$$

Thus

$$A^tY_0 \ge c$$

To show (c), consider the constant terms in the above equations. We must have

$$s_1b_1 + s_2b_2 + \cdots + s_mb_m = c \cdot X_0$$

that is,

$$b \cdot Y_0 = c \cdot X_0$$

Since $Y_0 \ge 0$ and $A^tY_0 \ge c$, the point Y_0 is a feasible solution to the dual problem. The value of the objective function v at Y_0, $c \cdot Y_0$, is equal to the value of the objective function z at X_0. Thus, from Corollary 1, the minimal value of v is $c \cdot Y_0$, and so the optimal values of both problems are equal.

Finally, suppose that we know initially that it is the dual problem that has the finite optimal solution. But in Section 4.1 it was shown that this problem is equivalent to a problem expressed in primal form. Thus we can apply what we have already proven to this equivalent problem, showing that its dual, the primal problem of the original problem, has the same optimal solution. ###

Corollary 3

If both the primal and dual problems have feasible solutions, then both objective functions have optimal solutions and maximum z = minimum v.

Proof Since both problems have feasible solutions, it follows from Theorem 1 that the objective function z is bounded above and the objective function v is bounded below. From the corollary in Section 3.8, both objective functions attain their optimal values and, from the Duality Theorem, these optimal values must be equal. ###

In summary, we have shown that there are exactly four different categories into which solutions to the primal and dual problems can fall.

1. Both problems have feasible solutions. Then the sets of possible values for the objective functions z and v relate on the real line in the following.

$$z = c \cdot X \quad | \quad v = b \cdot Y$$

optimal value for both

2. The objective function z is unbounded above and the dual problem has no feasible solutions.
3. The objective function v is unbounded below and the primal problem has no feasible solutions.
4. Both problems have no feasible solutions.

We conclude this section with a numerical example similar to the problem worked out in Example 1 of Section 4.2 and some of the problems of the two previous sections.

Example 2

$$\text{Maximize} \quad -5x_1 + 18x_2 + 6x_3 - 3x_4$$

subject to

$$\begin{aligned}
2x_1 \qquad\quad - x_3 + 3x_4 &\leq 20 \\
x_2 - 2x_3 - x_4 &\leq 30 \\
-3x_1 + 6x_2 + 3x_3 + 4x_4 &\leq 24 \\
x_1, x_2, x_3, x_4 &\geq 0
\end{aligned}$$

The dual to this problem is to

$$\text{Minimize} \quad 20y_1 + 30y_2 + 24y_3$$

subject to

$$\begin{aligned} 2y_1 \qquad\quad - 3y_3 &\geq -5 \\ y_2 + 6y_3 &\geq 18 \\ -y_1 - 2y_2 + 3y_3 &\geq 6 \\ 3y_1 - y_2 + 4y_3 &\geq -3 \\ y_1, y_2, y_3 &\geq 0 \end{aligned}$$

Table 4.3

	x_1	x_2	x_3	x_4	x_5	x_6	x_7	
x_5	2	0	-1	3	1	0	0	20
x_6	0	1	-2	-1	0	1	0	30
x_7	-3	(6)	3	4	0	0	1	24
	5	-18	-6	3	0	0	0	0
x_5	(2)	0	-1	3	1	0	0	20
x_6	$\frac{1}{2}$	0	$-\frac{5}{2}$	$-\frac{5}{3}$	0	1	$-\frac{1}{6}$	26
x_2	$-\frac{1}{2}$	1	$\frac{1}{2}$	$\frac{2}{3}$	0	0	$\frac{1}{6}$	4
	-4	0	3	15	0	0	3	72
x_1	1	0	$-\frac{1}{2}$	$\frac{3}{2}$	$\frac{1}{2}$	0	0	10
x_6	0	0	$-\frac{9}{4}$	$-\frac{29}{12}$	$-\frac{1}{4}$	1	$-\frac{1}{6}$	21
x_2	0	1	$\frac{1}{4}$	$\frac{17}{12}$	$\frac{1}{4}$	0	$\frac{1}{6}$	9
	0	0	1	21	2	0	3	112

Adding three slack variables to the primal problem and solving, we have the tableaux of Table 4.3. The maximum value of the function $-5x_1 + 18x_2 + 6x_3 - 3x_4$ is 112, and is attained at the point (10,9,0,0). The minimum value of the objective function of the dual problem is also 112 and, from the bottom row of the third tableau, this value is attained at the point (2,0,3).

PROBLEM SET 4.3

1. Prove Corollary 2.

2. Show that both the following linear programming problem and its dual do not have any feasible solutions.

$$\text{Maximize } x_1$$

subject to

$$\begin{aligned} x_1 - x_2 &\leq 1 \\ -x_1 + x_2 &\leq -2 \\ x_1, x_2 &\geq 0 \end{aligned}$$

3. Consider the linear programming problem of

$$\text{Maximizing } 4x_1 + 10x_2 - 3x_3 + 2x_4$$

subject to

$$\begin{aligned} 3x_1 - 2x_2 + 7x_3 + x_4 &\leq 26 \\ x_1 + 6x_2 - x_3 + 5x_4 &\leq 30 \\ -4x_1 + 8x_2 - 2x_3 - x_4 &\leq 10 \end{aligned}$$

   (a) Show that $(\frac{54}{5},\frac{16}{5},0,0)$ is a feasible solution to this problem. Compute the value of the objective function at this point.
   (b) Write out the dual problem. Show that $(\frac{7}{10},\frac{19}{10},0)$ is a feasible solution to the dual problem. What is the value of the objective function of the dual at this point.
   (c) Using Corollary 1, what can you conclude?

4. (a) Solve the following problem, which is in primal form, graphically.

$$\text{Maximize } 4x_1 + 3x_2$$

subject to

$$\begin{aligned} x_1 &\leq 6 \\ x_2 &\leq 8 \\ x_1 + x_2 &\leq 7 \\ 3x_1 + x_2 &\leq 15 \\ x_1, x_2 &\geq 0 \end{aligned}$$

   (b) Solve the dual problem using the simplex method. Compare the last entries in the two slack variable columns with the solution found in part a.

5. *Generalization of the proof of the Duality Theorem.* Suppose some of the constant terms b_i in (1) are negative. By rearranging the constraining equations if necessary, assume that

$b_i < 0$ for $1 \le i \le k$ and $b_i \ge 0$ for $k+1 \le i \le m$. Then, in order to apply the simplex method to (1), the first k equations must be multiplied by (-1), resulting in all non-negative terms in the right-hand column. However, now an initial basic feasible solution may not be apparent and, if not, artificial variables must be introduced and the simplex method initiated at stage one. Thus the initial tableau would look something like the following.

x_1		x_n	x_{n+1}		x_{n+k}	x_{n+k+1}		x_{n+m}	*Artificial variables*			
$-a_{11}$	$\cdots$	$-a_{1n}$	-1	$\cdots$	0	0	$\cdots$	0	1	$\cdots$	0	$-b_1$
$\vdots$												
$-a_{k1}$	$\cdots$	$-a_{kn}$	0	$\cdots$	-1	0	$\cdots$	0	0	$\cdots$	1	$-b_k$
$a_{k+1,1}$	$\cdots$	$a_{k+1,n}$	0	$\cdots$	0	1	$\cdots$	0	0	$\cdots$	0	b_{k+1}
$\vdots$												
a_{m1}	$\cdots$	a_{mn}	0	$\cdots$	0	0	$\cdots$	1	0	$\cdots$	0	b_m
$-c_1$	$\cdots$	$-c_n$	0	$\cdots$	0	0	$\cdots$	0				0

Since we have assumed that the problem of (1) has feasible solutions, the simplex method initiated on the above tableau will first drive the artificial variables from the basis and then second drive to the optimal value of the objective function. Let r_j, s_i, and t_i be defined just as in the proof of the Duality Theorem, for $1 \le j \le n$ and $1 \le i \le m$. Show that the proof given there can be extended to this case, with the only difference being that here $s_i = -t_i$ for $1 \le i \le k$.

6. Show that the r_j's as defined in the proof of the Duality Theorem measure the slack in the constraints of the dual problem at the $Y_0 = (s_1, s_2, \ldots, s_m)^t$ solution point.

7. *The Complementary Slackness Theorem.* Consider the primal and dual linear programming problems of (1) and (2) of Section 4.1. Suppose that $X^* = (x_1^*, x_2^*, \ldots, x_n^*)$ is a feasible solution to the primal, and $Y^* = (y_1^*, y_2^*, \ldots, y_m^*)$ is a feasible solution to the dual. Prove that X^* is an optimal solution to the primal and Y^* is an optimal solution to the dual if and only if

$$x_j^*\left(\sum_{i=1}^{m} a_{ij} y_i^* - c_j\right) = 0 \qquad \text{for } 1 \le j \le n$$

and

$$y_i^*\left(b_i - \sum_{j=1}^{n} a_{ij} x_j^*\right) = 0 \qquad \text{for } 1 \le i \le m$$

that is, if and only if, for each j, $x_j^* > 0$ implies that the slack variable of the corresponding dual constraint is 0, and similarly for each y_i^*.

(*Hint.* Expand on the proof of Theorem 1 of this section.)

8. Verify that the optimal solutions $X^* = (10,9,0,0)$ and $Y^* = (2,0,3)$ for the primal and dual problems, respectively, of Example 2 of this section satisfy the conditions of the Complementary Slackness Theorem of Problem 7.

9. Consider the linear programming problem in primal form of

$$\text{Maximizing} \quad x_1 + 2x_2$$

subject to

$$\begin{aligned} 2x_1 + x_2 &\leq 3 \\ x_1 + 2x_2 &\leq 3 \\ x_1, x_2 &\geq 0 \end{aligned}$$

(a) Determine the dual problem.
(b) Show that $X^* = (1,1)$ and $Y^* = (0,1)$ are optimal solutions for the primal and dual problems, respectively, by using the Complementary Slackness Theorem.
(c) Note that at these solution points, both y_1^* and the slack in the corresponding first constraint of the primal are zero.

10. Consider the linear programming problem in primal form of

$$\text{Maximizing} \quad 2x_1 + 2x_2$$

subject to

$$\begin{aligned} x_1 + + x_3 + x_4 &\leq 1 \\ x_2 + x_3 - x_4 &\leq 1 \\ x_1 + x_2 + 2x_3 \phantom{{}+x_4} &\leq 3 \\ x_1, x_2, x_3, x_4 &\geq 0 \end{aligned}$$

(a) Determine the dual problem.
(b) Show that $X^* = (1,1,0,0)$ and $Y^* = (1,1,1)$ are feasible solutions to the primal and dual problems, respectively.
(c) Show that for this pair of solutions, for each j, $x_j^* > 0$ implies that the slack variable in the corresponding dual constraint is zero.
(d) Show that Y^* is not an optimal solution to the dual.
(e) Does this contradict the Complementary Slackness Theorem?

5
Sensitivity Analysis

SECTION 5.1 EXAMPLES IN SENSITIVITY ANALYSIS

In some applications of linear programming there may be a need not only to optimize a given function under specified conditions, but also to evaluate the effects changes in the conditions of the problem have on the optimal solution. For example, it could be that some of the coefficients a_{ij} of the coefficient matrix A are just approximations, and it would be desirable to know how their variance effects the optimal solution. Or it could be that the results of purchasing raw materials from other sources, yielding altered cost coefficients c_j, or of expanding one's storage capacities, yielding altered constants b_i, are to be measured. Techniques to handle such problems fall into the categories of *sensitivity analysis* and *parametric programming*, depending on whether the changes in the constants of the problem are discrete or continuous.

In this section we introduce, through two examples, some concepts involved in such an analysis. In this first example we work with a linear programming problem with only two variables and two constraints. Our analysis of this problem is based on graphs in the plane, available because of the limited size of the problem, and the Duality Theorem. In the second example, a problem with three constraints, our analysis uses only the Duality Theorem, but is not as complete as that in the first example. Sensitivity analysis techniques for more general problems are developed in the subsequent sections of the chapter.

Example 1

Consider the problem of the poultry producer in Problem 7 of Section 2.2. His problem is to provide an adequate diet for his stock at minimal costs from two available feeds. Letting x_1 and x_2 denote the number of pounds of Feeds 1 and 2 in the diet, the linear programming problem is to

$$\text{Minimize} \quad 16x_1 + 14x_2$$

subject to

$$\begin{aligned} 10x_1 + 4x_2 &\geq 124 \\ 3x_1 + 5x_2 &\geq 60 \\ x_1, x_2 &\geq 0 \end{aligned}$$

The graph of the set of feasible solutions is sketched in Figure 5.1. Now the value of the objective function can be computed at the three vertices and its minimal value thus determined. Or, we can apply the argument presented in Section 2.2. The lines of the form $16x_1 + 14x_2 = c$, c a constant, all have slope $-\frac{8}{7}$, and the graph of one such line appears above. It is clear from the graph that the minimal valuc of $16x_1 + 14x_2$ is attained at the point (10,6), and thus is 244, since 244 is the smallest value of c for which a line in the family of lines $16x_1 + 14x_2 = c$ intersects the set of feasible solutions.

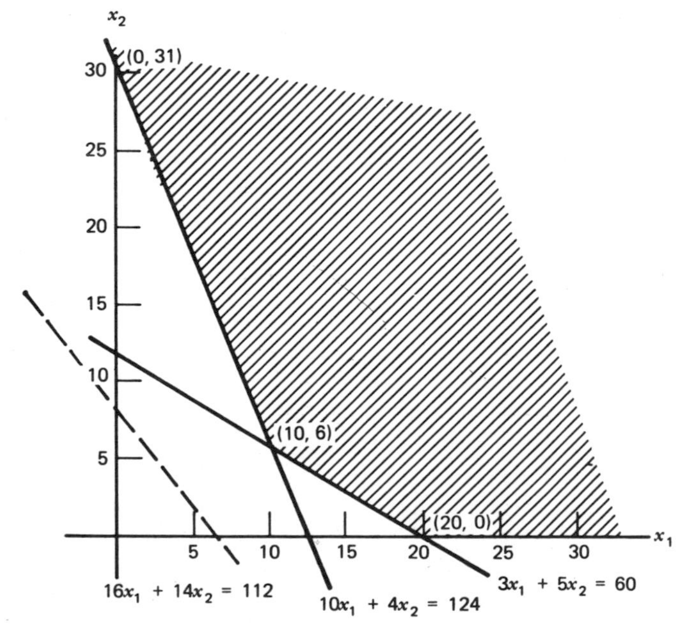

Figure 5.1

Now, suppose the costs of the two feeds vary due to market conditions, weather patterns, labor negotiations, and the like. The poultry producer would like to know when the (10,6) diet is still the minimal cost diet. To answer this question, consider Figure 5.1. It can be seen that as long as the slope of the lines in the family of lines determined by the objective function is between the slopes of the two lines intersecting at (10,6), then this point will provide the minimal cost diet. Algebraically, let c_1 and c_2 denote the costs of Feeds 1 and 2, respectively. The slope of the line $c_1x_1 + c_2x_2 = c$, where c is any constant, is $-c_1/c_2$. The slopes of the two lines intersecting at (10,6) are $-\frac{5}{2}$ and $-\frac{3}{5}$. Thus the condition on c_1 and c_2 is simply that

$$-\frac{5}{2} \leq -\frac{c_1}{c_2} \leq -\frac{3}{5}$$

or

$$\frac{3}{5} \leq \frac{c_1}{c_2} \leq \frac{5}{2}$$

Thus, as long as the ratio of the costs of Feed 1 to Feed 2 is between $\frac{3}{5}$ and $\frac{5}{2}$, 10 lb of Feed 1 and 6 lb of Feed 2 provide an adequate diet at minimal costs. (It is this type of consideration that is at the root of Problems 3 and 4 of Section 2.2.)

Suppose now that the poultry producer questions the effect of varying the daily nutritional requirements on the minimal cost of an adequate diet. It could be that the 124 units and 60 units required of elements A and B were estimates, and the producer would like to know what can be saved by decreasing these amounts. Or maybe the producer has discovered that by increasing the amounts of one or both of the nutrients, the stock has a higher market value, and he wonders how this increased value compares with the increased feeding costs.

To answer this question, consider the dual linear programming problem. Assuming that the costs of the two feeds are 16 and 14 cents/lb, as originally stated, the dual problem is to

$$\text{Maximize} \quad 124y_1 + 60y_2$$

subject to

$$\begin{aligned} 10y_1 + 3y_2 &\leq 16 \\ 4y_1 + 5y_2 &\leq 14 \\ y_1, y_2 &\geq 0 \end{aligned}$$

From the graph in Figure 5.2 we see that the maximum value of the function $124y_1 + 60y_2$ is attained at the point (1,2), and equals 244, the minimum value in the original problem. In fact, from the Duality Theorem, we know that the maximum value of the function $124y_1 + 60y_2$ on the hatched region above always equals the minimal cost of an adequate diet in the original problem.

We use this fact to estimate the effect of varying the required amounts of the nutritional elements. Let b_1 and b_2 denote the minimal amounts of elements A

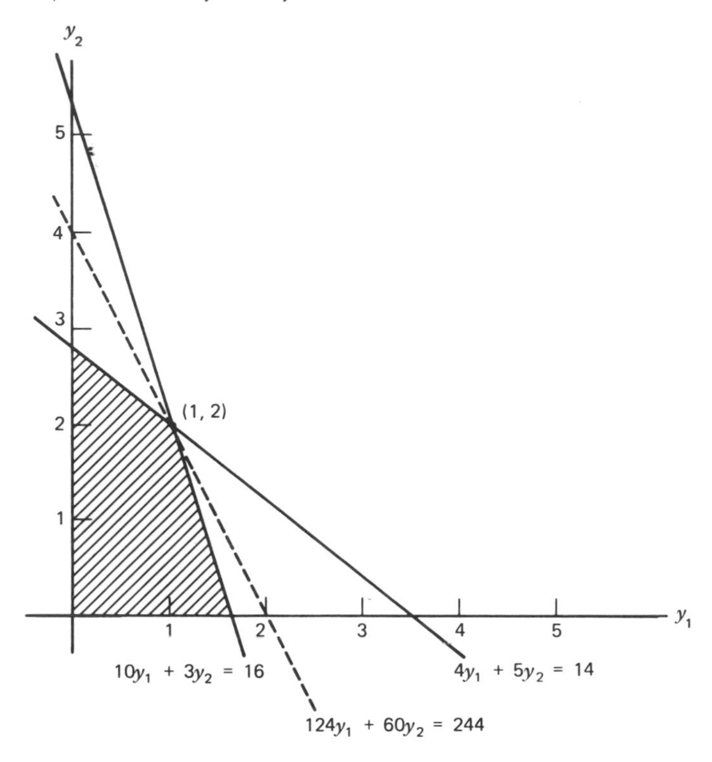

Figure 5.2

and B, respectively, required daily. Under these more general conditions, the only component changed in the dual problem is the objective function, which becomes $b_1y_1 + b_2y_2$. Now, as before, as long as the slope of the lines in the family of lines $b_1y_1 + b_2y_2 = c$, c a constant, remains between the slopes of the two lines intersecting at the point (1,2), the maximum value of the objective function will still be attained at this point. These two slopes are $-\frac{10}{3}$ and $-\frac{4}{5}$. Thus, as long as $-\frac{10}{3} \leq -b_1/b_2 \leq -\frac{4}{5}$, or $\frac{4}{5} \leq b_1/b_2 \leq \frac{10}{3}$, the maximum value of the dual objective function which, from the Duality Theorem, also equals the minimal cost of an adequate diet, will be $b_1 + 2b_2$. Thus the second question is answered. As long as the ratio of the number of required units of element A to element B is between $\frac{4}{5}$ and $\frac{10}{3}$, the minimal cost in cents to the producer is the number of required units of A plus twice the number of required units of *B*. Hence each required unit of A is costing the producer 1 cent, and each unit is costing 2 cents. Furthermore, it is clear that this analysis could be extended to include all possible ratios of the required units (see Problem 3).

In the second part of the above example we were able to measure the consequence of variations in the daily nutritional requirements for the stock on the minimal cost of an adequate diet; that is, we measured the effect of changes in the constant terms of the constraints of the problem on the optimal value of the objective function. We used the Duality Theorem along with the graph of the set of feasible solutions to the dual problem. Actually, for a given linear programming problem, even if the associated dual problem has more than two variables and graphing techniques are unavailable, we can still derive, using the Duality Theorem, useful information relating changes in the constant terms of the constraints of the original problem to the optimal value of the objective function. We illustrate with an example.

Example 2

Consider the problem of the boat manufacturer in Example 1 of Section 2.3. The problem is to determine a maximal profit production schedule in the manufacturing and selling of rowboats and canoes utilizing limited resources of aluminum, machine time, and labor. Letting R and C denote the number of rowboats and canoes, respectively, to be manufactured, the linear programming problem is to

$$\text{Maximize} \quad z = 50R + 60C$$

subject to

$$\begin{aligned} 50R + 30C &\leq 2000 \\ 6R + 5C &\leq 300 \\ 3R + 5C &\leq 200 \\ R,C &\geq 0 \end{aligned} \tag{1}$$

The constant terms of the constraints come from the limits on the resources of aluminum (2000 lb), machine time (300 min), and labor (200 hr). As determined in Problem 4 of Section 2.3, the maximum possible profit is \$2750 earned in the production of 25 rowboats and 25 canoes.

Suppose that the boat manufacturer has the opportunity to purchase more aluminum. With additional aluminum available, (possibly) more boats can be produced. But the manufacturer needs to know how the profits from the additional sales would compare with the cost of the extra aluminum. In other words, how much should he be willing to pay for more aluminum?

In terms of the original problem of (1), what we need is a measure of the effect of a change in the constant of the first inequality on the maximum value of the objective function. To determine this consider the dual to the problem of (1), and its solution. The dual is to

$$\text{Minimize } v = 2000y_1 + 300y_2 + 200y_3$$

subject to

$$\begin{aligned} 50y_1 + 6y_2 + 3y_3 &\geq 50 \\ 30y_1 + 5y_2 + 5y_3 &\geq 60 \\ y_1,y_2,y_3 &\geq 0 \end{aligned}$$

The optimal value of v is 2750 attained at $y_1 = \frac{7}{16}$, $y_2 = 0$, $y_3 = \frac{75}{8}$ (see Problem 2 of Section 4.2). Note that the coefficients of the dual objective function v are, by definition, the constant terms from the original constraints. In fact, from the Duality Theorem, we have that

$$\text{Max } z = \text{Min } v = 2000(\tfrac{7}{16}) + 300(0) + 200(\tfrac{75}{8})$$

It follows that as long as $(\frac{7}{16}, 0, \frac{75}{8})$ is an optimal solution point for the dual, the minimum of v and therefore the maximum profit will increase by $\$(\frac{7}{16}) \doteq 44$ cents for each available pound of aluminum above the original 2000 pounds. We can now answer the original question. Since the profit figures of \$50 for a rowboat and \$60 for a canoe would be determined by subtracting the cost of the required aluminum, machine time, and labor for each from the selling price, the manufacturer should be willing to pay for additional aluminum up to about 44 cents/lb more than he paid for the original 2000 lb of aluminum. For example, if he can purchase 48 lb for only \$15 more than the original cost of 48 lb of aluminum, he can increase his sales by $\$(\frac{7}{16})48 = \$21$ and, therefore, have a net gain of \$6.

Obviously we can extend this analysis to the other resources. Since $y_3 = \frac{75}{8} \doteq 9.38$ in the optimal solution to the dual, each additional hour of finishing labor would increase profits by \$9.38. Similarly, $y_2 = 0$ implies that an increase in available machine time over the original 300 min will provide no increase in profits. Actually, we could have reasoned to this by examining the original optimal solution to (1). As can be easily calculated, the production schedule of $R = C = 25$ utilizes all 2000 lb of aluminum and 200 hr of labor, but only 275 of the available 300 min of machine time. (The Complementary Slackness Theorem of Problem 7 of Section 4.3 is relevant here.) Thus production is restricted by the limited amounts of aluminum and labor available. Machine time is an underutilized resource, so increasing its availability has no effect on profits.

In sum, the solution $(\frac{7}{16}, 0, \frac{75}{8})$ to the dual has provided estimates on the effects changes in the constant terms of the problem have on the optimal value of the objective function. These numbers are sometimes called the *shadow prices* or *marginal values* of the constraints, since each provides some indication of the worth or cost on the optimal value of the objective function of a unit of the resource or demand generating the associated constraint. Certainly, however, there are limitations on their use. For example, the boat manufacturer would not be able to make unlimited profits even if there were available an unlimited supply of aluminum, because the other two constraints would still restrict the total production. Thus the estimate of an increased profit of 44 cents/lb for each additional pound of aluminum available is accurate only to some upper limit. (See Problem 7.) Once the change or changes in the constant terms of the original problem effect a change in the optimal solution point to the dual, these marginal values will change. In fact, in the first example of this section, by using geometry we were able to state precisely, in terms of the ratio of the two constant terms, when a change in these constant terms would alter the optimal solution point to the dual. For a general problem this question can be much more involved. Techniques for dealing with it are developed in the subsequent sections of this chapter.

PROBLEM SET 5.1

Problems 1–3 refer to the Example 1 of this section.

1. Show that if the ratio c_1/c_2 of the costs of the two feeds is less than $\frac{3}{5}$, the minimal cost diet is achieved using 20 lb of Feed 1 only; and that if this ratio is greater than $\frac{5}{2}$, the minimal cost diet uses 31 lb of Feed 2 only. What can be used if the ratio equals $\frac{3}{5}$ or $\frac{5}{2}$?

Assume for Problems 2 and 3 that the costs of the feeds are fixed at the original 16 and 14 cents.

2. Suppose that by doubling the number of units of nutritional element B for 2 weeks, the producer can realize $15 more from the sale of his stock. Is this worthwhile?

3. Show that if the ratio b_1/b_2 of the required daily amounts of the nutritional elements A and B is less than $\frac{4}{5}$, the minimal cost in cents of an adequate diet is $\frac{14}{5}$ times the number of required units of element B; and if this ratio is greater than $\frac{10}{3}$, the cost is $\frac{8}{5}$ times the number of required units of A.

4. Suppose the minimal value of the objective function $c_1x_1 + \cdots + c_nx_n$ of a linear programming problem is attained at the point X_0 with the first m variables as basic variables. True or false:
(a) If a c_j, $m + 1 \le j \le n$, is increased, the minimal value is still attained at X_0?
(b) If a c_j, $1 \le j \le m$, is decreased, the minimal value is still attained at X_0?

5. Consider the example in Problem 6 of Section 2.3. (See also Problem 6 of Section 3.5 and Problem 5 of Section 4.2.) Suppose the manager of the electronics firm wants some estimate on how the altering of the input of material and labor affects the maximum income earned from the sale of the circuits. Show that if b_1 and b_2 denote the number of units of material and labor, respectively, committed to the production of the circuits, then: if $b_1/b_2 \le \frac{3}{2}$, the maximum attainable income in dollars is $5b_1$; if $\frac{3}{2} \le b_1/b_2 \le 2$, the maximum income is $b_1 + 6b_2$; and if $b_1/b_2 \ge 2$, the maximum income is $8b_2$.

6. Using two raw materials, a firm can produce up to three different products. Inputs and profit per unit of production of each of the products are given by the following table.

	Input		
	Raw Material B_1	*Raw Material* B_2	*Profit* \$
Product 1	4	3	45
Product 2	2	1	17
Product 3	2	9	30

Suppose b_1 units of B_1 are available, and b_2 units of B_2.

(a) Formulate the mathematical model for the problem of determining a production schedule that maximizes profit.

(b) Express as a function of b_1 and b_2 the maximum profit function. (Your answer should resemble the answer suggested for Problem 5.)

7. This problem refers to Example 2 of this section.

(a) Show that an increase of 320 lb in the supply of aluminum increases the maximum profit by only \$116.67 and not $\$(\frac{7}{16})\ 320 = \$140$. (To do this you could solve the problem with the altered constant term by using the simplex algorithm. Or you might work with Figure 7.1 of Section 7.4. General and more efficient techniques will be developed later in this chapter.)

(b) Assuming that the amounts of the other two resources remain at their original amounts, use the graph of feasible solutions to the original problem (Figure 7.1 of Section 7.4) to show that once the amount of additional aluminum available exceeds $266\frac{2}{3}$ lb, the marginal value of 44 cents/lb for aluminum is no longer valid. What would be the marginal value of any additional aluminum over $266\frac{2}{3}$ lb?

(c) Assuming that the amounts of available aluminum is 2000 lb and machine time is 300 min, at what point will the marginal value of \$9.38 for additional hours of labor exceeding the original 200 hr become inoperative? Would additional hours of labor beyond this point have marginal value zero?

8. Reconsider Problem 5c of Section 4.2.

(a) Determine the marginal values for a unit of material and a unit of labor.

(b) Generalize these results. Suppose a total of b_1 units of material and b_2 units of labor are on hand. Determine the marginal values of these resources as a function of the ratio b_1/b_2.

9. Determine the marginal values for the resources of fabric, wood, and labor in Problem 11 of Section 2.3.

10. Consider the diet problem of the Example of Section 2.2. In Example 1 of Section 4.2 the shadow prices for the three nutritional requirements were determined to be 1 cent/unit for nutritional elements A and B, and 0 cents/unit for C. Suppose that because of inflation the cost per pound of each feed has been increased by 10 cents, so that now Feed 1 costs 20 cents/lb and Feed 2 costs 14 cents/lb. Determine the new shadow prices for the constraints.

11. Consider the transportation problem of Problem 2 of Section 2.4 and Example 3 of Section 4.2.

(a) Show that

$$x_{11} = 300,\ x_{12} = 300,\ x_{13} = 0,\ x_{14} = 0$$

$$x_{21} = 0,\ x_{22} = 50,\ x_{23} = 400,\ x_{24} = 450$$

and

$$y_1 = 24,\ y_2 = 0,\ y_3 = 44,\ y_4 = 40,\ y_5 = 50,\ y_6 = 44$$

are optimal solutions for the primal and dual problems. (You could use either Corollary 1 or Problem 7 of Section 4.3.)

(b) In the optimal shipping schedule Outlet 1 is supplied by Warehouse 1 at a cost of 20 cents/case. But an increase in demand at Outlet 1 would increase the total shipping cost by more than 20 cents/case for each additional case required. Explain why this follows from the optimal solution to the dual problem.

12. Consider the linear programming problem of (1) of Section 2.3.

(a) Show that (19.55,17.256,9.132,0) is a feasible solution.

(b) Determine the dual to the problem of (1).

(c) Show that (20.483,0,0,17.497,8.86) is a feasible solution to the dual.

(d) Show that these solutions are, in fact, optimal solution points for their respective problems. (Again, either Corollary 1 or Problem 7 of Section 4.3 can be used.)

(e) Referring to the last paragraph of Section 2.3, assist the division manager in responding to the vice president's questions.

(f) What does the marginal value for the first constraint of the original problem suggest about the use of overtime?

SECTION 5.2 MATRIX REPRESENTATION OF THE SIMPLEX TABLEAU

In this section we will develop formulas for the tableau presentation of a linear programming problem. In the ensuing sections we will use these formulas to analyze the effects of an alteration in the original data of a linear programming problem on an already determined optimal solution to the original problem. We will consider changes in the objective function coefficient vector c and in the constant column vector b, and will consider additions of a new variable (an additional column in the coefficient matrix A) and a new constraint (an additional row in A). Our primary goal is not to provide a complete catalog of sensitivity analysis tools (notice, for example, that we will not consider the effects of a change in the entries of the coefficient matrix A); we provide a deeper understanding of the questions that can be raised in postoptimality analysis and how the tools that we have at our disposal (and will have after this section and Section 5.6) can be used to answer these questions.

In matrix notation the standard form of the linear programming problem (Section 3.1) is to minimize $c \cdot X - z_0$ subject to $AX = b$, $X \geq 0$, where matrix A and column vectors c, b, and X are defined in the obvious manner (in fact, just as in Section 4.1). Let $A^{(j)}$ denote the jth column of the coefficient matrix A, for $j = 1,\ldots,n$. Then the matrix equation $AX = b$ is equivalent to the vector equation

$$\sum_{j=1}^{n} x_j A^{(j)} = b$$

in the m-dimensional column vectors $A^{(j)}$ and b. We have already discussed this equivalence in Example 3 of Section 3.2 (see pages 53–55).

We denote the initial tableau presentation of this problem in standard form as follows.

A	b
c	z_0

Suppose now the pivot operation is applied several times, with the resulting equivalent linear programming problem being in canonical form and having the following tableau representation.

A^*	b^*
c^*	z_0^*

Suppose the basic variables at this representation are $x_{j(1)},x_{j(2)},\ldots,x_{j(m)}$. Define the $m \times m$ matrix $B = [A^{j(1)},A^{j(2)},\ldots,A^{j(m)}]$, and the $1 \times m$ row vector $c_B = [c_{j(1)},c_{j(2)},\ldots,c_{j(m)}]$. In this section we will show that

$$A^* = B^{-1}A, \qquad b^* = B^{-1}b$$
$$c^* = c - c_B B^{-1}A, \qquad z_0^* = z_0 - c_B B^{-1}b$$

Thus the above tableau representation can be calculated from the original tableau data and these auxiliary terms B and c_B, with the resulting tableau given by

$B^{-1}A$	$B^{-1}b$
$c - c_B B^{-1}A$	$z_0 - c_B B^{-1}b$

Before these results are proven, we give an example.

Example

Consider the linear programming problem of pages 72–74. The initial and final tableaux from Tables 3.1 and 3.3 are given in Table 5.1. Here $j(1) = 5$ and $j(2) = 1$, and so $B = [A^{(5)},A^{(1)}] = \begin{bmatrix} 2 & -6 \\ 3 & -3 \end{bmatrix}$, and $c_B = [c_5,c_1] = [-2,5]$. Now

$$B^{-1} = \tfrac{1}{12}\begin{bmatrix} -3 & 6 \\ -3 & 2 \end{bmatrix} = \begin{bmatrix} -\frac{1}{4} & \frac{1}{2} \\ -\frac{1}{4} & \frac{1}{6} \end{bmatrix}$$

Table 5.1

	x_1	x_2	x_3	x_4	x_5	
x_3	-6	0	1	-2	2	6
x_2	-3	1	0	5	3	15
	5	0	0	3	-2	-21
x_5	0	$\frac{1}{2}$	$-\frac{1}{4}$	3	1	6
x_1	1	$\frac{1}{6}$	$-\frac{1}{4}$	$\frac{4}{3}$	0	1
	0	$\frac{1}{6}$	$\frac{3}{4}$	$\frac{7}{3}$	0	-14

and

$$B^{-1}A = \begin{bmatrix} -\frac{1}{4} & \frac{1}{2} \\ -\frac{1}{4} & \frac{1}{6} \end{bmatrix} \begin{bmatrix} -6 & 0 & 1 & -2 & 2 \\ -3 & 1 & 0 & 5 & 3 \end{bmatrix}$$

$$= \begin{bmatrix} 0 & \frac{1}{2} & -\frac{1}{4} & 3 & 1 \\ 1 & \frac{1}{6} & -\frac{1}{4} & \frac{4}{3} & 0 \end{bmatrix} = A^*$$

$$B^{-1}b = \begin{bmatrix} -\frac{1}{4} & \frac{1}{2} \\ -\frac{1}{4} & \frac{1}{6} \end{bmatrix} \begin{bmatrix} 6 \\ 15 \end{bmatrix} = \begin{bmatrix} 6 \\ 1 \end{bmatrix} = b^*$$

$$c - c_B B^{-1}A = c - c_B A^* = [5,0,0,3,-2] - [-2,5]A^*$$
$$= [5,0,0,3,-2] - [5,-\tfrac{1}{6},-\tfrac{3}{4},\tfrac{2}{3},-2]$$
$$= [0,\tfrac{1}{6},\tfrac{3}{4},\tfrac{7}{3},0] = c^*$$

$$z_0 - c_B B^{-1}b = z_0 - c_B b^* = -21 - (-7) = -14$$

In order to give some indication of how effective these formulas might be in sensitivity analysis, suppose that in this problem, the initial coefficient of x_4 in the objective function, $c_4 = 3$, is reduced from 3 to 1. Is the minimal value of z still 14 and attained at the point (1,0,0,0,6), or would a basic feasible solution using x_4 as a basic variable reduce the value of z? From the formulas for the final tableau, we see that this change in c_4 changes only the vector c and therefore only c^* (note that c_B remains fixed as c_1 and c_5 have not changed). In fact, since only the fourth component of c has changed, we need only calculate the new fourth component of c^*, which becomes

$$c_4 - c_B B^{-1}A^{(4)} = c_4 - c_B A^{*(4)}$$

$$= 1 - [-2,5]\begin{bmatrix} 3 \\ \frac{4}{3} \end{bmatrix} = \tfrac{1}{3}$$

Since $\frac{1}{3} \geq 0$, all the entries of the new c^* are nonnegative, the simplex process is terminated, and we have the same optimal solution as the original problem.

Now we begin the verification of these formulas.

Definition Let U_i denote the ith unit vector of R^m.

Lemma *The inverse of the matrix B always exists.*

Proof Define $Y = [y_1, y_2, \ldots, y_m]^t$. For fixed i, $i \leq i \leq m$, consider the system $BY = U_i$ of m equations in the m unknowns $y_1, \ldots, y_m$. By applying the same sequence of pivot steps to go from the original tableau

A	b
c	z_0

to the tableau

A^*	b^*
c^*	z_0^*

to this system of equations, one obtains an equivalent system through this application of elementary row operations. (At each pivot step of the original simplex operation, multiples of one fixed row of the system of equations $AX = b$ were added to the remaining rows. Here we apply these same elementary row operations to the system of equations $BY = U_i$.) Now, since in the final tableau, $x_{j(1)}, x_{j(2)}, \ldots, x_{j(m)}$ are the basic variables, this equivalent system will be in canonical form with the m basic variables $y_1, \ldots, y_m$. Thus this system of m equations and m unknowns has a solution, the results of these row operations on the column vector U_i. Denote this solution by the column vector $Y^{(i)}$ for each i, $1 \leq i \leq m$. Then

$$B[Y^{(1)}, Y^{(2)}, \ldots, Y^{(m)}] = [U_1, U_2, \ldots, U_m] = I$$

the $m \times m$ identity matrix. Therefore

$$B^{-1} = [Y^{(1)}, Y^{(2)}, \ldots, Y^{(m)}].$$

###

Example (continued)

Consider the system of equations $BY = U_1$; that is,

$$\begin{bmatrix} 2 & -6 \\ 3 & -3 \end{bmatrix} \begin{bmatrix} y_1 \\ y_2 \end{bmatrix} = \begin{bmatrix} 1 \\ 0 \end{bmatrix}$$

that is,

$$2y_1 - 6y_2 = 1$$
$$3y_1 - 3y_2 = 0$$

Using the row operations of the example on pages 72–74 (and the standard "detached coefficients" notation), we have

$$\left[\begin{array}{cc:c} 2 & -6 & 1 \\ 3 & -3 & 0 \end{array}\right] \sim \left[\begin{array}{cc:c} 1 & -3 & \frac{1}{2} \\ 0 & 6 & -\frac{3}{2} \end{array}\right] \sim \left[\begin{array}{cc:c} 1 & 0 & -\frac{1}{4} \\ 0 & 1 & -\frac{1}{4} \end{array}\right]$$

Therefore $y_1 = -\frac{1}{4}$, $y_2 = -\frac{1}{4}$, and $Y^{(1)} = \begin{bmatrix} -\frac{1}{4} \\ -\frac{1}{4} \end{bmatrix}$.

Theorem 1 $b^* = B^{-1}b$, *or*, $Bb^* = b$.

Proof The two systems of equations $AX = b$ and $A^*X = b^*$ are equivalent and so have the same solution set. Define $X^* = [x_1^*, x_2^*, \ldots, x_n^*]^t$, a column vector in R^n, by

$$x_j^* = \begin{cases} b_k^*, & x_j \text{ is the basic variable in the } k\text{th equation of the final tableau} \\ 0, & x_j \text{ is a nonbasic variable in the final tableau} \end{cases}$$

Then X^* is simply the basic feasible solution associated with the final tableau, and $A^*X^* = b^*$. Therefore

$$AX^* = b; \qquad \text{that is, } \sum_{k=1}^{m} A^{j(k)} b_k^* = b, \qquad \text{that is, } Bb^* = b \qquad \# \# \#$$

Thus we have the result discussed in Section 3.2 (pages 53–55), that each basic feasible solution corresponds to an expression for the vector b as a linear combination of m of the column vectors $A^{(j)}$, $1 \leq j \leq n$.

Theorem 2 *For any* j, $1 \leq j \leq n$, $A^{*(j)} = B^{-1}A^{(j)}$.

Proof Fix j, $1 \leq j \leq n$. Suppose the initial column vector b of Theorem 1 is $A^{(j)}$. Then the resulting b^* would simply be $A^{*(j)}$, the result of the sequence of pivot steps on $A^{(j)}$. Thus Theorem 1 implies that $A^{*(j)} = B^{-1}A^{(j)}$. [Note that some of the entries of $A^{(j)}$ and $A^{*(j)}$ may be negative, but that this has no affect on the proof of Theorem 1.]

# # #

Corollary

$A^* = B^{-1}A$.

Note that if $A^{(j)}$ is U_k, the kth unit vector in R^m, then $B^{-1}A^{(j)} = B^{-1}U_k =$ the kth column of B^{-1}. Thus, if the unit vectors $U_1, \ldots, U_m$ are present in the initial coefficient matrix A, the matrix B^{-1} can be read off immediately if the A^* coefficient matrix is available. For example, in the example of this section, $A^{(2)} = U_2$ and $A^{(3)} = U_1$. Hence the third column of A^* will be the first column of B^{-1}, and the second column of A^* will be the second column of B^{-1}. The reader should verify this.

Theorem 3 $z_0^* = z_0 - c_B B^{-1}b = z_0 - c_B b^*$.

Proof The original linear programming problem is to minimize z with $z = -z_0 + c_1x_1 + \cdots + c_nx_n$ subject to $AX = b$, $X \geq 0$. After applying a sequence of pivot operations, we have the equivalent problem of minimizing z with $z = -z_0^* + c_1^*x_1 + \cdots + c_n^*x_n$ subject to $A^*X = b^*$, $X \geq 0$. These two expressions for the objective function, say $F(X) = -z_0 + c_1x_1 + \cdots + c_nx_n$ and $G(X) = -z_0^* + c_1^*x_1 + \cdots + c_n^*x_n$, although different expressions, have the same value at any point that satisfies the equivalent systems of equations $AX = b$ or $A^*X = b^*$ (see the discussion on the representation of the objective function in Section 3.2). Now X^*, as defined in the proof of Theorem 1, is a solution to the system of constraints, and so $F(X^*) = G(X^*)$. But $F(X^*) = -z_0 + c_B b^*$, $G(X^*) = -z_0^*$. Therefore, $-z_0^* = -z_0 + c_B b^*$, or $z_0^* = z_0 - c_B b^*$. # # #

Theorem 4 *For any* j, $1 \leq j \leq n$, $c_j^* = c_j - c_B B^{-1}A^{(j)} = c_j - c_B A^{*(j)}$.

Proof Fix j, $1 \leq j \leq n$. Suppose the initial column vector b in Theorem 3 is $A^{(j)}$, and the initial constant z_0 equals c_j. Then the resulting b^* would be $A^{*(j)}$, and the resulting z_0^* would be c_j^*. Thus Theorem 3, applied to the linear programming problem of minimizing z with $z = -c_j + c_1x_1 + \cdots + c_nx_n$ subject to $AX = A^{(j)}$, $X \geq 0$, implies that $c_j^* = c_j - c_B A^{*(j)}$. (Note again that negative entries in either $A^{(j)}$ or $A^{*(j)}$ will not affect the proof of Theorem 3.) # # #

Corollary

$c^* = c - c_B A^* = c - c_B B^{-1}A$.

The formulas of this section show that at any step of the simplex algorithm all the relevant data can be calculated easily from the original data of the problem, as long as we know the corresponding matrix B^{-1}. This suggests another approach to the simplex process. Instead of using the full tableau as we do to record the results of each step of the algorithm, suppose we simply record the original data along with the B^{-1} matrix. In fact, note that at each step of the simplex algorithm exactly one column of the B matrix is altered as exactly one variable in the basis is replaced by another variable. Rules can be given that prescribe the effects of this alteration on the B^{-1} matrix, and this leads to a modification of the simplex process known as the revised simplex method. For large problems this technique for recording the simplex process has the advantage that it can require less computer time, calculation, and memory; therefore it enables one to handle larger problems with fixed computer facilities. We will not develop the revised simplex method in this text; refer to the books of Dantzig [6] or Hadley [16].

PROBLEM SET 5.2

1. Consider the problem of Example 1 of Section 3.5.
(a) Determine B, B^{-1}, and c_B for the final tableau. Note that B^{-1} is contained in the final tableau.
(b) Verify the formulas of this section for this tableau.
(c) By how many units can $c_3 = -3$ be decreased before the point $x_1 = 8$, $x_2 = 18$, $x_3 = 0$ is no longer optimal?

2. Consider the problem of Example 1 of Section 3.6.
(a) Determine the matrix B^{-1} for the third tableau of Table 3.8.
(b) Determine B^{-1} for the final tableau of Table 3.9.
(c) What does this suggest about the data in the artificial variable columns of the coefficient matrix?

3. Consider the linear programming problem of

$$\text{Minimizing} \quad z = 3x_1 + 2x_2 + 5x_3 - 4x_4$$

subject to

$$\begin{aligned} x_1 - x_2 + 3x_3 + 2x_4 + 7x_5 &= 31 \\ -2x_1 + 3x_2 - 6x_3 + 4x_4 - x_5 &= 2 \\ x_1, x_2, x_3, x_4, x_5 &\geq 0 \end{aligned}$$

(a) Determine the B, B^{-1}, and c_B for the tableau presentation of this problem with basic variables x_3 and x_4.
(b) Express $b = [31,2]^t$ as a linear combination of $A^{(3)} = [3,-6]^t$ and $A^{(4)} = [2,4]^t$ using Theorem 1.
(c) Show that this solution to the system of constraints is an optimal solution to the linear programming problem. (Compute c^* using the corollary to Theorem 4.)

4. In the linear programming problem of Problem 3, suppose we select x_1 and x_3 as potential basic variables. Determine the corresponding matrix B. Does B have an inverse? Why does this not contradict the lemma of this section?

SECTION 5.3 CHANGES IN THE OBJECTIVE FUNCTION

Suppose that the linear programming problem of minimizing $z = c \cdot X - z_0$ subject to $AX = b$, $X \geq 0$, has been solved, with the final tableau given by

A^*	b^*
c^*	z_0^*

Suppose that X^* is the associated basic feasible solution, with $X^* = [x_1^*, \ldots, x_n^*]^t$ defined by

$$x_j^* = \begin{cases} b_k^*, \text{ if } x_j \text{ is the basic variable of the } k\text{th row of the final tableau} \\ 0, \text{ if } x_j \text{ is a nonbasic variable in the final tableau} \end{cases}$$

Now suppose that an initial cost c_j is altered. How will this affect the solution to the problem? To answer this question, suppose that the coefficient c_s is increased by an amount q, so that the new coefficient $\bar{c}_s = c_s + q$. (Note that a negative q would simply mean that $\bar{c}_s$ represents a decrease over c_s.) The only data that will be affected by this change is the bottom row of the final tableau, so X^* remains a basic feasible solution. Denote the new entries in this bottom row by $\bar{c}^*$ and $\bar{z}_0^*$. Then X^* will still be an optimal solution if $\bar{c}^* \geq 0$.

To determine if $\bar{c}^* \geq 0$, consider two cases. Suppose first that x_s is a nonbasic variable in the final tableau. Then the vector c_B defined in Section 5.2 is unchanged by the change in c_s. Since, from Section 5.2, $c^* = c - c_B B^{-1} A = c - c_B A^*$, the only change in $\bar{c}^*$ from c^* will be in the sth component, with

$$\bar{c}_s^* = c_s + q - c_B A^{*(s)} = c_s - c_B A^{*(s)} + q = c_s^* + q$$

Therefore $\bar{c}_s^* \geq 0$ if and only if $c_s^* + q \geq 0$; that is, $q \geq -c_s^*$. Thus, if q is positive, this inequality will always be satisfied, so the cost coefficient of a nonbasic variable can be increased without bound without affecting the optimal solution. (Intuitively, why should this be so?) And this cost coefficient can be decreased up to the amount c_s^* without changing the optimal solution. However, once c_s is reduced by an amount more than c_s^*, more iterations of the simplex method may be necessary to drive the problem to completion. Note that the simplex method would be initiated with the

A^*	b^*
$\bar{c}^*$	z_0^*

tableau, with $\bar{c}_s^*$ determined by the formula $\bar{c}_s^* = c_s^* + q$.

Consider now the case in which x_s is a basic variable. Suppose x_s is the basic variable isolated in the rth row of the final tableau. Then c_B and $\bar{c}_B$ differ only in their rth components, with the rth component of $\bar{c}_B$ being q more than the corresponding component of c_B.

Using $\bar{c}^* = \bar{c} - \bar{c}_B A^*$, we have, for $j \neq s$,

$$\bar{c}_j^* = c_j - c_B A^{*(j)} - q a_{rj}^* = c_j^* - q a_{rj}^*$$

and

$$\begin{aligned} \bar{c}_s^* &= \bar{c}_s - c_B A^{*(s)} - q a_{rs}^* \\ &= c_s + q - c_B A^{*(s)} - q \\ &= c_s^* = 0 \end{aligned}$$

since $a_{rs}^* = 1$ and $c_s^* = 0$ since x_s is a basic variable isolated in the rth row. Thus, in order that all $\bar{c}_j^* \geq 0$, we must have

$$0 \leq c_j^* - q a_{rj}^*, \quad \text{or } q a_{rj}^* \leq c_j^*, \qquad \text{for all } j,\ 1 \leq j \leq n,\ j \neq s$$

If $a_{rj}^* > 0$, we must have that $q \leq c_j^*/a_{rj}^*$. Since this must hold for all $a_{rj}^* > 0$, we must have $q \leq \text{Min}\{c_j^*/a_{rj}^* | a_{rj}^* > 0, j \neq s\}$. Similarly, if $a_{rj}^* < 0$, then q must satisfy $q \geq c_j^*/a_{rj}^*$, and so we must have $q \geq \text{Max}\{c_j^*/a_{rj}^* | a_{rj}^* < 0\}$.

Example

Suppose the problem of minimizing z, $z = 3x_1 - 2x_2 + 7x_3 + 10x_4 - 5x_5$, subject to two constraints, results in the final tableau of Table 5.2. Since x_3, x_4, and x_5 are nonbasic variables, c_3^*, c_4^*, and c_5^* determine the maximum amounts that c_3, c_4, or c_5 could be reduced before more pivot operations would be necessary to bring the problem to completion. Thus, if no other components of c are altered, c_3 could be reduced by up to 9 units before the optimal value of z is no longer attained at (7,3,0,0,0). Similarly, c_4 could be reduced by up to 5 units, or c_5 could be reduced by up to 3 units.

Consider now a change only in c_1, say, to $\bar{c}_1 = c_1 + q_1$. Since x_1 is the basic variable of the second row, the limits on q_1 are

$$q_1 \leq \text{Min}\left\{\frac{c_j^*}{a_{2j}^*}\middle| a_{2j}^* > 0, j \neq 1\right\} = \text{Min}\left\{\frac{9}{2}, \frac{5}{1}\right\} = \frac{9}{2}$$

Table 5.2

	x_1	x_2	x_3	x_4	x_5	
x_2	0	1	4	−1	−2	3
x_1	1	0	2	1	−4	7
	0	0	9	5	3	−15

and

$$q_1 \geq \text{Max}\left\{\frac{c_j^*}{a_{2j}^*}\middle| a_{2j}^* < 0\right\} = \text{Max}\left\{-\frac{3}{4}\right\} = -\frac{3}{4}$$

Therefore, if $-\frac{3}{4} \leq q_1 \leq \frac{9}{2}$, that is, if $\frac{9}{4} \leq \bar{c}_1 \leq \frac{15}{2}$, the point (7,3,0,0,0) remains an optimal solution.

Similarly, (7,3,0,0,0) remains optimal if $-\frac{7}{2} \leq \bar{c}_2 \leq \frac{1}{4}$ and all the other c_j's remain fixed. (See Problem 1.)

If, on the other hand, more than one c_j is changed, the new $\bar{c}^*$ can simply be calculated from the data of the final tableau to determine if the original solution remains optimal.

Example (continued)

Suppose in the above example c_1 is increased from 3 to 5, and c_4 is decreased from 10 to 6. Notice that while these new values for c_1 and c_4 are both within their corresponding ranges found above, we can no longer be certain that (7,3,0,0,0) remains an optimal solution as more than one coefficient has been changed. In fact, we have

$$\begin{aligned}\bar{c}^* &= \bar{c} - \bar{c}_B A^* \\ &= [5, -2, 7, 6, -5] - [-2, 5]\begin{bmatrix} 0 & 1 & 4 & -1 & -2 \\ 1 & 0 & 2 & 1 & -4 \end{bmatrix} \\ &= [0, 0, 5, -1, 11]\end{aligned}$$

Table 5.3

	x_1	x_2	x_3	x_4	x_5	
x_2	0	1	4	−1	−2	3
x_1	1	0	2	(1)	−4	7
	0	0	5	−1	11	−29
x_2	1	1	6	0	−6	10
x_4	1	0	2	1	−4	7
	1	0	7	0	7	−22

Thus $\bar{c}_4^* < 0$, and more iterations are necessary to complete the problem. Now

$$\bar{z}_0^* = z_0 - \bar{c}_B b^* = 0 - [-2, 5]\begin{bmatrix} 3 \\ 7 \end{bmatrix} = -29$$

and so, picking up the action at the modified, previously final tableau of Table 5.2, we have the tableaux of Table 5.3. Hence, with $\bar{c}_1 = 5$ and $\bar{c}_4 = 6$, the minimal value of z is 22 and is attained at the point (0,10,0,7,0).

PROBLEM SET 5.3

1. For the example of this section, verify that (7,3,0,0,0) remains optimal if $-\frac{7}{2} \leq \bar{c}_2 \leq \frac{1}{4}$ and all the other c_j's remain fixed.

2. Complete the original problem of the example if:
(a) c_3 is reduced to 0.
(b) c_3 is reduced to -2.
(c) c_3 is reduced to -3.
(d) c_5 is reduced to -9.
(e) c_1 is increased to 8.
(f) c_1 is reduced to 2.
(g) c_1 is reduced to 2 and c_5 is increased to -4.
(h) c_2 is increased to 0 and c_3 is reduced to 4.

3. Consider the linear programming problem of (1) of Section 3.3, with $c_1 = -4$, $c_2 = 1$, $c_3 = 1$, $c_4 = 6$, $c_5 = 3$. Determine the range of each c_j such that if all the other coefficients of c remain fixed at their original values, the point (1,0,0,0,6), as determined from the tableau of Table 3.3 of Section 3.5, remains an optimal solution point.

4. Consider the poultry producer problem developed in Example 1 of Section 5.1.
(a) Using the techniques of this section, determine the range of costs of the two feeds so that a singular change in one of the costs will not alter the optimal solution point (10,6).
(b) The answers to the above are $\frac{42}{5} \leq c_1 \leq 35$ and $\frac{32}{5} \leq c_2 \leq \frac{80}{3}$. Relate these results to the results of page 129, that is, that the point (10,6) remains optimal if $\frac{3}{5} \leq c_1/c_2 \leq \frac{5}{2}$.

5. Show that an optimal solution remains optimal even if several c_j's are changed simultaneously, as long as the only altered c_j's correspond to nonbasic variables of the final optimal tableau, and none of these are reduced by more than their respective bounds c_j^*.

6. True or false: Suppose that for a completed linear programming problem an initial cost c_j is altered, but that the previously determined optimal solution remains optimal. Then the previously determined optimal value z_0^* for the objective function does not change?

7. Reconsider Problem 7 of Section 5.1 using the ideas developed in this section (and the tableaux you calculated to answer Problem 2c of Section 4.2.)

8. Consider the diet problem of the example of Section 2.2 and Example 1 of Section 4.2. The shadow prices for the three nutritional requirements, from the final tableau of Table 4.2 (page 112), are 1 cent/unit for nutritional elements A and B, and 0 cents/unit for C. Determine, using the final tableau of Table 4.2, the range on the required number of units of A such that for any nutritional requirement on A within this range, with the requirements for B and C fixed at 84 and 72, respectively, the above shadow prices remain accurate. Similarly, determine the corresponding ranges for the requirements for B and for C.

SECTION 5.4 ADDITION OF A NEW VARIABLE

Suppose that the simplex process has been used to solve the problem of minimizing $z = c \cdot X - z_0$ subject to $AX = b$, $X \geq 0$, with the final tableau given by

A^*	b^*
c^*	z_0^*

and with X^* the associated basic feasible solution. However, suppose that now we wish to add another variable in the formulation of the original problem. For example, this variable could measure the amount of goods to be shipped by means of a newly opened shipping line, or it could measure the amount of production of a certain new side product for which a market has just been developed. Let x_{n+1} be this new variable, with cost coefficient c_{n+1} and column vector of coefficients for the constraining equations $A^{(n+1)}$. Then the expanded, modified problem is to minimize z, with $z = c \cdot X + c_{n+1}x_{n+1}$, subject to

$$[A, A^{(n+1)}]\begin{bmatrix} X \\ x_{n+1} \end{bmatrix} = b, \qquad X \geq 0, x_{n+1} \geq 0 \tag{1}$$

Now X^* is still a basic feasible solution to (1) if we simply set the value of the nonbasic variable x_{n+1} equal to 0. Moreover, this point will provide an optimal solution if $c_{n+1}^* \geq 0$. And $c_{n+1}^* = c_{n+1} - c_B B^{-1} A^{(n+1)}$, from Theorem 4 of Section 5.2. Thus this quantity can be easily calculated. If $c_{n+1}^* \geq 0$, the original optimal solution remains optimal. If $c_{n+1}^* < 0$, the data for the new $(n + 1)$ column in the system of constraints, $A^{*(n+1)}$, is needed in order to proceed with the simplex process. But $A^{*(n+1)} = B^{-1}A^{(n+1)}$, from Theorem 2 of Section 5.2, and so this information can also be easily calculated.

Example

Consider the example of pages 72–74 and Section 5.2. Suppose we wish to introduce a new variable x_6 with the problem becoming the following.

Minimize z with $x_i \geq 0, 1 \leq i \leq 6$

$$\begin{aligned}
-6x_1 \qquad\quad + x_3 - 2x_4 + 2x_5 - 4x_6 &= 6 \\
-3x_1 + x_2 \qquad\quad + 5x_4 + 3x_5 + 6x_6 &= 15 \\
5x_1 \qquad\qquad\quad + 3x_4 - 2x_5 + x_6 &= -21 + z
\end{aligned}$$

Since the optimal value for the original problem is attained at the point (1,0,0,0,6), a basic feasible solution for this expanded problem is (1,0,0,0,6,0). This solution remains optimal if $c_6^* \geq 0$. Now,

$$\begin{aligned}
c_6^* &= c_6 - c_B B^{-1} A^{(6)} \\
&= 1 - [-2,5]\begin{bmatrix} -\frac{1}{4} & \frac{1}{2} \\ -\frac{1}{4} & \frac{1}{6} \end{bmatrix}\begin{bmatrix} -4 \\ 6 \end{bmatrix} = -1
\end{aligned}$$

Thus (1,0,0,0,6,0) is no longer optimal, and $A^{*(6)}$ must be calculated. Now

$$\begin{aligned}
A^{*(6)} &= B^{-1} A^{(6)} \\
&= \begin{bmatrix} -\frac{1}{4} & \frac{1}{2} \\ -\frac{1}{4} & \frac{1}{6} \end{bmatrix}\begin{bmatrix} -4 \\ 6 \end{bmatrix} = \begin{bmatrix} 4 \\ 2 \end{bmatrix}
\end{aligned}$$

To complete the problem, we expand the final tableau of Table 5.1 to include this new column, and use the simplex algorithm. The resulting tableaux are given in Table 5.4. The minimum value of z is now $13\frac{1}{2}$, and is attained at the point $(0,0,0,0,4,\frac{1}{2})$.

Table 5.4

	x_1	x_2	x_3	x_4	x_5	x_6	
x_5	0	$\frac{1}{2}$	$-\frac{1}{4}$	3	1	4	6
x_1	1	$\frac{1}{6}$	$-\frac{1}{4}$	$\frac{4}{3}$	0	(2)	1
	0	$\frac{1}{6}$	$\frac{3}{4}$	$\frac{7}{3}$	0	-1	-14
x_5	-2	$\frac{1}{6}$	$\frac{1}{4}$	$\frac{1}{3}$	1	0	4
x_6	$\frac{1}{2}$	$\frac{1}{12}$	$-\frac{1}{8}$	$\frac{2}{3}$	0	1	$\frac{1}{2}$
	$\frac{1}{2}$	$\frac{1}{4}$	$\frac{5}{8}$	3	0	0	$-13\frac{1}{2}$

PROBLEM SET 5.4

1. Solve the linear programming problem of the example in this section if a variable $x_6 \geq 0$ is added, with:

(a) $A^{(6)} = \begin{bmatrix} 2 \\ 1 \end{bmatrix}$, $c_6 = -1$

(b) $A^{(6)} = \begin{bmatrix} 2 \\ 1 \end{bmatrix}$, $c_6 = -2$

(c) $A^{(6)} = \begin{bmatrix} 8 \\ 6 \end{bmatrix}$, $c_6 = 9$

(d) $A^{(6)} = \begin{bmatrix} 8 \\ 6 \end{bmatrix}$, $c_6 = -9$

(e) $A^{(6)} = \begin{bmatrix} 8 \\ 6 \end{bmatrix}$, $c_6 = -11$

2. Suppose a variable $x_5 \geq 0$ is added to the problem of Example 1 of Section 3.6, with the problem becoming:

$$\text{Minimize} \quad 2x_1 - 3x_2 + x_3 + x_4 + 5x_5$$

subject to

$$\begin{aligned} x_1 - 2x_2 - 3x_3 - 2x_4 + 4x_5 &= 3 \\ x_1 - x_2 + 2x_3 + x_4 + 3x_5 &= 11 \\ x_1,x_2,x_3,x_4,x_5 &\geq 0 \end{aligned}$$

Solve this expanded problem.

SECTION 5.5 CHANGES IN THE CONSTANT COLUMN VECTOR

Suppose that we have solved the problem of minimizing z, $z = c \cdot X - z_0$, subject to $AX = b$, $X \geq 0$, but that now we must alter the constant column vector b. In fact, suppose that the quantities q_i are to be added to each b_i, $1 \leq i \leq$ m. Using bars to again denote the terms of this new problem, we have then that

$$\bar{b} = b + q$$

where $q = (q_1, \ldots, q_m)^t$.

However, changing the entries in the constant term column will have no affect on the c_j^*'s of the final tableau of the original problem. Thus the solution of this new problem will have the same basic variables as the solution to the original problem if the altered entries $\bar{b}^*$ in the constant term column of the final tableau remain nonnegative. To determine $\bar{b}^*$, we can use, from Theorem 1 of Section 5.2,

$$\bar{b}^* = B^{-1}\bar{b} = B^{-1}(b + q) = B^{-1}b + B^{-1}q = b^* + B^{-1}q \tag{1}$$

Example

Consider Example 3 of Section 3.5. Suppose that b_1 is decreased by 2 units and b_2 increased by 3; that is,

$$b = \begin{bmatrix} 20 \\ 10 \\ 60 \end{bmatrix}, \quad q = \begin{bmatrix} -2 \\ 3 \\ 0 \end{bmatrix}, \quad \text{and } \bar{b} = b + q = \begin{bmatrix} 18 \\ 13 \\ 60 \end{bmatrix}$$

From the final tableau of Table 3.6 (page 77), using the entries of the seventh, second, and fifth columns, we have

$$B^{-1} = \begin{bmatrix} \frac{1}{12} & -\frac{7}{24} & \frac{1}{24} \\ -\frac{1}{6} & \frac{1}{12} & \frac{5}{12} \\ -\frac{1}{2} & \frac{1}{4} & \frac{1}{4} \end{bmatrix}$$

and so

$$\begin{aligned} \bar{b}^* &= b^* + B^{-1}q \\ &= \begin{bmatrix} \frac{5}{4} \\ \frac{45}{2} \\ \frac{15}{2} \end{bmatrix} + \begin{bmatrix} \frac{1}{12} & -\frac{7}{24} & \frac{1}{24} \\ -\frac{1}{6} & \frac{1}{12} & \frac{5}{12} \\ -\frac{1}{2} & \frac{1}{4} & \frac{1}{4} \end{bmatrix} \begin{bmatrix} -2 \\ 3 \\ 0 \end{bmatrix} = \begin{bmatrix} \frac{5}{24} \\ \frac{277}{12} \\ \frac{37}{4} \end{bmatrix} \geq \begin{bmatrix} 0 \\ 0 \\ 0 \end{bmatrix} \end{aligned}$$

Therefore the final tableau of Table 3.6, with the b^* entries replaced with these $\bar{b}^*$ entries and the z_0^* value corrected, provides the final tableau for the modified problem. The optimal value of the objective function is now attained at the point $(\frac{5}{24},0,0,\frac{277}{12},0,\frac{37}{4},0)$. The new optimal value can be easily determined by simply evaluating the original objective function at this solution point. This value is -227.

On the other hand, if b_2 were to be increased by 4 units, we would have

$$\bar{b}^* = b^* + B^{-1}\begin{bmatrix} -2 \\ 4 \\ 0 \end{bmatrix} = \begin{bmatrix} -\frac{1}{12} \\ \frac{139}{6} \\ \frac{19}{2} \end{bmatrix}$$

Thus the point $(-\frac{1}{12},0,0,\frac{139}{6},0,\frac{19}{2},0)$ is a solution to this new system of constraints, but not a feasible solution. Note that now how to proceed is not at all clear. Since the simplex method as we have developed it must move from feasible solution to feasible solution, it would seem that we are forced to initiate the simplex algorithm back on the original problem but with this new constant term column of 18, 14, 60, and that we cannot make use of the data of the final tableau of Table 3.6. However, there is a variation of the simplex algorithm that could be used here with negative entries in the constant term column, and that, in this case, would save considerable effort. In the next section we will develop the algorithm; then, with this new tool at our disposal, we will come back to the above problem.

PROBLEM SET 5.5

1. Consider the problem of Example 1 of Section 3.6. Determine both the minimal value of z and a point at which this value is attained, if:
   (a) b_1 is increased from 3 to 7 and b_2 is decreased from 11 to 8.
   (b) b_1 is increased from 3 to 8 and b_2 is decreased from 11 to 8.

2. Consider the linear programming problem of Example 1 of Section 3.5. Determine the maximum value of the objective function and a point at which this value is attained if
   (a) b_2 is increased from 10 to 30 units, b_1 and b_3 remaining unchanged.
   (b) b_1, b_2, and b_3 are each decreased by 10 units from their original values.

3. Consider the problem of Example 2 of Section 3.5, a linear programming problem with an unbounded objective function. Using equation (1), solve this problem if b_2 is increased from 3 to 7.

4. Consider the problem of Example 2 of Section 4.3.
   (a) Show that if the quantity q_1 is added to $b_1 = 20$ while b_2 and b_3 remain at their original values of 30 and 24, respectively; then, as long as $-20 \leq q_1 \leq 84$, the optimal value of the objective function is $112 + 2q_1$. [Use equation (1) to determine the limits on q_1 so that $\bar{b}^* \geq 0$. Then calculate the value of the objective function at the corresponding basic feasible solution.]
   (b) Show that if the quantity q_2 is added to b_2 while b_1 and b_3 remain at their original values, then, as long as $q_2 \geq -21$, the optimal value of the objective function is still 112.
   (c) Make a similar analysis assuming that q_3 is added to b_3 while b_1 and b_2 remain at their original values.
   (d) What are the marginal values for the three constraints? How are these numbers related to the above results?

SECTION 5.6 THE DUAL SIMPLEX ALGORITHM

In the example considered at the end of the last section, we saw that changes in the initial values of the constant term column vector b can bring about a linear programming problem in the following form. The system of constraints is in canonical form with a specified set of basic variables, the objective function is expressed in terms of the nonbasic variables, and the corresponding coefficients c_j are nonnegative, but the associated basic solution is not feasible; that is, the constant term column contains negative entries. With the simplex method as we have developed it, the only way we have of handling such a problem would be to multiply those equations with negative constant terms by (-1), add artificial variables if necessary to put the problem into canonical form, and proceed with the two stage simplex process.

Example 1

$$\text{Minimize} \quad z = 10x_1 + 5x_2 + 4x_3$$

subject to

$$\begin{aligned} &3x_1 + 2x_2 - 3x_3 \geq 3 \\ &4x_1 \qquad\quad + 2x_3 \geq 10 \\ &x_1, x_2, x_3 \geq 0 \end{aligned} \tag{1}$$

Adding slack variables x_4 and x_5, we have

$$\begin{aligned} 3x_1 + 2x_2 - 3x_3 - x_4 \qquad &= 3 \\ 4x_1 \qquad\quad + 2x_3 \qquad - x_5 &= 10 \\ 10x_1 + 5x_2 + 4x_3 \qquad\qquad &= z \end{aligned}$$

To apply the simplex process to this problem, we would now add two artificial variables and proceed on. On the other hand, by multiplying the two constraints by (-1), we have the following.

$$\begin{aligned} -3x_1 - 2x_2 + 3x_3 + x_4 \qquad &= -3 \\ -4x_1 \qquad\quad - 2x_3 \qquad + x_5 &= -10 \\ 10x_1 + 5x_2 + 4x_3 \qquad\qquad &= z \end{aligned} \tag{2}$$

Notice that the problem expressed in this way is in the form described above; that is, the system of constraints is in canonical form with basic variables x_4 and x_5, the objective function is expressed in terms of the nonbasic variables x_1, x_2, and x_3, and the associated coefficients 10, 5, and 4 are nonnegative, but that the associated basic solution $x_1 = x_2 = x_3 = 0$, $x_4 = -3$, $x_5 = -10$ is not feasible.

In this section we will develop an algorithm for resolving problems in this form. The algorithm, called the Dual Simplex Algorithm, is intimately related to the dual problem of the linear programming problem under consideration. We will develop this relationship after describing the steps of the algorithm by means of the above example.

The basic step of the Dual Simplex Algorithm is the pivot operation that we have already seen. However, this algorithm differs from the standard simplex process by the rules used to determine the pivot term at each step. In this algorithm, at each step first the row to pivot in is determined and then the column is determined. Thus here we determine first what variable to extract from the basis, and then we determine what variable to enter into the basis. We demonstrate the technique with the following example.

Example 1 (continued)

Consider the tableau presentation in Table 5.5 of the problem as stated in (2).

Table 5.5

	x_1	x_2	x_3	x_4	x_5	
x_4	-3	-2	3	1	0	-3
x_5	-4	0	-2	0	1	-10
	10	5	4	0	0	0

To apply the Dual Simplex Algorithm, we determine first the row to pivot in. According to the algorithm, the pivot term can be in any row with a negative constant term. In this tableau, $b_1 = -3$ and $b_2 = -10$; therefore the pivot term can come from either row. An arbitrary rule to use in such a case is to pivot in that row with the smallest b_i term, and so, here, to pivot in the second row, extracting x_5 from the basis.

Next we determine what column to pivot in. The algorithm dictates that the pivot term be at a negative a_{ij} entry, and so, here, the pivot term will be either at $a_{21} = -4$ or $a_{23} = -2$. To determine at which entry we pivot, the ratios c_j/a_{rj} must be considered for those $a_{rj} < 0$ (where r is the pivoting row), and the pivot term be in that column, say column s, for which

$$\frac{c_s}{a_{rs}} = \text{Max}\left\{\frac{c_j}{a_{rj}} \middle| a_{rj} < 0\right\}$$

In this case we compare $c_1/a_{21} = \frac{10}{-4} = -\frac{5}{2}$ with $c_3/a_{23} = \frac{4}{-2} = -2$. The maximum occurs in the third column, and therefore we pivot at $a_{23} = -2$. (Note that here we are comparing two nonpositive ratios and seeking the maximum, and therefore are actually seeking that ratio of minimum absolute value. By nature of the algorithm, this will always be the case.)

Pivoting here, we have the tableaux of Table 5.6.

Table 5.6

	x_1	x_2	x_3	x_4	x_5	
x_4	-3	-2	3	1	0	-3
x_5	-4	0	$\boxed{-2}$	0	1	-10
	10	5	4	0	0	0
x_4	-9	-2	0	1	$\frac{3}{2}$	-18
x_3	2	0	1	0	$-\frac{1}{2}$	5
	2	5	0	0	2	-20

Notice that the c_j^* entries, the 2, 5, 0, 0, and 2, remain nonnegative. Our choice of pivoting column guarantees this. Now we proceed on. In the second tableau, $b_1^* = -18$ is the only negative constant term, so we must pivot in the first row. Comparing those ratios corresponding to negative a_{rj}^* terms, we have $c_1^*/a_{11}^* = -\frac{2}{9} > c_2^*/a_{12}^* = -\frac{5}{2}$, and so we pivot at the $a_{11}^* = -9$ term. The resulting tableau is in Table 5.7.

Note that after this step, the constant term column entries are nonnegative. In fact, with the original problem presented in this form, we see that we have reached the resolution of the problem as Theorem 1 of Section 3.4 applies. The minimum value of the objective function is 24 and is attained at the point (2,0,1,0,0).

Table 5.7

	x_1	x_2	x_3	x_4	x_5	
x_4	$\boxed{-9}$	-2	0	1	$\frac{3}{2}$	-18
x_3	2	0	1	0	$-\frac{1}{2}$	5
	2	5	0	0	2	-20
x_1	1	$\frac{2}{9}$	0	$-\frac{1}{9}$	$-\frac{1}{6}$	2
x_3	0	$-\frac{4}{9}$	1	$\frac{2}{9}$	$-\frac{1}{6}$	1
	0	$\frac{41}{9}$	0	$\frac{2}{9}$	$\frac{7}{3}$	-24

We summarize now the steps of the *Dual Simplex Algorithm*: Consider the linear programming problem of minimizing $z = c \cdot X - z_0$ subject to $AX = b$, $X \geq 0$ (with b not necessarily ≥ 0).

0. Assume that
   (a) The system of constraints is in canonical form with a specified set of basic variables.
   (b) The objective function z is expressed in terms of the nonbasic variables only, and the corresponding coefficients c_j are all nonnegative.

1. If all the $b_i \geq 0$, the minimum value of the objective function has been attained (Theorem 1 of Section 3.4 applies).

2. If there exists an r such that $b_r < 0$ and $a_{rj} \geq 0$ for all j, the system of constraints has no feasible solutions.

3. Otherwise pivot. To determine the pivot term:
   (a) Pivot in any row with a negative b_i term. If there are several negative b_i terms, pivoting in that row with the smallest b_i may reduce the total number of steps necessary to complete the problem. Assume we pivot in row r.
   (b) To determine the column of the pivot term, find that column, say column s, such that

$$\frac{c_s}{a_{rs}} = \text{Max}\left\{\frac{c_j}{a_{rj}} \,\middle|\, a_{rj} < 0\right\}$$

4. After pivoting, the problem will remain in the form described in Step 0. Thus now return to Step 1.

There are some obvious questions associated with the algorithm. We list some here. (See also Problems 1 and 2.)

1. If the problem initially is not in the form described in Step 0, is there a systematic way of driving the problem into this form?

2. If the pivoting term is chosen as in Step 3, why will the problem remain in the form described in Step 0?

3. Will this algorithm always terminate; that is, will we always reach a point where either Step 1 or Step 2 applies?

The answer to the first question is yes. However, for our purposes, we will need and use the algorithm only for problems already in the form described in Step 0, and we will not develop here any of the techniques for putting a problem into the desired form. Refer to Dantzig [6] and Lemke [21].

The answers to the other two questions follow from the relationship between the Dual Simplex Algorithm and the standard simplex algorithm applied to the dual of the original problem. To see this, consider the linear programming problem in the form described above in Step 0. To determine its dual we do the following.

1. Consider the basic variables of the system of constraints as slack variables with coefficients + 1. Drop them from the problem, replacing the equations of the constraints with ($\leq$) inequalities.
2. Multiply each constraint by (-1), creating a minimization problem with ($\geq$) inequalities for constraints.

The dual of this equivalent problem is readily determined, since the problem is now in the form of the dual problem described on page 104. In fact, the dual would be a maximization problem, with the coefficients of the objective function corresponding to the negative of the b_i terms of the original problem, and the constant terms of the dual corresponding to the c_j terms of the original.

To apply the simplex process to this dual problem, it must first be put into canonical form. The slack variables added to the ($\leq$) inequalities in the system of constraints can serve as the basic variables, and the associated basic solution is feasible as the coefficients c_j of the original objective function were assumed to be nonnegative. Finally, multiplication of the coefficient of the dual objective function by -1 produces the required minimization problem.

The simplex algorithm can now be applied. Notice that the coefficients of the nonbasic variables of the objective function are precisely the b_i terms of the original problem, so that determining the pivoting column corresponds directly to Step 3a of the Dual Simplex Algorithm. And the a_{ij} entries in this pivoting column correspond to the negative of the a_{ij} entries in the pivoting row determined by Step 3a of the Dual Simplex Algorithm. Thus determination of the pivoting row here corresponds directly to Step 3b of the Dual Simplex Algorithm.

Instead of attempting to write out in precise terms the relationship between the algorithms for a general linear programming problem and its dual, we will demonstrate the relationships by means of the example of this section.

Example 1 (continued)

The problem as stated in (2) is to minimize z with

$$\begin{aligned}
-3x_1 - 2x_2 + 3x_3 + x_4 \quad &= -3 \\
-4x_1 \quad\quad - 2x_3 \quad + x_5 &= -10 \\
10x_1 + 5x_2 + 4x_3 \quad\quad &= z
\end{aligned}$$

Dropping the basic variables, the constraints become

$$\begin{aligned}
-3x_1 - 2x_2 + 3x_3 &\leq -3 \\
-4x_1 \quad\quad - 2x_3 &\leq -10
\end{aligned}$$

Multiplying each by (-1), we have the equivalent problem of minimizing z with

$$
\begin{aligned}
3x_1 + 2x_2 - 3x_2 &\geq 3 \\
4x_1 \qquad\quad + 2x_3 &\geq 10 \\
10x_1 + 5x_2 + 4x_3 &= z
\end{aligned}
$$

[Note that this is simply the problem in (1).]

The dual problem is to maximize v with

$$
\begin{aligned}
3y_1 + 4y_2 &\leq 10 \\
2y_1 \qquad &\leq 5 \\
-3y_1 + 2y_2 &\leq 4 \\
3y_1 + 10y_2 &= v \\
y_1, y_2 &\geq 0
\end{aligned}
$$

Table 5.8

	y_1	y_2	y_3	y_4	y_5	
y_3	3	4	1	0	0	10
y_4	2	0	0	1	0	5
y_5	-3	(2)	0	0	1	4
	-3	-10	0	0	0	0
y_3	(9)	0	1	0	-2	2
y_4	2	0	0	1	0	5
y_2	$-\frac{3}{2}$	1	0	0	$\frac{1}{2}$	2
	-18	0	0	0	5	20
y_1	1	0	$\frac{1}{9}$	0	$-\frac{2}{9}$	$\frac{2}{9}$
y_4	0	0	$-\frac{2}{9}$	1	$\frac{4}{9}$	$\frac{41}{9}$
y_2	0	1	$\frac{1}{6}$	0	$\frac{1}{6}$	$\frac{7}{3}$
	0	0	2	0	1	24

The equivalent problem in canonical form is to minimize $-v$ with

$$\begin{aligned}
3y_1 + 4y_2 + y_3 &= 10 \\
2y_1 + y_4 &= 5 \\
-3y_1 + 2y_2 + y_5 &= 4 \\
-3y_1 - 10y_2 &= (-v) \\
y_1, y_2, y_3, y_4, y_5 &\geq 0
\end{aligned}$$

Compare now the steps of the simplex algorithm applied to this problem (Table 5.8) with the steps of the Dual Simplex Algorithm applied to the original problem.

Example 2

We conclude this section by applying the Dual Simplex Algorithm to the unresolved problem of the example in Section 5.5. In that example we started with the problem of Example 3 of Section 3.5, but then altered the constant vector b from $\begin{bmatrix}20\\10\\60\end{bmatrix}$ to $\begin{bmatrix}18\\14\\60\end{bmatrix}$. Using the data from the final tableau of Table 3.6 (page 77), an equivalent form of this problem is represented by the tableau in Table 5.9. [The

Table 5.9

	x_1	x_2	x_3	x_4	x_5	x_6	x_7	
x_1	1	$-\frac{7}{24}$	$-\frac{7}{4}$	0	$\frac{1}{24}$	0	$\frac{1}{12}$	$-\frac{1}{12}$
x_4	0	$\frac{1}{12}$	$-\frac{5}{2}$	1	$\frac{5}{12}$	0	$-\frac{1}{6}$	$\frac{139}{6}$
x_6	0	$\frac{1}{4}$	$-\frac{5}{2}$	0	$\frac{1}{4}$	1	$-\frac{1}{2}$	$\frac{19}{2}$
	0	0	3	0	2	0	0	226

constant term column entries $\bar{b}^*$ were computed in the last section. The $\bar{z}_0^* = 226$ is simply the negative of the value of the objective function at the point $(-\frac{1}{12},0,0,\frac{139}{6},0,\frac{19}{2},0)$.] Applying the Dual Simplex Algorithm, we have the tableaux of Table 5.10.

Thus our problem is resolved after only one iteration. The minimal value of the objective function is -226 and is attained at the point $(0,\frac{2}{7},0,\frac{162}{7},0,\frac{66}{7},0)$.

Table 5.10

	x_1	x_2	x_3	x_4	x_5	x_6	x_7	
x_1	1	$\boxed{-\frac{7}{24}}$	$-\frac{7}{4}$	0	$\frac{1}{24}$	0	$\frac{1}{12}$	$-\frac{1}{12}$
x_4	0	$\frac{1}{12}$	$-\frac{5}{2}$	1	$\frac{5}{12}$	0	$-\frac{1}{6}$	$\frac{139}{6}$
x_6	0	$\frac{1}{4}$	$-\frac{5}{2}$	0	$\frac{1}{4}$	1	$-\frac{1}{2}$	$\frac{19}{2}$
	0	0	3	0	2	0	0	226
x_2	$-\frac{24}{7}$	1	6	0	$-\frac{1}{7}$	0	$-\frac{2}{7}$	$\frac{2}{7}$
x_4	$\frac{2}{7}$	0	-3	1	$\frac{3}{7}$	0	$-\frac{1}{7}$	$\frac{162}{7}$
x_6	$\frac{6}{7}$	0	-4	0	$\frac{2}{7}$	1	$-\frac{3}{7}$	$\frac{66}{7}$
	0	0	3	0	2	0	0	226

PROBLEM SET 5.6

1. Explain the conclusion of Step 2 in the Dual Simplex Algorithm.

2. If a linear programming problem is presented in the form described in Step 0 of the Dual Simplex Algorithm, there cannot exist a set of feasible solutions to the system of constraints on which the objective function is unbounded below. Prove this.

In Problems 3–5, solve the new problem, finding both the optimal value of the objective function and a point at which this optimal value is attained.

3. Consider the problem of Example 3 of Section 3.5 (see also the example of Section 5.5 and Example 2 of this section). Suppose the constant term vector b is changed from $\begin{bmatrix} 20 \\ 10 \\ 60 \end{bmatrix}$ to $\begin{bmatrix} 28 \\ 2 \\ 48 \end{bmatrix}$.

4. Consider the problem of Example 1 of Section 3.6 (see also Problem 2 of Section 5.5). Suppose b_1 is increased from 3 to 9 and b_2 is decreased from 11 to 8.

5. Consider the problem of Example 1 of Section 3.5 (see also Problem 2 of Section 5.5). Suppose the column vector b is changed from $\begin{bmatrix} 60 \\ 10 \\ 50 \end{bmatrix}$ to $\begin{bmatrix} 40 \\ 25 \\ 50 \end{bmatrix}$.

6. Solve Problem 10 of Section 5.1 using the ideas developed in the last two sections and the tableaux of Table 4.2 (page 112).

SECTION 5.7 ADDITION OF A CONSTRAINT

Suppose that after the optimal value for a linear programming problem has been found by means of the simplex method, we wish to alter the original problem by the addition of a new constraint. Now it could be that the optimal solution found previously satisfies this new constraint. If this is the case, this solution is also optimal for the expanded problem, because clearly, by this addition of a constraint, we have not changed the objective function nor increased the set of feasible solutions to the system of constraints. On the other hand, if the previous optimal solution does not satisfy this new constraint, we must find a new optimal solution. Under certain circumstances, however, this problem may be resolved quite easily by the application of the Dual Simplex Algorithm to data determined from the final tableau solution to the original problem. We demonstrate the technique by means of an example.

Example

Consider again the example of pages 72–74. Using the final tableau of Table 3.3 (page 74), the problem is to minimize z with

$$\begin{aligned}
\tfrac{1}{2}x_2 - \tfrac{1}{4}x_3 + 3x_4 + x_5 &= 6\\
x_1 + \tfrac{1}{6}x_2 - \tfrac{1}{4}x_3 + \tfrac{4}{3}x_4 &= 1\\
\tfrac{1}{6}x_2 + \tfrac{3}{4}x_3 + \tfrac{7}{3}x_4 &= -14 + z\\
x_1, x_2, x_3, x_4, x_5 &\geq 0
\end{aligned}$$

The minimal value is 14 and is attained at the point (1,0,0,0,6). Suppose now that we also demand that the solution satisfy the constraint

$$x_3 + 6x_4 + 4x_5 \leq 8$$

The point (1,0,0,0,6) does not satisfy this constraint. Thus we add the constraint to the problem and attempt to put this expanded problem into a form to which we can apply a solution algorithm. Adding the slack variable x_6 to the new constraint, we first seek a set of basic variables for this system of three equations and six unknowns. An obvious choice, as long as the added constraint introduces a slack variable, is this slack variable along with the basic variables from the final tableau of the original problem. Here then we can use as basic variables x_5, x_1, and x_6. The only operation that must be considered is the removal of the basic variables of the previous problem from the new constraint; clearly, this can always be done by simply adding appropriate multiples of the original constraints to the new constraint. Here by subtracting four times the first equation from the new constraint, the x_5 variable will be eliminated from this equation, and the resulting equivalent system of equations will be in canonical form. And the expression for the objective function from the last tableau still contains no basic variables, since the added basic variable x_6 is a slack variable with a zero cost coefficient. Moreover, the associated coefficients of the objective function are nonnegative, corresponding to a terminating tableau in the simplex algorithm, and so this new problem is now in a form amenable to the Dual Simplex Algorithm. The modified problem is to

Minimize z with

$$\begin{aligned}
\tfrac{1}{2}x_2 - \tfrac{1}{4}x_3 + 3x_4 + x_5 \qquad &= 6 \\
x_1 + \tfrac{1}{6}x_2 - \tfrac{1}{4}x_3 + \tfrac{4}{3}x_4 \qquad &= 1 \\
-2x_2 + 2x_3 - 6x_4 \qquad + x_6 &= -16 \\
\tfrac{1}{6}x_2 + \tfrac{3}{4}x_3 + \tfrac{7}{3}x_4 \qquad &= -14 + z \\
x_1, x_2, x_3, x_4, x_5, x_6 \geq 0 &
\end{aligned}$$

Table 5.11

	x_1	x_2	x_3	x_4	x_5	x_6	
x_5	0	$\frac{1}{2}$	$-\frac{1}{4}$	3	1	0	6
x_1	1	$\frac{1}{6}$	$-\frac{1}{4}$	$\frac{4}{3}$	0	0	1
x_6	0	$\boxed{-2}$	2	-6	0	1	-16
	0	$\frac{1}{6}$	$\frac{3}{4}$	$\frac{7}{3}$	0	0	-14
x_5	0	0	$\frac{1}{4}$	$\frac{3}{2}$	1	$\frac{1}{4}$	2
x_1	1	0	$\boxed{-\frac{1}{12}}$	$\frac{5}{6}$	0	$\frac{1}{12}$	$-\frac{1}{3}$
x_2	0	1	-1	3	0	$-\frac{1}{2}$	8
	0	0	$\frac{11}{12}$	$\frac{11}{6}$	0	$\frac{1}{12}$	$-\frac{46}{3}$
x_5	-3	0	0	4	1	$\frac{1}{2}$	1
x_3	-12	0	1	-10	0	-1	4
x_2	-12	1	0	-7	0	$-\frac{3}{2}$	12
	11	0	0	11	0	1	-19

Table 5.11 displays the results of the Dual Simplex Algorithm. The minimal value of the objective function with the added constraint is 19 and is attained at the point (0,12,4,0,1).

PROBLEM SET 5.7

1. Solve the problem of Example 1 of Section 3.6, if the following constraints are added.
(a) $x_1 + x_2 + 2x_3 + 2x_4 \leq 37$
(b) $x_1 + x_2 + 2x_3 + 2x_4 \leq 17$
(c) $x_1 - x_2 \leq 7$
(d) $x_1 + 2x_2 + 20x_3 - 2x_4 \geq 38$

2. Consider the linear programming problem in the example of this section, that is, the problem of minimizing z with

$$\begin{aligned} \tfrac{1}{2}x_2 - \tfrac{1}{4}x_3 + 3x_4 + x_5 &= 6 \\ x_1 + \tfrac{1}{6}x_2 - \tfrac{1}{4}x_3 + \tfrac{4}{3}x_4 \qquad &= 1 \\ \tfrac{1}{6}x_2 + \tfrac{3}{4}x_3 + \tfrac{7}{3}x_4 \qquad &= -14 + z \\ x_1, x_2, x_3, x_4, x_5 \geq 0 & \end{aligned}$$

Solve the problem when the following constraints are added.
(a) $x_4 + x_5 = 5$
(b) $x_4 + x_5 = 7$
Hint. In each, first eliminate the basic variable x_5 from the new constraint. Now, since there is no slack variable to be added, we must add an artificial variable to this new constraint to complete the basis. However, we do not have to introduce the w function and use the two stage simplex method to solve the problem. We can drive this artificial variable out of the basis with one iteration of the Dual Simplex Algorithm, as long as the constant term in the new constraint is negative. In part a, this will be the case. In part b, this will be the case if we simply multiply the new constraint, modified by the elimination of the x_5 variable, by -1 before adding the artificial variable.

3. Solve the problem of Example 1 of Section 3.5, if the following constraints are added.
(a) $2x_1 - x_2 - 5x_3 \geq 12$
(b) $x_3 \geq 6$
(c) $2x_1 + x_2 \geq 39$
(d) $x_1 + x_2 + x_3 = 20$
(e) $x_1 + x_2 + x_3 = 27$

4. In the example of this section, a $(\leq)$ inequality was added to the set of constraints of a linear programming problem. The solution point to the original problem did not satisfy this new constraint, and so the expanded problem was modified first by the elimination of the basic variable (the x_5 variable) from the new constraint, resulting in a constraint with a negative constant term (here -16). Since this term was negative, we were able to initiate the Dual Simplex Algorithm after the introduction of a slack variable. What would we have done if this resulting constant term were nonnegative?

6 Integer Programming

SECTION 6.1 INTRODUCTION TO INTEGER PROGRAMMING

The term "integer programming" (or "mixed integer programming") refers to the study of linear programming problems for which the domains of all (or some) of the variables of the problem are restricted to be integral. Models leading to such problems readily occur. For example, suppose we are considering a transportation problem, where the units to be shipped are automobiles, refrigerators, or soldiers, or a production problem, where the units to be produced are homes, swimming pools, or submarines. Certainly the optimal solution to such a problem cannot contain fractional values. (How would you attempt to explain to your Vice-President of Profits that the Superstructure and Girder Division that you manage is planning to optimize revenue by constructing $1\frac{5}{7}$ radio broadcast towers and $2\frac{3}{5}$ bridges?) There are various other optimization problems that can be successfully formulated as integer programming problems, problems that initially do not suggest the use of variables restricted to be integral. We will see some examples of such problems in Section 6.2.

In this section a numerical example of an integer programming problem will be discussed in order to demonstrate some of the difficulties involved with such problems. Then, in Sections 6.3 and 6.4, we will develop two different algorithms that can be used to calculate optimal solutions to integrally restricted problems. The first algorithm will use a cutting plane method of solution; the second will use a branch and bound approach. Both algorithms will make use of the Dual Simplex Algorithm.

Suppose we have a linear programming problem with integral restrictions on some or all of the variables. The most obvious possible solution technique would be initially to ignore the integral restrictions, solve the corresponding linear programming problem using whatever algorithm or means is suitable, and round off this calculated solution to integral values. However, this method fails, as the following example will demonstrate.

Example

$$\text{Maximize } f(x_1,x_2) = 3x_1 + 13x_2$$

subject to

$$2x_1 + 9x_2 \le 40$$
$$11x_1 - 8x_2 \le 82$$
$$x_1,x_2 \ge 0 \text{ and integral}$$

Table 6.1

	x_1	x_2	x_3	x_4	
x_3	2	(9)	1	0	40
x_4	11	-8	0	1	82
	-3	-13	0	0	0
x_2	$\frac{2}{9}$	1	$\frac{1}{9}$	0	$4\frac{4}{9}$
x_4	($\frac{115}{9}$)	0	$\frac{8}{9}$	1	$117\frac{5}{9}$
	$-\frac{1}{9}$	0	$\frac{13}{9}$	0	$57\frac{7}{9}$
x_2	0	1	$\frac{11}{115}$	$-\frac{2}{115}$	$2\frac{2}{5}$
x_1	1	0	$\frac{8}{115}$	$\frac{9}{115}$	$9\frac{1}{5}$
	0	0	$\frac{167}{115}$	$\frac{1}{115}$	$58\frac{4}{5}$

If we ignore the integral restrictions, we can apply the simplex algorithm (after adding slack variables x_3 and x_4). From the tableaux of Table 6.1, the maximum value of f for the simple linear programming problem is $58\frac{4}{5}$ and is attained at the point $(9\frac{1}{5},2\frac{2}{5})$. Now, for the original problem with the integrally restricted variables, it would seem reasonable that we should simply round off the above solution point $(9\frac{1}{5},2\frac{2}{5})$ to (9,2), or maybe (10,2), or (10,3), or (9,3). However, none of these four points are feasible; the first three do not satisfy the second inequality, and the last two do not satisfy the first. Thus the simplex algorithm has provided us with no useful information, and it is not at all clear how one could proceed in general.

Actually, for this simple example with only two variables, we can graph the solution set to the system of constraints and work from there. This graph is sketched in Figure 6.1. There are 36 lattice points (points with both coordinates integral) in the region bounded by the constraints. Since the coefficients of the objective function are positive, there can be no lattice points in the feasible region that are to the right or above the point at which the maximal value of the objective function is attained, and so the optimal value of f must occur at either (2,4), (6,3), or (8,2). Now $f(2,4) = 58$, $f(6,3) = 57$, and $f(8,2) = 50$, so the maximal value of f is 58 and is attained at the point (2,4). Contrast here the proximities of the feasible lattice point (8,2) and the actual solution point (2,4) to the solution point of the nonintegrally restricted problem, $(9\frac{1}{5}, 2\frac{2}{5})$, and the difference in values of f at these two points, 50 and 58. Clearly, integer programming problems may require special techniques.

Figure 6.1

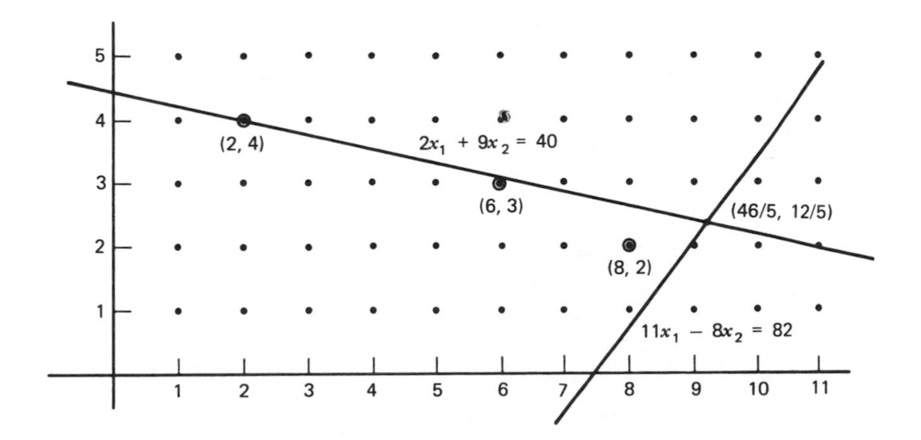

PROBLEM SET 6.1

1. Consider the problem of

$$\text{Maximizing} \quad 5x_1 + 2x_2$$

subject to

$$\begin{aligned} 6x_1 + 2x_2 &\leq 13 \\ -6x_1 + 7x_2 &\leq 14 \\ x_1, x_2 &\geq 0 \text{ and integral} \end{aligned}$$

(a) Using the simplex algorithm, show that the optimal value of the objective function of the problem with the integral restrictions ignored is $11\frac{5}{6}$ and is attained at the point $(1\frac{1}{6},3)$.

(b) Show graphically that the optimal solution to the original restricted variable problem is at (2,0).

Conclusion. Given an integer programming problem, the fact that some (but not all of the restricted) coordinates of the solution point of the corresponding linear programming problem are integral is of little value in solving the integer programming problem.

2. Consider the problem of

$$\text{Maximizing} \quad 2x_1 + 9x_2$$

subject to

$$\begin{aligned} 2x_1 + x_2 &\leq 20 \\ x_1 + 5x_2 &\leq 24 \\ x_1, x_2 &\geq 0 \text{ and integral} \end{aligned}$$

(a) Using the simplex algorithm, show that the optimal value of the problem without the integral constraints is attained at $(8\frac{4}{9},3\frac{1}{9})$.

(b) Show that the lattice point (8,3) is a feasible solution to the system of constraints.

(c) Show graphically (or algebraically) that the optimal solution to the original integrally restricted variable problem is at (4,4).

Conclusion. Given an integer programming problem, the fact that the coordinates of the solution point of the corresponding linear programming problem round off to a feasible solution to the system of constraints is of little value in solving the original problem.

3. Consider the problem of

$$\text{Maximizing} \quad x_1 + x_2$$

subject to

$$\begin{aligned} 3x_2 &\geq 1 \\ 3x_2 &\leq 2 \\ x_1, x_2 &\geq 0 \text{ and integral} \end{aligned}$$

(a) Solve the problem ignoring the integral restrictions.
(b) Solve the problem with the integral restrictions.
(c) Draw a conclusion.

4. Given an integer programming problem, suppose the optimal value of the corresponding linear programming problem is attained at a point with all coordinates integral. Is this point the solution point to the original problem?

SECTION 6.2 MODELS WITH INTEGER PROGRAMMING FORMULATIONS

In this section we will discuss various problem situations that can be formulated as integer (or mixed integer) programming models. Three general types will be discussed: the capital budget problem, the fixed charge problem, and the problem of alternative constraints. We want to stress that restricting the domains of some of the variables of a problem to a discrete set can be an effective tool in the formulating of mathematical models. However, although we are extending the range of our subject, the lack of efficiency of the presently known algorithms for solving integer programming problems sometimes makes the resolutions of large-scale problems of these types impractical using integer programming techniques.

The Capital Budget Problem

Suppose we have a fixed amount of a resource, such as capital, space, or machine time, that we wish to utilize in a way that maximizes profit or gain. Suppose also that the alternative methods through which our resource can be utilized are such that only multiples of fixed sized lots can be allocated to each. By introducing integrally restricted variables to correspond to the allotment made to each of the alternatives, such problems can be formulated as integer programming problems.

Example 1

An investment firm, wishing to maximize profit, has \$100,000 to invest in a construction project requiring an investment of \$48,000 and providing a profit of \$2900, in any number of units of a portfolio of stocks requiring an investment of \$19,000/unit and yielding a profit of \$1100/unit, and in any number of shares of a certain stock costing \$1750/share and yielding a profit of \$95/share.

To formulate a mathematical model, introduce three variables x_1, x_2, and x_3, defined as follows: x_1 will be 1 if the construction project is utilized and 0 if not, x_2

will be the number of units of the portfolio utilized, and x_3 will be the number of units of the stock utilized. The integer programming model is:

$$\text{Maximize} \quad 2900x_1 + 1100x_2 + 95x_3$$

subject to

$$48000x_1 + 19000x_2 + 1750x_3 \leq 100000$$
$$0 \leq x_1 \leq 1, x_2, x_3 \geq 0 \text{ and } x_1, x_2, x_3 \text{ integral}$$

Example 2 ***The Knapsack Problem***

An airplane can carry up to W lb of extra cargo on a scheduled flight. There are n different items that could be transported, with item i weighing a_i lb and providing a profit of c_i dollars if transported. What items should be shipped so as to maximize profit?

Define n variables x_i,

$$x_i = \begin{cases} 1, & \text{the } i\text{th item is to be shipped} \\ 0, & \text{if not} \end{cases}$$

The corresponding integer programming problem is:

$$\text{Maximize} \sum_{i=1}^{n} c_i x_i$$

subject to

$$\sum_{i=1}^{n} a_i x_i \leq W$$
$$0 \leq x_i \leq 1 \text{ and integral}, i = 1, \ldots, n$$

Note that the final models in these two examples are simple integer programming problems. This is our goal for each of the examples of this section, that is, to translate the problem at hand into a representative linear programming problem with integral restrictions on (possibly some of) the variables of the problem, so that the problem could be resolved by the use of available integer programming solution techniques. Remember that the introduction and interpretation of the variables for the problem is just an intermediate (but often the most difficult) step, and that it is contingent on the problem formulator to ascertain whether the final integer programming problem adequately reflects the situation at hand and, in particular, the desired interpretation of the variables.

The Fixed Charge Problem

Consider an operation in which the costs or profits associated can involve lump sum quantities, depending on whether or not certain processes or facilities or whatever are utilized, that is, situations in which there is a fixed charge or cost for the use of a process only if that process is used. For example, if a machine is to be used in the manufacture of a product, there could be a setup cost to prepare the equipment for the run. Or, if a new market is to be opened for the selling of a product, there could be an initial market development cost totally independent of the number of units eventually sold in that market. Or, in a transportation problem, there could be the rental costs of the warehouses utilized.

To formulate models for such operations, we introduce, for each fixed charge, an auxiliary variable, restricted to be either 0 or 1, with the interpretation that the variable will be 1 only if the corresponding fixed charge is to be assessed. Then, to the objective function of the model the term—the product of the fixed charge times the variable—is added, and to the constraints inequalities are added so that this "on or off" role of the auxiliary variable is maintained.

Example 3

Suppose a machine shop has three processes that it can use to manufacture two different parts, with each process combining various amounts of raw material and labor to produce different quantities of the two parts. Suppose the shop has weekly limits on its raw material and labor and must meet weekly fixed demands for the two parts. Assume that with each process there is both a setup and maintenance charge d_i if the process is utilized, and also a cost c_i for every hour of its operation. We wish to determine that weekly operation of the shop that minimizes total cost while meeting the given demands for the parts.

Define variables:

$$x_i = \text{number of hours per week that Process } i \text{ is used}$$

$$y_i = \begin{cases} 0, & x_i = 0 \\ 1, & x_i > 0 \end{cases}$$

Then the restrictions on the available raw material and labor would be reflected in ($\leq$) inequalities involving the x_i's, the weekly demands for the parts would correspond to ($\geq$) inequalities in the x_i's, and the objective function for the model would be

$$c_1x_1 + c_2x_2 + c_3x_3 + d_1y_1 + d_2y_2 + d_3y_3$$

Now all we need do is add constraints to regulate the y_i's. Suppose that there are known upper bounds on the x_i's independent of the final operation of the plant. In this problem, let us assume that each x_i must be less than or equal to 40. Then, to the constraints we add the inequalities

$$y_i \geq \frac{x_i}{40}$$

$$0 \leq y_i \leq 1 \text{ and integral, } i = 1,2,3$$

Notice that if the optimal operation calls for the use of Process i, the restrictions on y_i force it to be 1, since $x_i/40$ will be greater than 0, whereas if Process i is not to be used, the constraints would permit y_i to be 0 or 1, but the minimization of the objective function would lead to $y_i = 0$. Thus the desired role of these auxiliary variables is achieved.

Example 4

A distributor supplies n retail outlets, with Outlet j requiring d_j units monthly. The distributor can rent storage facilities in up to m warehouses, with Warehouse i having a storage capacity of s_i units and a monthly rental fee of r_i dollars. There is a cost of c_{ij} dollars to ship 1 unit from Warehouse i to Outlet j. Determine what warehouses are to be utilized in the implementation of a feasible shipping schedule that minimizes total costs. Notice that this problem would be a standard transportation problem if it were not for the rental fees of the occupied warehouses.

Define variables:

$$x_{ij} = \text{number of units shipped monthly from Warehouse } i \text{ to Outlet } j$$

$$y_i = \begin{cases} 0, \text{ if Warehouse } i \text{ is not utilized} \\ 1, \text{ if Warehouse } i \text{ is utilized} \end{cases}$$

Then an integer programming model for this problem is to:

$$\text{Minimize} \quad \sum_{i,j} c_{ij}x_{ij} + \sum_i r_i y_i$$

subject to

$$\sum_j x_{ij} \leq s_i, \qquad i = 1, \ldots, m$$

$$\sum_i x_{ij} \geq d_j, \qquad j = 1, \ldots, n$$

$$y_i \geq \sum_j \frac{x_{ij}}{s_i}, \qquad i = 1, \ldots, m$$

$$0 \leq x_{ij}, 0 \leq y_i \leq 1 \text{ and integral}$$

Notice that in a solution to this problem, the y_i's would indicate what warehouses are to be used, and the x_{ij}'s would provide a corresponding minimal-cost shipping schedule.

The Problem of Alternative Constraints

Consider a situation in which the amounts of the quantities involved, represented by the variables of the problem, must satisfy one (or more) set(s) from two (or more) alternative sets of constraints. For example, in the manufacture of a certain product, it may be economically feasible to produce either none at all or an amount exceeding some minimal batch size. Or, in the bidding for various contracts, those bids submitted must exceed certain minimal operating costs, and so the bids are either 0 or greater than or equal to some lower bounds. Or, in the utilization of the resources of a plant, one of several methods may be employed, and associated with each method is a corresponding system of constraints reflecting the limited supplies and required demands.

To formulate models for problems of this type, we again introduce auxiliary on or off variables equal to either 0 or 1, as in the fixed charge problem. Here, however, we use the variables to modify the system of constraints and not the terms of the objective function.

Example 5

Suppose a variable x_i, representing the amount of an item to be produced, must be either 0 or greater than or equal to a minimal batch size b_i. Introduce a variable y_i to be 0 if $x_i = 0$ and 1 if $x_i \geq b_i$, and suppose there exists an upper bound, say m_i, on x_i. Then, to the constraints we add

$$\begin{aligned} &x_i \leq m_i y_i \\ &x_i \geq b_i y_i \\ &0 \leq y_i \leq 1 \text{ and integral} \end{aligned}$$

Now y_i can equal only 0 or 1. If $y_i = 0$, the two inequalities force $x_i = 0$, and if $y_i = 1$, the first inequality has no significance but the second demands $x_i \geq b_i$. Thus x_i is properly restricted.

Example 6

Suppose nonnegative variables x_1 and x_2 must satisfy either $2x_1 + 9x_2 \leq 18$ or $x_1 + x_2 \leq 6$. The graph of the set of feasible points (x_1,x_2) is illustrated in Figure 6.2. (Notice that this set is not convex.) Introduce variables y_1 and y_2, with

$$y_1 = \begin{cases} 0, \text{ first inequality is not satisfied} \\ 1, \text{ first inequality is satisfied} \end{cases}$$

and similarly for y_2 and the second inequality.

Now, to formulate a model to reflect these interpretations, we again need upper bounds in the constraints. Notice that no matter what inequality a feasible point (x_1,x_2) satisfies, we must have $x_1 \leq 9$ and $x_2 \leq 6$. Thus, for any feasible point (x_1,x_2).

$$2x_1 + 9x_2 - 18 \leq 18 + 54 - 18 = 54$$

and

$$x_1 + x_2 - 6 \leq 9 + 6 - 6 = 9$$

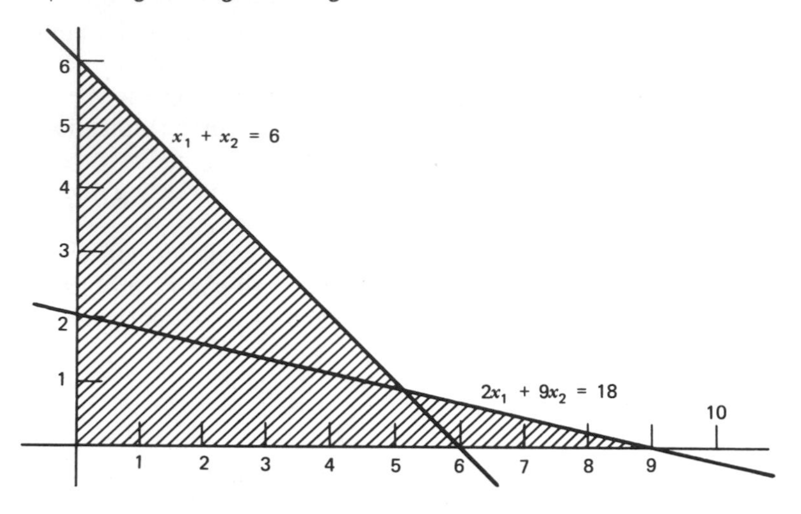

Figure 6.2

Using these bounds, we can express the system of constraints as follows.

$$\begin{aligned} 2x_1 + 9x_2 - 18 &\le 54(1 - y_1) \\ x_1 + x_2 - 6 &\le 9(1 - y_2) \\ y_1 + y_2 &\ge 1 \\ x_1, x_2 &\ge 0 \\ 0 \le y_1, y_2 &\le 1 \text{ and integral} \end{aligned} \tag{1}$$

To ascertain that this system properly reflects the original conditions of the problem, notice first that the inequality $y_1 + y_2 \ge 1$ demands that at least one of the y_i's $= 1$. And if $y_1 = 1$, the first inequality reduces to the original $2x_1 + 9x_2 \le 18$, and so if (x_1, x_2, y_1, y_2) is a solution to (1) with $y_1 = 1$, the point (x_1, x_2) must satisfy this inequality. Similarly, if $y_2 = 1$, a solution point must satisfy the original second inequality. Finally, if $y_1 = 0$, the first constraint places no meaningful restrictions on the solution set, because all feasible points (x_1, x_2) must satisfy $2x_1 + 9x_2 - 18 \le 54$. Similarly if $y_2 = 0$. In summary, if (x_1, x_2, y_1, y_2) is a solution to (1), then either $y_1 = 1$ or $y_2 = 1$ (or both), and the point (x_1, x_2) must satisfy the first original constraint if $y_1 = 1$ and the second if $y_2 = 1$.

Notice the nature of the end product in the above example. We have a simple collection of linear inequalities involving nonnegative variables, with some of the variables restricted to be integral. Thus, unless there were present other difficulties, integer programming techniques could be used with these constraints. Of course, if the optimization problem were simply to maximize, say $f(x_1, x_2) = 4x_1 + x_2$ subject to either $x_1 + x_2 \le 6$ or $2x_1 + 9x_2 \le 18$, with x_1 and $x_2 \ge 0$, one could consider each constraint separately, solve the corresponding linear programming problem, and compare the two optimal values. The larger value would be the solution to the problem with the either/or constraints. However, the techniques of the example do show some of the range of integer programming and also generalize quite easily to encompass more complicated situations.

Example 7

Consider the problem of determining all points (x_1,x_2,x_3) satisfying

$$x_1 + 2x_2 + 3x_3 \leq 600, \qquad x_1,x_2,x_3 \geq 0$$

and at least three of the following five alternative sets of inequalities:

$$\{5x_1 + 10x_2 \leq 500\}, \qquad \{x_1 - x_2 + x_3 \leq 450\}, \qquad \{x_2 + x_3 \geq 100\},$$

$$\begin{Bmatrix} x_1 - x_3 \leq 250 \\ x_3 \geq 50 \end{Bmatrix}, \qquad \begin{Bmatrix} x_1 - x_3 \leq 250 \\ x_2 + x_3 \leq 400 \end{Bmatrix}$$

Introduce discrete variables y_i, with

$$y_i = \begin{cases} 0, & \text{the } i\text{th set of inequality is not satisfied} \\ 1, & \text{the } i\text{th set is satisfied} \end{cases}$$

To determine bounds to render the constraints redundant when the y_i's equal 0, note first that $x_1 + 2x_2 + 3x_3 \leq 600$ implies that $x_1 \leq 600$, $x_2 \leq 300$, $x_3 \leq 200$ for any solution point. Thus, for any feasible point (x_1,x_2,x_3), we have

$$\begin{aligned}
5x_1 + 10x_2 - 500 &\leq 5(600) + 10(300) - 500 = 5500 \\
x_1 - x_2 + x_3 - 450 &\leq 600 - 0 + 200 - 450 = 350 \\
x_2 + x_3 - 100 &\geq 0 + 0 - 100 = -100 \\
x_1 - x_3 - 250 &\leq 600 - 0 - 250 = 350 \\
x_3 - 50 &\geq 0 - 50 = -50 \\
x_2 + x_3 - 400 &\leq 300 + 200 - 400 = 100
\end{aligned}$$

The problem then becomes one of determining all (x_1,x_2,x_3) and (y_1,y_2,y_3,y_4,y_5) satisfying

$$\begin{aligned}
x_1 + 2x_2 + 3x_3 &\leq 600 \\
5x_1 + 10x_2 - 500 &\leq 5500\,(1 - y_1) \\
x_1 - x_2 + x_3 - 450 &\leq 350\,(1 - y_2) \\
x_2 + x_3 - 100 &\geq -100\,(1 - y_3) \\
x_1 - x_3 - 250 &\leq 350\,(1 - y_4) \\
x_3 - 50 &\geq -50\,(1 - y_4) \\
x_1 - x_3 - 250 &\leq 350\,(1 - y_5) \\
x_2 + x_3 - 400 &\leq 100\,(1 - y_5)
\end{aligned}$$

$$y_1 + y_2 + y_3 + y_4 + y_5 \geq 3$$

$$0 \leq y_1,y_2,y_3,y_4,y_5 \leq 1 \text{ and integral}$$

$$x_1,x_2,x_3 \geq 0$$

PROBLEM SET 6.2

1. In working with auxiliary variables in problems with fixed charges or alternative constraints, we used upper and lower bounds in the inequalities associated with these variables so that these inequalities were rendered redundant when the corresponding auxiliary variables were "off" (i.e., usually equal to 0). To show the need for these bounds, consider Example 6. Suppose we attempt to replace the either/or constraints of that example with the following.

$$\begin{aligned} 2x_1 + 9x_2 - 18 &\le 1 - y_1 \\ x_1 + \ \ x_2 - \ \ 6 &\le 1 - y_2 \\ y_1 + y_2 &\ge 1 \\ x_1,x_2 &\ge 0 \\ 0 \le y_1,y_2 \le 1 &\text{ and integral} \end{aligned} \tag{2}$$

[This is simply (1), with the upper bounds of 54 and 9 omitted from the first two inequalities, respectively.]

(a) Show that if (x_1,x_2,y_1,y_2) satisfies (2), then either $2x_1 + 9x_2 \le 18$ or $x_1 + x_2 \le 6$.

(b) Show that for some (x_1,x_2) that satisfy either $2x_1 + 9x_2 \le 18$ or $x_1 + x_2 \le 6$, we cannot find y_1 and y_2 such that (x_1,x_2,y_1,y_2) is a solution to (2). [Try finding y_1 and y_2 for the point (3,3), say.]

(c) In fact, show that if (x_1,x_2,x_1,y_2) is a solution to (2), then (x_1,x_2) satisfies

$$\left\{\begin{aligned} 2x_1 + 9x_2 &\le 18 \\ x_1 + \ \ x_2 &\le 7 \end{aligned}\right\} \quad \text{or} \quad \left\{\begin{aligned} 2x_1 + 9x_2 &\le 19 \\ x_1 + \ \ x_2 &\le 6 \end{aligned}\right\}$$

(d) If you still are not convinced, graph the solution set to the either/or system of constraints in part c, and then compare this set with the feasible set graphed in Figure 6.2.

Formulate integer (or mixed integer) programming models for the following.

2. An investment firm, wanting to maximize profit, has \$500,000 to invest among the following.

A development project, requiring a lump sum investment of \$390,000 and yielding a return of 6.7%.

A construction project, requiring a lump sum investment of \$220,000 and yielding a return of 6.5%.

Any number of units in a portfolio of stocks, costing \$25,000/unit and yielding a return of 6.3%.

Any number of units of a certain stock, costing \$1,300/share and yielding a return of 5.7%.

3. A construction firm has available M dollars in capital to be used for the development of up to n different sites, with Site i costing a_i dollars for development and returning an expected profit of c_i dollars upon completion. However, $\sum a_i > M$. What sites should be selected to optimize expected profit?

4. A backpacker's knapsack has a volume of V in.^{3} and can hold up to W lb of gear. The backpacker has a choice of n items to carry in it, with the ith item requiring a_i in.^{3} of space, weighing w_i lb, and providing c_i units of value for the trip. What items should be taken in the knapsack?

5. A road construction firm seeks to assign its force optimally over the next 28-week period. They can be assigned to any combination of the following.

For any number of 10-week periods, working for the state, and earning a profit of $3200/week.

For any number of 6-week periods, working for the county, and earning a profit of $2900/week.

For any number of 3-week periods, working for a private land developer, and earning $2750/week.

For any number of weeks, working on parking lot construction, and earning $2550/week.

However, if the firm does any work at all for either the state or county (or both), it is expected to contribute $7500 to the campaign funds of a certain anonymous political figure.

6. A manufacturer supplies six outlets and has a choice of renting space in up to three warehouses to maintain stocks to deliver to the outlets. Determine the minimal cost renting and shipping schedule, using the following data.

Shipping Costs Per Unit		***Outlets***						*Storage Capacity*	*Monthly Rent*
		1	2	3	4	5	6		
Warehouse	1	12	9	16	13	11	23	150	300
	2	10	13	12	7	12	26	200	500
	3	13	12	14	10	17	21	300	700
Monthly Demands		70	45	35	50	75	60		

7. A firm has M units of a new product to be sold in up to n different new market areas. To develop the ith market for sales, there is an initial research and advertising cost of d_i dollars. Once opened, the ith market can seell up to u_i units at a profit of c_i dollars/unit. What areas should be developed and how many units should each of these areas receive so as to maximize profit?

8. A small division of an automobile plant manufactures two parts to meet the monthly demands of the major assembly plant. Three different machines can be used in the process, each having varying input and output capacities and setup and maintenance costs if used.

Determine the most economical monthly operation of the division, using the following data.

	One hr of Operation				
	Input		*Output*		
	Labor (*man-hr*)	*Raw Material* (*lb*)	*Part A* (*units*)	*Part B* (*units*)	*Monthly Maintenance Costs*
Machine 1	12	95	16	15	\$ 900
Machine 2	9	70	10	12	\$ 625
Machine 3	14	75	5	17	\$1050

Monthly demands: Part A—400 units
Part B—500 units
Monthly supplies: Labor—550 man-hr
Raw material—unlimited
Costs: Man-hours—\$7.80/hr
Raw material—\$5/lb

9. An ice cream plant can make up to 28 different flavors of ice cream each month. Flavor i requires a_i lb of sugar/gal and earns c_i dollars/gal sold, but at least u_i gal must be made per month if the flavor is to sell. With M lb of sugar available for the month, what flavors should be made to optimize profit?

10. Maximize $5x_1 + 12x_2$

subject to

$$7x_1 + 3x_2 \leq 16 \quad \text{or} \quad 3x_1 + 10x_2 \leq 20 \quad \text{and} \quad x_1,x_2 \geq 0$$

11. Maximize $3x_1 + 5x_2 + 7x_3$

subject to

$5x_1 + 4x_2 + 2x_3 \leq 300,\ x_1,x_2,x_3 \geq 0,$ and

$$\begin{Bmatrix} x_1 + x_3 \leq 100 \\ x_1 - x_2 \geq \quad 0 \end{Bmatrix} \quad \text{or} \quad \{2x_1 + 4x_2 + 5x_3 \leq 250\}$$

12. Maximize $9x_1 + 8x_2 + 7x_3$

subject to

$x_1 + x_2 + x_3 \leq 500$
$x_1,x_2,x_3 \geq 0$

and such that x_1,x_2,x_3 satisfies at least two of the following three constraints.

$$\begin{aligned} 3x_1 + 3x_2 + 4x_3 &\leq 1000 \\ x_1 \qquad\quad + x_3 &\leq 400 \\ x_1 + x_2 \qquad\quad &\leq 600 \end{aligned}$$

13. Consider the situation in Problem 7, but suppose also that because of personnel limitations, at most k market areas ($k < n$) can be developed.

SECTION 6.3 GOMORY'S CUTTING PLANE ALGORITHM

There are various algorithms available for the resolution of integer programming problems. The reason for this abundance is that no one algorithm has proved to be computationally efficient for all problems, and thus the search continues for more effective algorithms. To introduce some of the methods presently used for solving these problems, we will present in this section and the next two integer programming algorithms. Specifically, in this section we will develop one version of Gomory's Cutting Plane Algorithm. This algorithm was one of the first of its kind, published in 1958 ([14]), and is still an effective tool for solving certain integer programming problems. (For a more thorough development of integer programming techniques, see the literature on the subject, such as the books by Garfinkel and Nemhauser [13] or Taha [31].)

Consider a pure integer programming problem, that is, a standard linear programming problem with integral restrictions on all of the variables. The fundamental idea underlying Gomory's Cutting Plane Algorithm is to add constraints to the problem one at a time so that we eventually have a linear programming problem with an optimal solution with integral coordinates. The algorithm works as follows. First we solve the original linear programming problem, ignoring the integral restrictions. Then, if this solution has all integral coordinates, it is also an optimal solution to the integer programming problem, and we are done. If not, we generate a new constraint to be added to the problem. This constraint will have two fundamental properties: first, the nonintegral optimal solution to the original linear programming problem will not satisfy this constraint; and second, all integral feasible solutions to the original problem will satisfy the new constraint. Thus this constraint essentially cuts off a subset of the set of feasible solutions to the linear programming problem, but a subset that contains no feasible integral solutions. We add this constraint to the problem and proceed to solve the expanded problem as before, first ignoring the integral restrictions, and continuing on. We demonstrate the algorithm first by an example.

Example 1

$$\text{Minimize} \quad x_1 - 3x_2$$

subject to

$$\begin{aligned} x_1 - x_2 &\leq 2 \\ 2x_1 + 4x_2 &\leq 15 \\ x_1, x_2 &\geq 0 \text{ and integral} \end{aligned}$$

Adding integrally restricted slack variables x_3 and x_4 and using the simplex algorithm on the associated linear programming problem, we have the tableaux of Table 6.2. The minimal value of the objective function, ignoring the integral constraints, is attained at $x_1 = 0$, $x_2 = \frac{15}{4} = 3\frac{3}{4}$ (and $x_3 = 2\frac{3}{4}, x_4 = 0$). Since this point has nonintegral coordinates, we need to generate a new constraint. To do this we can work with any constraining equation from the final tableau that has a nonintegral constant term, and so, in this case, we could use either equation. Let us consider the equation defined by the first row of the final tableau.

$$\tfrac{3}{2}x_1 + x_3 + \tfrac{1}{4}x_4 = \tfrac{23}{4}$$

Separating all constants into their integral and fractional parts, we have

$$(1 + \tfrac{1}{2})x_1 + x_3 + \tfrac{1}{4}x_4 = 5 + \tfrac{3}{4}$$

Table 6.2

	x_1	x_2	x_3	x_4	
x_3	1	−1	1	0	2
x_4	2	(4)	0	1	15
	1	−3	0	0	0
x_3	$\frac{3}{2}$	0	1	$\frac{1}{4}$	$\frac{23}{4}$
x_2	$\frac{1}{2}$	1	0	$\frac{1}{4}$	$\frac{15}{4}$
	$\frac{5}{2}$	0	0	$\frac{3}{4}$	$\frac{45}{4}$

Therefore we must have

$$\tfrac{1}{2}x_1 + \tfrac{1}{4}x_4 - \tfrac{3}{4} = 5 - x_1 - x_3$$

Now, since we want only integral solutions, the right side and therefore the left side of this equation must be integral. And, since all variables are nonnegative, the left side must be greater than or equal to $-\frac{3}{4}$. Combining, we want solutions such that

$$\tfrac{1}{2}x_1 + \tfrac{1}{4}x_4 - \tfrac{3}{4}$$

is a nonnegative integer, say x_5. It is this constraint that we add to the two original constraints, giving us the appended problem of

$$\text{Minimizing} \quad \tfrac{5}{2}x_1 + \tfrac{3}{4}x_4 - \tfrac{45}{4}$$

subject to

$$\begin{aligned}
\tfrac{3}{2}x_1 \qquad\quad + x_3 + \tfrac{1}{4}x_4 \qquad\quad &= \tfrac{23}{4} \\
\tfrac{1}{2}x_1 + x_2 \qquad\quad + \tfrac{1}{4}x_4 \qquad\quad &= \tfrac{15}{4} \\
\tfrac{1}{2}x_1 \qquad\qquad\quad + \tfrac{1}{4}x_4 - x_5 &= \tfrac{3}{4} \\
x_1, x_2, x_3, x_4, x_5 \geq 0 \text{ and integral}
\end{aligned}$$

Here we have used the final tableau data of Table 6.2 in expressing the original problem. Notice that the optimal solution found above, $x_1 = x_4 = 0$, $x_2 = \frac{15}{4}$, $x_3 = \frac{23}{4}$, does not satisfy the new constraint.

Now we proceed as before, solving the corresponding linear programming problem. To do this, since we have simply added a constraint to a completed problem, we can use the Dual Simplex Algorithm (Sections 5.6 and 5.7). After multiplying the new constraint by (-1), we have as basic variables for the first tableau the basic variables of the previous final tableau (x_3 and x_2) along with the new variable x_5. The tableaux are presented in Table 6.3. The solution point here has all integral coordinates, and therefore we have the solution to the original integer programming problem. The minimal value of the objective function is -9 and is attained at the point $x_1 = 0$, $x_2 = 3$.

Geometrically, what has happened is the following. The feasible points for the original system of constraints are the lattice points of the hatched region of Figure 6.3. Now the added constraint can be expressed as $\frac{1}{2}x_1 + \frac{1}{4}x_4 - \frac{3}{4} \geq 0$. Using $x_4 = 15 - 2x_1 - 4x_2$, this inequality reduces to

$$2x_1 + (15 - 2x_1 - 4x_2) \geq 3, \qquad \text{or} \qquad 4x_2 \leq 12, \qquad \text{or} \qquad x_2 \leq 3$$

As can be seen in Figure 6.4, this new constraint is equivalent to an inequality that cuts off from the feasible set the original nonintegral optimal solution $(0,3\frac{3}{4})$, but does not exclude from consideration any feasible lattice points.

Now we describe in detail how these new constraints are generated. Suppose after solving the associated linear programming problem, the constant term of the

Table 6.3

	x_1	x_2	x_3	x_4	x_5	
x_3	$\frac{3}{2}$	0	1	$\frac{1}{4}$	0	$\frac{23}{4}$
x_2	$\frac{1}{2}$	1	0	$\frac{1}{4}$	0	$\frac{15}{4}$
x_5	$-\frac{1}{2}$	0	0	$\boxed{-\frac{1}{4}}$	1	$-\frac{3}{4}$
	$\frac{5}{2}$	0	0	$\frac{3}{4}$	0	$\frac{45}{4}$
x_3	1	0	1	0	1	5
x_2	0	1	0	0	1	3
x_4	2	0	0	1	-4	3
	1	0	0	0	3	9

ith row of the final tableau is not an integer. Then, in the optimal basic solution corresponding to this final tableau, the value of the basic variable isolated in the ith row will not be integral, and so we need to add a new constraint.

Attaching back the variables, suppose this ith constraint is

$$\sum_j a_{ij}x_j = b_i \tag{1}$$

Letting $[a]$ denote the greatest integer in a (i.e., the greatest integer less than or equal to a, and so $[3\frac{4}{5}] = 3$, $[1] = 1$, $[-3\frac{2}{5}] = -4$), define the fractional part of a

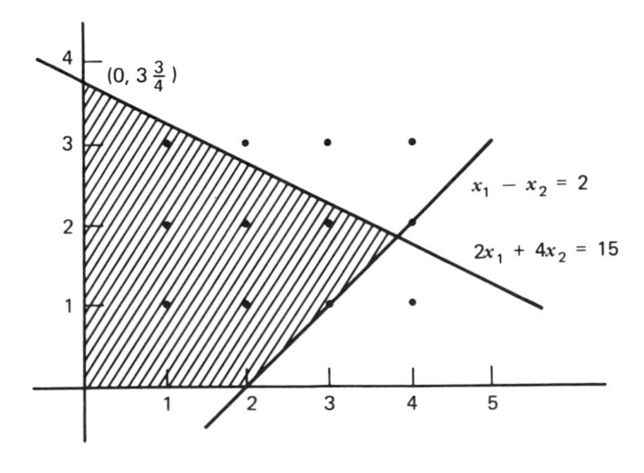

Figure 6.3

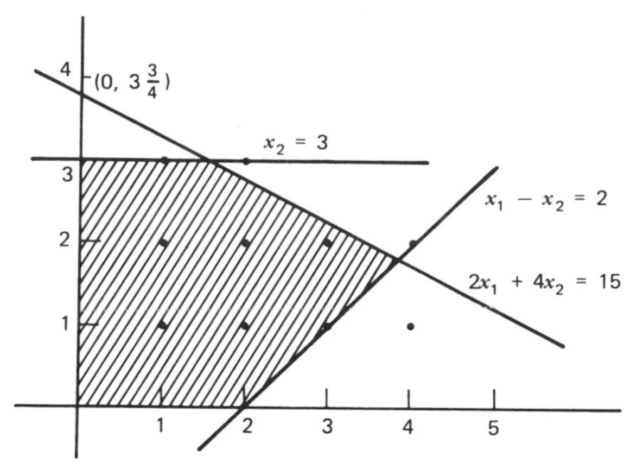

Figure 6.4

number a by $a - [a]$. Thus the fractional part of $3\frac{4}{5}$ is $\frac{4}{5}$, of 1 is 0, and of $-3\frac{2}{5}$ is $-3\frac{2}{5} - (-4) = \frac{3}{5}$. Notice that the fractional part of a number must be nonnegative and less than 1. Now let f_{ij} and f_i denote the fractional parts of a_{ij} and b_i, respectively, that is,

$$f_{ij} = a_{ij} - [a_{ij}]$$
$$f_i = b_i - [b_i]$$

Then we can rewrite (1) as

$$\sum_j ([a_{ij}] + f_{ij})x_j = [b_i] + f_i \quad \text{or} \quad \sum_j f_{ij}x_j - f_i = [b_i] - \sum_j [a_{ij}]x_j \quad (2)$$

Notice that all the constant terms on the right side of (2) are integral. Thus, for any integral solution to the original system of constraints, the right side, and therefore the left side of (2), must be integral. Moreover, since all variables are nonnegative and f_i is less than 1, the left side of (2) must be greater than or equal to the integer 0. Hence the new constraint:

$$\sum_j f_{ij}x_j - f_i \geq 0 \text{ and integral} \quad (3)$$

We have developed (3) so that any feasible integral solution to the original system of constraints will satisfy this new constraint, and that by adding this constraint, we still have a pure integer programming problem. Furthermore, the optimal basic feasible solution from the final tableau of the corresponding linear programming problem does not satisfy this constraint. We have chosen i such that b_i was not an integer, and so $f_i > 0$. Now the only variables x_j that can appear in (3) are the nonbasic variables of the final tableau; the coefficients of the basic variables are either 0 or 1, and so have fractional part 0. (In Example 1, the variables of the added constraint were x_1 and x_4, the nonbasic variables of the final tableau resolution of the original problem.) Hence the corresponding basic feasible solution, with $x_j = 0$ for all nonbasic variables x_j, does not satisfy (3).

If, in the final tableau for the corresponding linear programming problem, several of the constant terms b_i are not integral, we have a choice of what row to use to generate the new constraint. In fact, rules governing the choice of row to use can be given (Gomory [15] or Hadley [17]) that will guarantee in theory at least the convergence of the algorithm to an optimal integral solution in a finite number of steps. However, after many iterations and several hours of computer time with no feasible solution in sight, your Vice-President of Computer Affairs would probably be somewhat unimpressed with theoretical convergence arguments. In practice, the simple rule of using that row containing the constant term b_i of largest fractional value is easy to apply and usually quite effective.

We summarize now the steps of Gomory's Cutting Plane Algorithm for pure integer programming problems. Consider the integer programming problem of optimizing $c \cdot X$ subject to $AX = b$, $X \geq 0$ and integral.

1. Solve the corresponding linear programming problem simply ignoring the integral restrictions on X. If this solution has all integral coordinates, then it is an optimal solution to the original problem.

2. Otherwise a new constraint is added to the problem.

(a) To construct this constraint, select any row from the final optimal tableau solution of the linear programming problem with a non-integral constant term b_i. (Using that row containing the constant term with largest fractional value may reduce the total number of iterations necessary for convergence.)

(b) Suppose the ith row is selected and the corresponding equation is

$$\sum_j a_{ij} x_j = b_i$$

Then form the constraint

$$-\sum_j f_{ij} x_j + x = -f_i$$

where $f_{ij} = a_{ij} - [a_{ij}]$ = fractional part of a_{ij}

$f_i = b_i - [b_i]$ = fractional part of b_i

x = a new slack variable, restricted to be nonnegative and integral

(c) Add this constraint to the problem and return to Step 1. Note that now when solving the corresponding linear programming problem, the Dual Simplex Algorithm can be used.

Example 2

$$\text{Minimize} \quad x_1 - 2x_2$$

subject to

$$\begin{aligned} 2x_1 + x_2 &\leq 5 \\ -4x_1 + 4x_2 &\leq 5 \\ x_1, x_2 \geq 0 &\text{ and integral} \end{aligned}$$

Adding integrally restricted slack variables x_3 and x_4 and applying the simplex algorithm, we have the tableaux of Table 6.4. Both constant terms of the final tableau are nonintegral, but $\frac{5}{2}$ has the larger fractional value. The second row of this tableau generates the constraint

$$-\tfrac{1}{3}x_3 - \tfrac{1}{6}x_4 + x_5 = -\tfrac{1}{2}$$

where x_5 is a new slack variable. Adding this equation and using the Dual Simplex Algorithm leads to the tableaux of Table 6.5. The first row of the final tableau of this table generates the constraint

$$-\tfrac{3}{4}x_4 + x_6 = -\tfrac{3}{4}$$

where x_6 is a new slack variable. (Note that $-\frac{1}{4} - [-\frac{1}{4}] = -\frac{1}{4} - (-1) = \frac{3}{4}$. In general, be careful when working with the fractional part of a negative number.)

Table 6.4

	x_1	x_2	x_3	x_4	
x_3	2	1	1	0	5
x_4	-4	(4)	0	1	5
	1	-2	0	0	0
x_3	(3)	0	1	$-\frac{1}{4}$	$\frac{15}{4}$
x_2	-1	1	0	$\frac{1}{4}$	$\frac{5}{4}$
	-1	0	0	$\frac{1}{2}$	$\frac{5}{2}$
x_1	1	0	$\frac{1}{3}$	$-\frac{1}{12}$	$\frac{5}{4}$
x_2	0	1	$\frac{1}{3}$	$\frac{1}{6}$	$\frac{5}{2}$
	0	0	$\frac{1}{3}$	$\frac{5}{12}$	$\frac{15}{4}$

Table 6.5

	x_1	x_2	x_3	x_4	x_5	
x_1	1	0	$\frac{1}{3}$	$-\frac{1}{12}$	0	$\frac{5}{4}$
x_2	0	1	$\frac{1}{3}$	$\frac{1}{6}$	0	$\frac{5}{2}$
x_5	0	0	$\boxed{-\frac{1}{3}}$	$-\frac{1}{6}$	1	$-\frac{1}{2}$
	0	0	$\frac{1}{3}$	$\frac{5}{12}$	0	$\frac{15}{4}$
x_1	1	0	0	$-\frac{1}{4}$	1	$\frac{3}{4}$
x_2	0	1	0	0	1	2
x_3	0	0	1	$\frac{1}{2}$	-3	$\frac{3}{2}$
	0	0	0	$\frac{1}{4}$	1	$\frac{13}{4}$

Table 6.6

	x_1	x_2	x_3	x_4	x_5	x_6	
x_1	1	0	0	$-\frac{1}{4}$	1	0	$\frac{3}{4}$
x_2	0	1	0	0	1	0	2
x_3	0	0	1	$\frac{1}{2}$	-3	0	$\frac{3}{2}$
x_6	0	0	0	$\boxed{-\frac{3}{4}}$	0	1	$-\frac{3}{4}$
	0	0	0	$\frac{1}{4}$	1	0	$\frac{13}{4}$
x_1	1	0	0	0	1	$-\frac{1}{3}$	1
x_2	0	1	0	0	1	0	2
x_3	0	0	1	0	-3	$\frac{2}{3}$	1
x_4	0	0	0	1	0	$-\frac{4}{3}$	1
	0	0	0	0	1	$\frac{1}{3}$	3

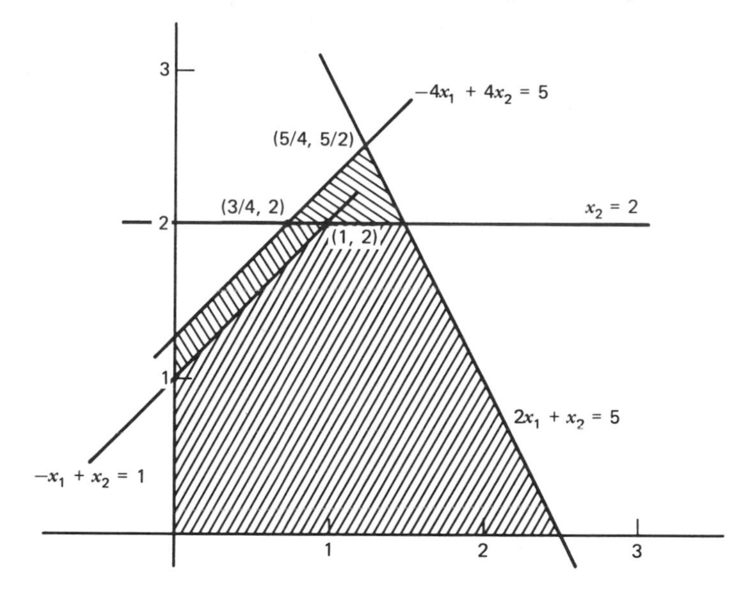

Figure 6.5

Adding this constraint, we have the tableaux of Table 6.6. Thus the minimal value of the objective function is -3 and is attained at $x_1 = 1, x_2 = 2$.

Although our problem is completed, we will also show in this example the action of the cutting planes geometrically. The first constraint added corresponds to

$$\tfrac{1}{3}x_3 + \tfrac{1}{6}x_4 - \tfrac{1}{2} \geq 0$$

That is,

$$\tfrac{1}{3}(5 - 2x_1 - x_2) + \tfrac{1}{6}(5 + 4x_1 - 4x_2) - \tfrac{1}{2} \geq 0$$

This reduces to $x_2 \leq 2$. Similarly, the second additional constraint reduces from $\frac{3}{4}x_4 - \frac{3}{4} \geq 0$ to $-x_1 + x_2 \leq 1$. The graph is sketched in Figure 6.5.

PROBLEM SET 6.3

1. In each of the following, solve using the Cutting Plane Algorithm and then sketch the graph of the original feasible region and the cutting plane. (In each, it should be necessary to add only one additional constraint before reaching an optimal integral solution.)

   (a) Minimize $x_1 - x_2$

   subject to

   $3x_1 + 4x_2 \leq 6$

   $x_1 - x_2 \leq 1$

   $x_1, x_2 \geq 0$ and integral

(b) Maximize $x_1 + 2x_2$

subject to

$$\begin{aligned} x_1 + 3x_2 &\leq 13 \\ 2x_1 - x_2 &\leq 6 \\ x_1, x_2 &\geq 0 \text{ and integral} \end{aligned}$$

2. Solve the following using the Cutting Plane Algorithm.

(a) Maximize $x_1 + 3x_2$

subject to

$$\begin{aligned} -x_1 + 3x_2 &\leq 6 \\ 2x_1 + x_2 &\leq 12 \\ x_1, x_2 &\geq 0 \text{ and integral} \end{aligned}$$

(b) Maximize $3x_1 + 8x_2$

subject to

$$\begin{aligned} x_1 + 2x_2 &\leq 9 \\ 2x_2 &\leq 5 \\ x_1, x_2 &\geq 0 \text{ and integral} \end{aligned}$$

(c) Maximize $2x_1 - 4x_2 + x_3$

subject to

$$\begin{aligned} x_1 - x_2 &\leq 12 \\ 2x_2 + 3x_3 &\leq 28 \\ x_1, x_2, x_3 &\geq 0 \text{ and integral} \end{aligned}$$

(d) Maximize $2x_1 - 4x_2 + x_3$

subject to

$$\begin{aligned} x_1 - x_2 &= 12 \\ 2x_2 + 3x_3 &= 28 \\ x_1, x_2, x_3 &\geq 0 \text{ and integral} \end{aligned}$$

3. Is it possible, after several iterations of the Cutting Plane Algorithm, to arrive at an optimal solution to the corresponding linear programming problem with the property that only the slack variables defined by the appended constraints assume nonintegral values, and that all the variables of the original problem assume integer values?

4. What do you suppose would happen if the Cutting Plane Algorithm were applied to an integer programming problem with the property that the corresponding linear programming problem had feasible (nonintegral) solutions but no feasible integral solutions?

5. Test your answer to Problem 4 on the following.

Maximize $x_1 + 2x_2$

subject to

$$
\begin{aligned}
3x_1 + 3x_2 &\leq 2 \\
3x_1 \quad\quad &\geq 1 \\
x_1, x_2 &\geq 0 \text{ and integral}
\end{aligned}
$$

6. Solve using the Cutting Plane Algorithm.

Maximize $x_1 + 2x_2 + x_3$

subject to

$$
\begin{aligned}
x_1 + 4x_2 + 2x_3 &\leq 7 \\
-x_1 + 3x_2 \quad\quad &\geq 4 \\
x_1, x_2, x_3 &\geq 0 \text{ and integral}
\end{aligned}
$$

SECTION 6.4 A BRANCH AND BOUND ALGORITHM

In the last section we developed a version of Gomory's Cutting Plane Algorithm. In this section we will demonstrate the basic idea underlying another important class of algorithms, branch and bound algorithms, used to solve integer programming problems. These algorithms originate from the work of Land and Doig [21], published in 1960, and the version we demonstrate is a modification due to Dakin [3]. Branch and bound algorithms, along with their refinements and extensions, form a constructive and promising set of solution techniques for integer programming problems.

Given an integer programming problem, the first step in the branch and bound approach is to ignore the integral restrictions and solve the corresponding linear programming problem. If this problem does not have an integral optimal solution then, as in the Cutting Plane Algorithm, new constraints are generated to cut off this optimal nonintegral solution. But here, instead of expanding the original problem by the addition of a single constraint, we create two distinct problems, each coming about by the addition of a new constraint to the original set of constraints. These two new constraints are generated from the nonintegral optimal solution to the original problem as follows. Select a variable, say x_j, that assumes a nonintegral value in this optimal solution. Suppose $x_j = b_i$ in this solution. Then the two new problems are created by adding to the original constraint set for one problem the constraint $x_j \leq [b_i]$ and for the other the constraint $x_j \geq [b_i] + 1$. For example, if in the original problem the optimal solution has the variable $x_3 = 8\frac{1}{4}$, the constraint sets for the two new problems would contain the original set of constraints plus, for one, the constraint $x_3 \leq 8$, and for the other, the constraint $x_3 \geq 9$. Note that the original nonintegral optimal solution is not a feasible solution to either new problem, but that any integral feasible solution to the original problem would be a solution to one of these new problems. However, in contrast to the Cutting Plane Algorithm, now we have two integer programming problems to deal with, and the integral optimal solution to the original problem could be contained in either problem. We continue, considering the two new problems just as before. For each we initially ignore the integral restrictions, solve and, if the problem has a nonintegral optimal solution, we again branch from that problem, formulating two new problems using the above method.

It may seem that with this branching process we are compounding our difficulties by continually expanding the set of problems to be solved. However, this in not quite the case, for two reasons. First, some of the newly formed problems may have no feasible solutions as a result of the increased restrictions from the additional constraints. Second, some of these problems may have integral optimal solutions. Such a solution would certainly satisfy the constraints of the original problem, and would provide a bound for the optimal value to the original integer programming problem. This bound would allow us to eliminate from consideration any problems generated through the branching process from a problem with an optimal value not better than this bound. By making use of such arguments based on bounds, we can eliminate problems to consider and will eventually be left with the optimal integral solution to the original problem.

We illustrate with an example.

Example 1

Consider the integer programming problem of Example 2 of Section 6.3. The problem is to

$$\text{Minimize} \quad z = x_1 - 2x_2$$

subject to

$$\begin{aligned} 2x_1 + x_2 &\leq 5 \\ -4x_1 + 4x_2 &\leq 5 \\ x_1, x_2 \geq 0 &\text{ and integral} \end{aligned}$$

The solution found in the last section to the problem with the integral restrictions ignored is Min $z = -3\frac{3}{4}$ attained at $x_1 = 1\frac{1}{4}, x_2 = 2\frac{1}{2}$. Using the branch and bound algorithm, we formulate two new problems here by restricting either the x_1 or the x_2 variable. We arbitrarily select the x_1 variable and use the constraints $x_1 \leq 1$ and $x_1 \geq 2$ to form two new problems. We denote this branching process in Figure 6.6. The first box in the figure corresponds to the original problem: its solution, with the integral restrictions ignored, is inside. The boxes labeled 2A and 2B correspond to the two new problems, with the new constraints indicated on the branches leading to

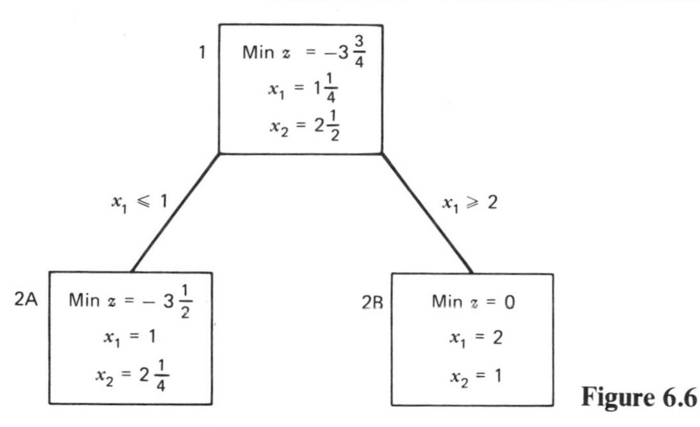

Figure 6.6

the respective boxes. The solutions to these problems, again with the integral restrictions ignored, could be determined by using the Dual Simplex Algorithm in conjunction with the final tableau solution to the original problem, because in each we have simply added a constraint to a completed problem. These solutions are listed in the appropriate boxes. (Note that finding these solutions can require some effort. This is an instance where complicated details have been left to and provided by the interested author.) Problem 2A has an optimal solution of Min $z = -3\frac{1}{2}$

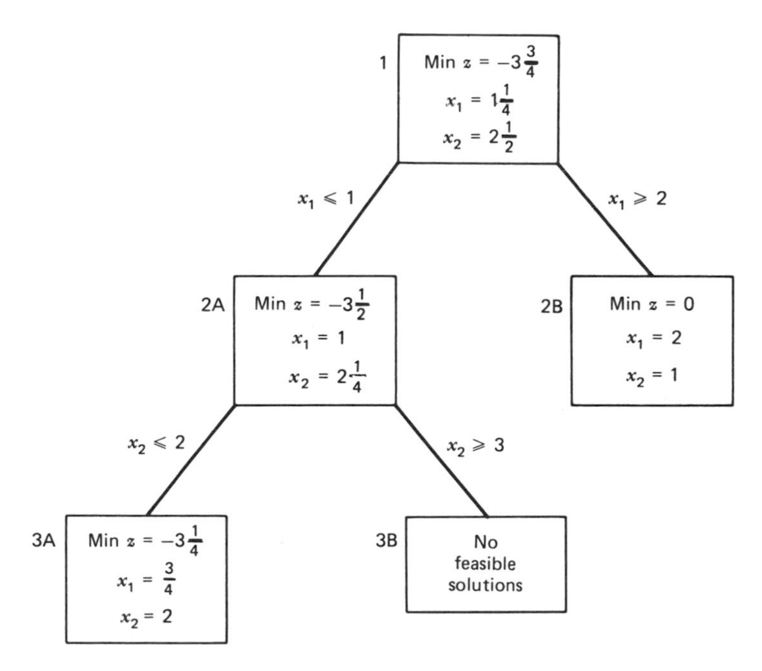

Figure 6.7

at $x_1 = 1, x_2 = 2\frac{1}{4}$. The optimal value for this problem is greater than the optimal value for Problem 1, as would be expected, because we have added a constraint and therefore reduced the solution set of feasible points on which to minimize the objective function. Problem 2B provides a feasible integral solution $x_1 = 2$, $x_2 = 1$ to the constraints of the original problem and an upper bound of 0 for the optimal value of the integrally restricted problem. Since better integral solutions may be contained in Problem 2A, we branch again off Problem 2A using the x_2 variable, as illustrated in Figure 6.7. The branch to Problem 3B terminates here, since this problem has no feasible solutions. However, we must continue by branching at Problem 3A. See Figure 6.8. Problem 4A yields the integral feasible solution of $x_1 = 1, x_2 = 2$ with the optimal value of $z = -3$. The optimal value for Problem 4B is $-2\frac{1}{2}$, and so further branching here can lead only to solutions with value greater than $-2\frac{1}{2}$. Since we already have an integral solution to the constraints of

Figure 6.8

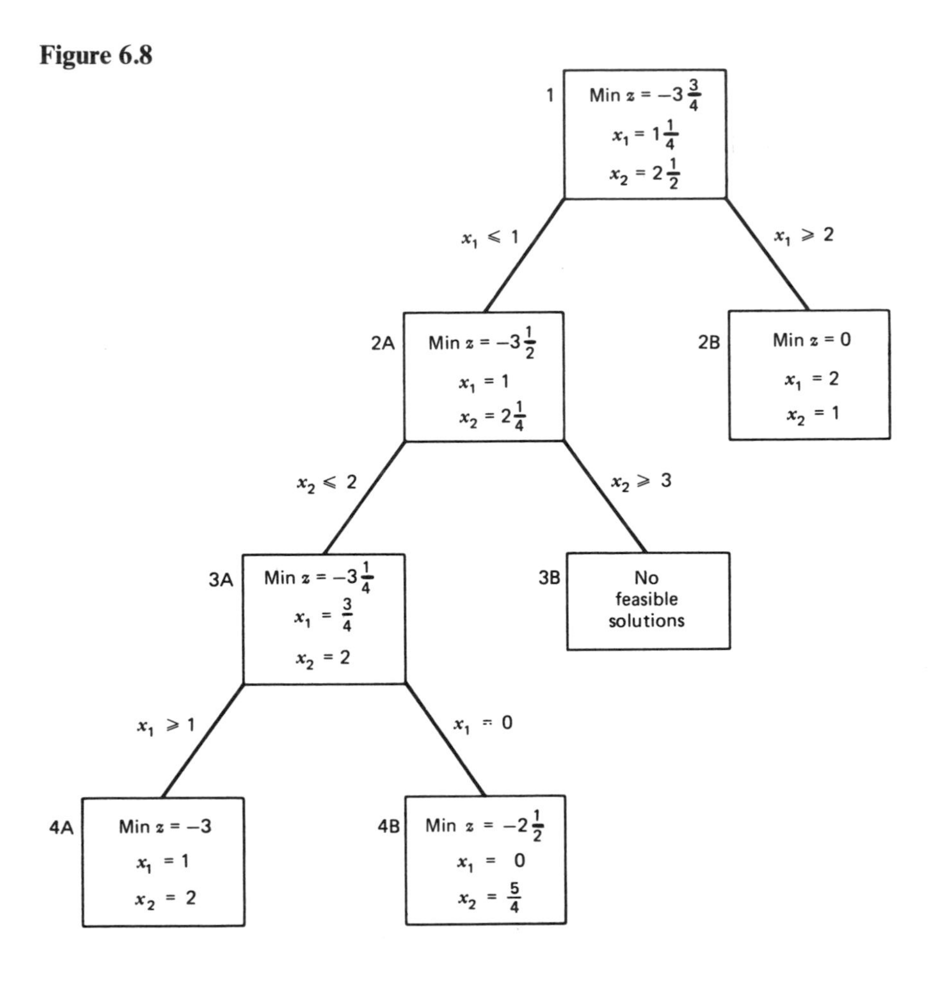

the original problem at which the value of the objective function is -3, there is no need to branch at Problem 4B. No other problems remain to be considered, so we are done. The original integer programming problem must have an optimal value of -3 attained at $x_1 = 1, x_2 = 2$, the optimal solution to Problem 4A. Note that here, in fact, we could have ceased calculations once the solution to Problem 4A had been found, because the objective function has integral coefficients; it follows from the optimal solution to Problem 1 that the smallest value the objective function can possibly attain at a feasible integral solution is -3, and the solution to Problem 4A provides a lattice point at which this value is attained.

In using this branch and bound algorithm, notice that there are three possible reasons for not continuing a branch at a particular problem. The problem under consideration may have no feasible solutions, or it may have an optimal integral solution. Finally, bounds from previously determined integral solutions may render further consideration of the problem unnecessary. In fact, one advantage of this algorithm over the Cutting Plane Algorithm of the last section is that this algorithm generates feasible integral solutions to the constraints of the problem as it proceeds, along with estimates on how close to the optimal value these solutions might be. Thus, in a large and complicated problem, if we are unable to persevere to the completion of the problem using the branch and bound algorithm, maybe because of limited computer capacities, the algorithm may still provide an integral feasible solution not necessarily optimal but adequate for our purposes.

In the quest for increased efficiency of the algorithm, questions such as what problem to examine next if two or more branches remain open and what variable to use when defining the new constraints if more than one variable takes on non-integral values in the optimal solution to the problem at hand must be considered. There are no universally accepted answers to these questions. Rules to follow when selecting the variable to use in defining the new constraints can be rather involved. When selecting the problem to branch from, one possible rule is to select the problem with the most favorable optimal value; another frequently used rule is to select the problem most recently generated. (See the following Example 3 for an application of this first rule.)

We conclude this section with two more examples.

Example 2

$$\text{Maximize} \quad z = 8x_1 + 15x_2$$

subject to

$$\begin{aligned} 10x_1 + 21x_2 &\le 156 \\ 2x_1 + x_2 &\le 22 \\ x_1, x_2 &\ge 0 \text{ and integral} \end{aligned}$$

Figure 6.9 contains the completed branch and bound diagram. Upon solving the original problem, Problem 1 of the diagram, we know that the optimal value of z

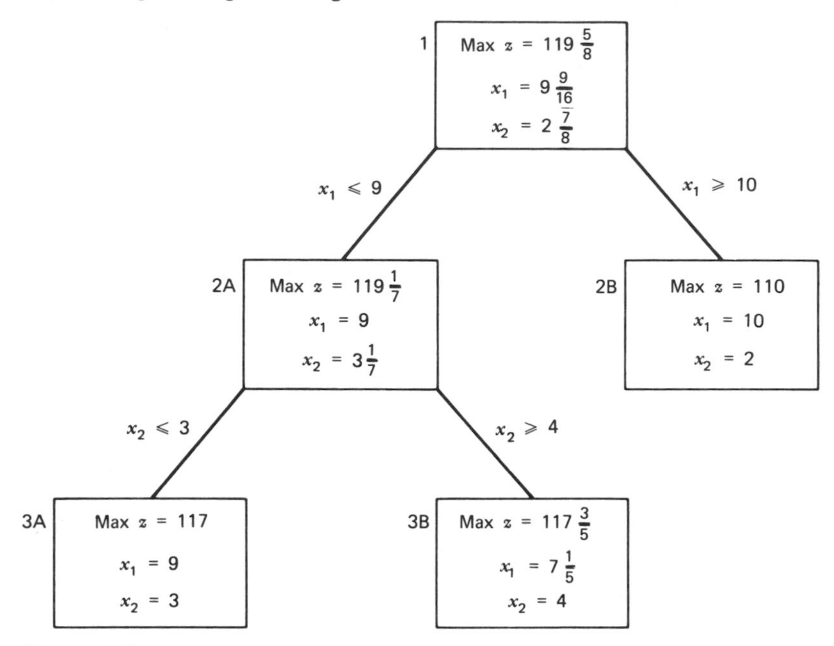

Figure 6.9

restricted to integral feasible solutions is at most 119. Arbitrarily selecting the x_1 variable, we create Problems 2A and 2B. Problem 2B provides a feasible integral solution to the original constraints and a lower bound of 110 for the final maximal value of z. Restrictions on x_2 lead to Problems 3A and 3B from Problem 2A. The integral solution to Problem 3A yields the improved lower bound of 117 for the optimal value of z. The optimal value for Problem 3B exceeds 117, but only by a fraction, and so the value of z at any integral solution to the constraints of Problem 3B cannot exceed 117. Thus the algorithm terminates. The optimal value for the objective function for the integrally restricted problem is 117, and one point at which this value is attained is $x_1 = 9$, $x_2 = 3$.

Example 3

Consider the problem of the Example of Section 6.1. The problem is to

$$\text{Maximize} \quad z = 3x_1 + 13x_2$$

subject to

$$\begin{aligned} 2x_1 + 9x_2 &\leq 40 \\ 11x_1 - 8x_2 &\leq 82 \\ x_1, x_2 &\geq 0 \text{ and integral} \end{aligned}$$

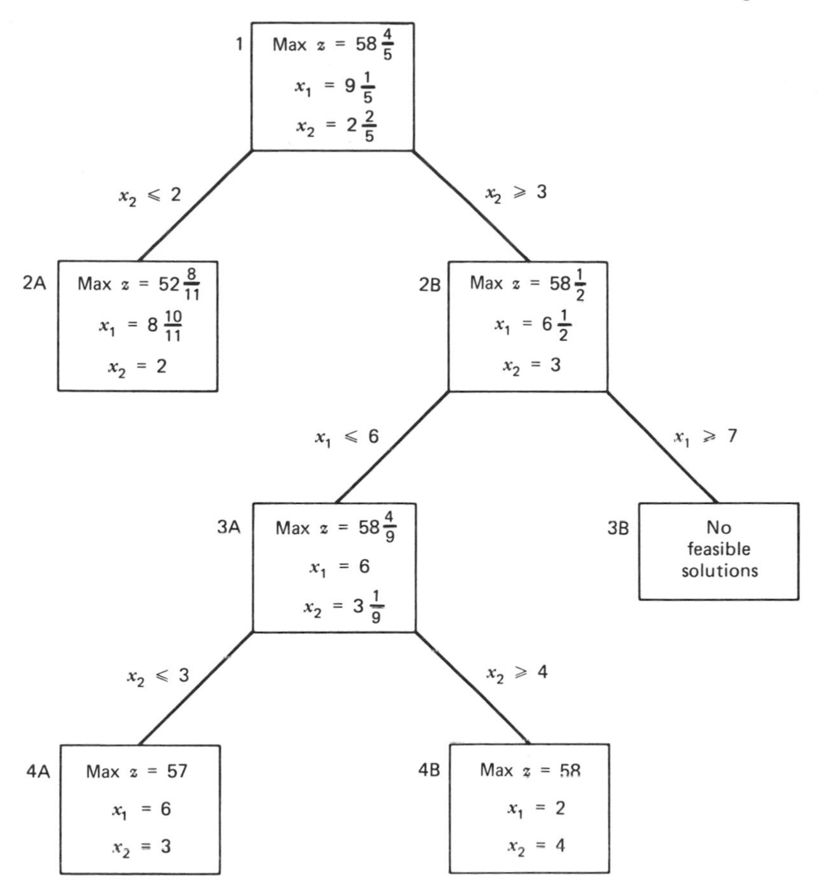

Figure 6.10

Figure 6.10 contains the completed branch and bound diagram. Note that after branching from Problem 1 to Problems 2A and 2B, and solving these problems, we have an option of which problem to work from. Here, however, the choice seems obvious. The maximal value of the objective function under the constraints of Problem 2B is $58\frac{1}{2}$, but under the constraints of Problem 2A is only $52\frac{8}{11}$. Thus, at this time we work from Problem 2B and hold Problem 2A in abeyance for future consideration. But then the optimal value for z of 58 in Problem 4B attained at the feasible integral solution $x_1 = 2$, $x_2 = 4$ makes further consideration of Problem 2A unnecessary, and the algorithm terminates with this solution as the optimal solution to the integer programming problem.

PROBLEM SET 6.4

1. Solve the integer programming problems of the following using the branch and bound algorithm.

   (a) Section 6.1, Problem 1.
   (b) Section 6.1, Problem 2.
   (c) Section 6.3, Problem 2a.
   (d) Section 6.3, Problem 2c.
   (e) Section 6.3, Problem 6.

2. Solve the following using the branch and bound approach.

   (a) Maximize $4x_1 + 5x_2 + 3x_3$

   subject to

   $$3x_1 - 2x_2 + x_3 \leq 14$$
   $$x_1 + 2x_2 + x_3 \leq 8$$
   $$x_1, x_2, x_3 \geq 0 \text{ and integral}$$

   (b) Maximize $9x_1 + 2x_2 + 3x_3$

   subject to

   $$x_1 + x_2 - x_3 \leq 5$$
   $$2x_1 - x_2 + 3x_3 \leq 8$$
   $$x_1, x_2, x_3 \geq 0 \text{ and integral}$$

   (c) Maximize $32x_1 + 21x_2 + 12x_3$

   subject to

   $$3x_1 + 7x_2 + 3x_3 \leq 14$$
   $$9x_1 + 5x_2 + 3x_3 \leq 37$$
   $$x_1, x_2, x_3 \geq 0$$
   $$x_1, x_2 \text{ integral}$$

3. Construct a flow chart for the branch and bound algorithm.

7
The Transportation Problem and Other Topics

SECTION 7.1 A DISTRIBUTION PROBLEM

In Section 2.4 we formulated a mathematical model for the standard transportation problem, the problem of determining a minimal cost shipping schedule between sources and destinations. The model is a linear programming problem, and so the simplex method of Chapter 3 can be used on the problem. However, one would be faced with a formidable linear programming problem if one was about to determine an optimal shipping schedule between, say 100 warehouses and 300 retail outlets. There would be 30,000 variables and 400 constraints. Fortunately, other much more efficient algorithms exist for the resolution of the transportation problem, algorithms that allow for the determination of the solutions to large-scale problems with only moderate amounts of computer effort. We will develop one such algorithm in Section 7.2. Then, in Section 7.3, we will apply apply this technique to determine the resolution of the assignment problem.

In this section we will develop an algorithm to solve the following distribution problem. Suppose a single commodity is produced in varying amounts at a set of plants or origins and is in demand at a set of markets or destinations, with varying demands at the different destinations. To meet the demands, the commodity must be shipped through links connecting the sources with the destinations, but each link has an upper limit on the amount of the commodity that can be shipped through the link. Is it possible to meet the demands of the destinations with the supplies at the sources with a shipping schedule that does not exceed the capacities of the links? Although this problem is of interest in its own right, the computational technique we develop here will be used in the transportation problem algorithm we develop in Section 7.2, and it is primarily for this reason that we consider the above problem.

Specifically, suppose there are m origins or plants and n destinations or markets. Let a_i denote the supply at the ith origin, b_j the demand at the jth destination, and k_{ij} the maximum amount that can be shipped from the ith origin to the jth destination, where k_{ij} is a nonnegative integer, $1 \leq i \leq m$, $1 \leq j \leq n$. Letting x_{ij} denote the amount to be shipped from the ith origin to the jth destination, we want to determine if there exists a shipping flow $\{x_{ij}\}$ satisfying

$$\sum_{j=1}^{n} x_{ij} \leq a_i, \qquad \text{for each } i,\ 1 \leq i \leq m$$

$$\sum_{i=1}^{m} x_{ij} \geq b_j, \qquad \text{for each } j,\ 1 \leq j \leq n$$

$$0 \leq x_{ij} \leq k_{ij}, \qquad \text{for } 1 \leq i \leq m,\ 1 \leq j \leq n$$

and if such a flow exists, a method of finding it. The first set of inequalities requires that the flow out of each origin does not exceed its supply, and the second set requires that the demand at each destination is met. Clearly, in order to have a *feasible solution* to the problem, we must have

$$\sum_{i=1}^{m} a_i \geq \sum_{j=1}^{n} b_j$$

The approach we use to solve this distribution problem is straightforward. We begin at the first origin and assign flow values x_{1j} within the link capacities and meeting the demands, as much as possible, at the destinations. Then we move on to the next origin, and so on. If these initial assignments do not meet all the demands, and they probably will not, then we simply rearrange the flow, "returning" to origins units from destinations that can be supplied by surplus from other origins, and then shipping these returned units to points where there is still demand. We illustrate with a simple example.

Example 1

Consider the distribution problem with three origins, four destinations, and data given by Table 7.1. In the table, the entries in the right-hand column represent the supplies, the entries in the bottom row the demands, and the ijth entry in the main body of the table is the capacity k_{ij}. We have for this problem the total supply equal to the total demand equal to 18, and so it is possible that feasible solutions exist.

Table 7.1

		Destinations				
		1	2	3	4	
	1	4	1	5	1	5
Origins	2	2	2	6	5	7 *Origin Supplies*
	3	5	1	4	0	7
		3	3	9	4	
		Destination Demands				

Starting with the first row and moving from left to right, we ship units to the destinations, the amounts shipped limited only by the destination demands, the link capacities, and the supply. Thus, from the first origin we send 3 units to the first destination, 1 to the second, and 1 to the third. By expanding our table, we denote in Table 7.2 these initial x_{ij} values alongside the corresponding k_{ij}.

Table 7.2

4	3	1	1	5	1	1		5
2		2		6		5		7
5		1		4		0		7
3		3		9		4		

Now we continue in this manner, shipping the 7 units from the second row and 3 units from the third. See Table 7.3.

Table 7.3

4	3	1	1	5	1	1		5	
2		2	2	6	5	5		7	
5		1		4	3	0		7	*S*
3		3		9		4	4		

In this simple manner we have constructed an initial shipping flow. However, the flow is not feasible, since a demand of 4 units remains in the fourth column. We denote these unmet demands with the auxiliary entries in the bottom row. This demand cannot be met directly with the surplus from the third row, the only row with surplus. We label this row with an "*S*." We now begin the more complicated part of the solution process, the readjusting of this flow in an attempt to increase it. For example, 1 unit of the demand for 3 units in the second column could be met by the third row. This would free the 1 unit presently assigned to the second column from the first row, and this unit could then be reassigned to the fourth column. We illustrate in Table 7.4.

Table 7.4

4	3	1	~~1~~0	5	1	1	~~0~~1	5		
2		2	2	6	5	5		7		
5		1	~~0~~1	4	3	0		7		*S*
3		3		9		4	~~4~~3			

Similarly, we increase the flow to the fourth column through the third row, the third column, and the second row. See Table 7.5.

Table 7.5

4	3	1	0	5	1	1	1	5		
2		2	2	6	~~5~~4	5	~~0~~1	7		
5		1	1	4	~~3~~4	0		7		*S*
3		3		9		4	~~3~~2			

Now we still have an unmet demand of 2 units in the fourth column and a surplus of 2 units in the third row. However, these 2 units in the third row can be shipped only to the first column, since all the other links from this source are used to full capacity. Shipping these units to the first column frees 2 units in the first row. Although these cannot be shipped directly to the fourth column, they can be shipped to the third column, freeing 2 units from the second row that can be used to meet the fourth column demand. Thus we have the following chain:

row 3 → column 1 → row 1 → column 3 → row 2 → column 4

With this adjustment, illustrated in Table 7.6, we have constructed a feasible shipping schedule, and the problem is completed.

Table 7.6

4	~~3~~1	1		5	~~1~~3	1	1	5	
2		2	2	6	~~4~~2	5	~~1~~3	7	
5	~~0~~2	1	1	4	4	0		7	~~S~~
3		3		9		4	~~2~~0		

In larger problems the construction of such chains that enable the flow to be readjusted and thereby increased might be rather complicated. Thus we will use a labeling scheme on the rows and columns of the table to facilitate the finding and tracing back of these chains. When units can be shipped to a column, we will record in that column the row number from which the units can come. Similarly, if a row has units that can be reassigned, we will record in that row the column number that allows the reassignment. Then, if we reach a column with an unmet demand, we can use these numbers to trace the chain back to the source with surplus.

We illustrate this procedure with the above example. Consider the problem with the flow given in Table 7.5 (see Table 7.7). Now the surplus in row 3 can be sent to column 1, and so we place a 3 at the bottom of column 1, indicating the

Table 7.7

4	3	1		5	1	1	1	5	
2		2	2	6	4	5	1	7	
5		1	1	4	4	0		7	S
3		3		9		4	2		

source of supply. Continuing, column 1 is receiving units from row 1, so these units in row 1 can be reassigned. We label row 1 with a 1, indicating that from this row units can be reassigned, and that this is possible because of a readjustment in column 1. The augmented table is illustrated in Table 7.8. Now the units of row 1 could be sent to columns 2 or 3, so we append columns 2 and 3 with a 1. From either of these columns row 2 units can be reassigned, so we append row 2 with a 2 or 3. But now column 4 could be sent reassigned units from row 2, so column 4 is

Table 7.8

4	3	1		5	1	1	1	5	1
2		2	2	6	4	5	1	7	
5		1	1	4	4	0		7	S
3		3		9		4	2		
	3								

labeled with a 2. See Table 7.9. Now we have labeled a column with an unmet demand, so we readjust and increase the flow. We can use the row and column labels to trace back from column 4 the route of supply.

column 4 ← row 2 ← columns 2 or 3 ← row 1 ← column 1 ← row 3

If we use column 3, we see that we can adjust the flow by 2 units, and thus we can complete the problem, as previously done.

Table 7.9

4	3	1		5	1	1	1	5	1
2		2	2	6	4	5	1	7	2 or 3
5		1	1	4	4	0		7	S
3		3		9		4	2		
	3		1		1		2		

Notice the double role of the numbers used in the labeling scheme. The fact that a row is labeled indicates that there are free units at that row that can be assigned to any column for which the corresponding shipping link is not being used to full capacity. And the source of these free units is the column number used as the label. Similarly, a labeled column indicates that any row shipping units to that column can reassign these units elsewhere. And the alternate row source for this column is the number used as the label. This scheme is a compromise between simplicity and completeness; we could introduce into the scheme a second index indicating at each step the maximum number of units that could flow to that point through the chain already constructed. For our textbook problems such an index is not necessary. However, if this algorithm were to be programmed to handle large-scale problems, this second index should be incorporated into the program in order to systematize the determination of the flow increase and the readjustment of the flow. For a discussion of this expanded scheme, see Ford and Fulkerson [11].

We now summarize the steps of the *distribution problem algorithm*. Consider a distribution problem with m sources, n destinations, supplies a_i, demands b_j, and capacities k_{ij} and $\sum a_i \geq \sum b_j$.

1. Construct an initial flow $\{x_{ij}\}$ by shipping as much as possible without exceeding demands or capacities first from row 1, then row 2, and so on.
2. Calculate the unmet demand $b_j - \sum_i x_{ij}$ for each column. If this is 0 for all columns, we have a feasible flow and the problem is completed.
3. Otherwise initiate the labeling procedure.
   (a) Label all rows with surplus units, that is, rows with $\sum_j x_{ij} < a_i$, with an "S." Let I denote this set of rows.
   (b) For each $i \in I$, determine all unlabeled columns j for which $x_{ij} < k_{ij}$. Label these columns with the corresponding row number $i \in I$. Let J denote this set of columns.
   (c) For each $j \in J$, determine all unlabeled rows i for which $x_{ij} > 0$. Label these rows with the corresponding column number $j \in J$. Let I denote this new set of labeled rows. Return to Step b.
4. Continue this labeling procedure, moving from rows to columns to rows to columns, and so on, until either:
   (a) In Step 3b, a column with an unmet demand is labeled. Then increase the flow into this column by readjusting the flow values x_{ij}. Erase all labels and return to Step 2.

   or

   (b) In either Step 3b or 3c, no previously unlabeled column or row, respectively, is labeled. Then the problem has no feasible solution.

Before discussing the convergence of the algorithm and the claim of non-feasibility made in Step 4b, we illustrate the algorithm with two additional examples.

Example 2

Consider the distribution problem of Table 7.10.

Table 7.10

4	0	10	10	50	15
8	7	10	8	15	20
10	9	0	6	20	40
5	10	15	20	22	

Here the total supply of 75 exceeds the demand of 72, so we begin the algorithm by constructing the initial flow, determining all unmet demands, and initiating the

Table 7.11

4	4	0		10	10	10	1	50		15	1
8	1	7	7	10	5	8	7	15		20	1 or 2
10		9	3	0		6	6	20	20	40	S
5		10		15		20	6	22	2		
3		3				1 or 2		1 or 2			

labeling procedure, as depicted in Table 7.11. Now there are various chains we can use to ship units from row 3 to columns 4 and 5. For example, we can send 2 units through row 3 → column 1 → row 1 → column 5, 2 units through row 3 → column 1 → row 1 → column 4, and 1 unit through row 3 → column 2 → row 2 → column 4. Recording these changes, erasing the old labels, noting that unmet demands remain in column 4, and labeling again, we have Table 7.12. From this table we see that a chain can be constructed to column 4 and, working backward, we have

$$\text{column 4} \leftarrow \text{row 1} \leftarrow \text{column 3 (or column 5)} \leftarrow \text{row 2}$$
$$\leftarrow \text{column 2 (or column 1)} \leftarrow \text{row 3}$$

Using column 3 and column 2, we can, in fact, increase the flow by the required 3 units. Table 7.13 presents this feasible flow.

Table 7.12

4	0	0	0	10	10	10	3	50	2	15	3 or 5
8	1	7	6	10	5	8	8	15		20	1 or 2
10	4	9	4	0		6	6	20	20	40	S
5		10		15		20	3	22			
3		3		2		1		2			

Table 7.13

4	0	0	0	10	~~10~~7	10	~~3~~6	50	2	15
8	1	7	~~6~~3	10	~~5~~8	8	8	15		20
10	4	9	~~4~~7	0		6	6	20	20	40
5		10		15		20		22		

Example 3

Consider the distribution problem of Table 7.14. Supply equals demands, and so we construct the initial flow, determine umnet demands, and label (Table 7.15).

Table 7.14

4	3	1	4	0	8
5	3	2	0	4	6
3	0	1	10	0	13
0	2	4	6	4	5
6	6	6	9	5	

Table 7.15

4	4	3	3	1	1	4		0		8	1
5	2	3	3	2	1	0		4		6	1
3		0		1	1	10	9	0		13	S
0		2		4	3	6		4	2	5	
6		6		6		9		5	3		
	3				2		3		2		

Thus 2 units can flow through row 3 → column 1 → row 2 → column 4. Recording this adjustment, deleting old labels, and labeling anew, we are led to Table 7.16.

Table 7.16

4	4	3	3	1	1	4		0		8	1
5	0	3	3	2	1	0		4	2	6	
3	2	0		1	1	10	9	0		13	S
0		2		4	3	6		4	2	5	
6		6		6		9		5	1		
	3						3				

Now, however, we reach a point where Step 4b of the algorithm applies. The surplus in row 3 can be sent to columns 1 or 4, and so the units assigned to column 1 from row 1 can be reassigned. But the only link out of row 1 not being used to capacity is to column 4, an already labeled column, Thus our labeling procedure terminates and, since we have been unable to label the column with the unmet demand, we can conclude that the problem has no feasible shipping schedule.

Actually, we do not need to apply to the algorithm to demonstrate that this problem has no feasible solution. Consider the total demand of columns 2, 3, and 5 (the unlabeled columns), that is, 17. The demand can be partially met with the supply from rows 2 and 4, (the unlabeled rows). However, the total supply of these two rows is 11, and so, in any feasible solution to the problem, at least $6 = 17 - 11$ units must flow from rows 1 and 3 (the labeled rows) to columns 2, 3, and 5. But the sum of the capacities of the six links connecting these two rows with these three columns is only $5 = 3 + 0 + 1 + 1 + 0 + 0$. Thus there can be no feasible solutions.

It is precisely the generalization of this argument that we will use to prove the nonfeasibility claim of Step 4b of the algorithm. Clearly, however, the rows and columns to consider will come from the labeling procedure. Before we do this we will prove a theorem stated primarily for its application to the transportation problem algorithm of the next section. The theoretical questions of convergence and nonfeasibility of the distribution problem algorithm will then be discussed.

Theorem 1 *Suppose the distribution problem algorithm is applied to a distribution problem, generating a flow $\{x_{ij}\}$ but that, with this flow, Step 4b of the algorithm is reached. Let R denote the set of all the labeled rows and C the set of all the labeled columns. Then*

(a) $i \in R, j \notin C$ *implies that* $x_{ij} = k_{ij}$.

(b) *For any* $j \in C$, $x_{ij} > 0$ *implies that* $i \in R$.

(c) $\sum_{j \in C} b_j < \sum_{i \in R} a_i$.

Proof (a) If $i \in R$, either row i has a surplus or units in row i can be reassigned. If $x_{ij} < k_{ij}$, this shipping link could be used to send units from row i to column j, and j would be in C. Thus $j \notin C$ implies $x_{ij} = k_{ij}$.

(b) If $j \in C$, column j has an alternate source of supply, and units assigned to column j can be reassigned. Thus if $x_{ij} > 0$, row i has such units and so $i \in R$.

(c) For any $j \in C$, $\sum_{\text{all } i} x_{ij} = b_j$; otherwise Step 4a of the algorithm would have been implemented. Therefore

$$\sum_{j \in C} b_j = \sum_{j \in C}\left(\sum_{\text{all } i} x_{ij}\right) = \sum_{j \in C}\left(\sum_{i \in R} x_{ij}\right) \qquad \text{(from part } b\text{)}$$

$$= \sum_{i \in R}\left(\sum_{j \in C} x_{ij}\right) \le \sum_{i \in R}\left(\sum_{j \in C} x_{ij}\right) + \sum_{i \in R}\left(\sum_{j \notin C} x_{ij}\right)$$

$$= \sum_{i \in R}\left(\sum_{\text{all } j} x_{ij}\right)$$

But for any i, $\sum_{\text{all } j} x_{ij} \le a_i$, and there is at least one row in R with surplus, that is, at least one $i \in R$ with $\sum_{\text{all } j} x_{ij} < a_i$. Therefore $\sum_{i \in R} (\sum_{\text{all } j} x_{ij}) < \sum_{i \in R} a_i$. ###

Theorem 2 *Let R and C be as defined as in Theorem 1, and let R' and C' denote their complements. Then*

$$\sum_{j \in C'} b_j > \sum_{i \in R'} a_i + \sum_{j \in C'} \left(\sum_{i \in R} k_{ij} \right)$$

That is, the total demand in the C' columns is strictly greater than the total supply in the R' rows plus the sum of the capacities of the links from the remaining rows to the C' columns.

Proof For any j, $\sum_{\text{all } i} x_{ij} \le b_j$, and for at least one $j \in C'$, $\sum_{\text{all } i} x_{ij} < b_j$. Note also that from Part b of Theorem 1, $i \in R'$ and $j \in C$ implies that $x_{ij} = 0$. Thus

$$\begin{aligned}
\sum_{j \in C'} b_j &> \sum_{j \in C'} \left(\sum_{\text{all } i} x_{ij} \right) \\
&= \sum_{j \in C'} \sum_{i \in R'} x_{ij} + \sum_{j \in C'} \sum_{i \in R} x_{ij} \\
&= \sum_{i \in R'} \sum_{j \in C'} x_{ij} + \sum_{j \in C'} \sum_{i \in R} k_{ij} \\
&= \sum_{i \in R'} \sum_{\text{all } j} x_{ij} + \sum_{j \in C'} \sum_{i \in R} k_{ij} \\
&= \sum_{i \in R'} a_i \quad + \sum_{j \in C'} \sum_{i \in R} k_{ij}
\end{aligned}$$

since $i \in R'$ implies that row i has no surplus. ###

Corollary 1

If Step 4b is reached in the application of the distribution problem algorithm, the associated problem has no feasible solutions.

Corollary 2

If the distribution problem algorithm is applied to a distribution problem with integral supplies, demands, and capacities, the solution process must terminate in a finite number of iterations either at Step 2 with the construction of a feasible integral solution or at Step 4b with the determination of nonfeasibility.

Proof Notice first that the initial flow and all subsequent modified flows will be integral. In fact, every readjustment of the flow must increase the total flow by at least 1 unit. Since the total demand is finite, the algorithm must therefore terminate after a finite number of iterations. ###

PROBLEM SET 7.1

1. For each of the following distribution problems, determine either a feasible shipping schedule or the set of rows R' and columns C' of Theorem 2, and verify that these rows and columns satisfy the inequality of that theorem.

(a)

20	5	8	0	12
15	1	8	4	23
2	12	10	5	15
10	13	21	6	

(b)

3	2	0	1	5	4
1	0	3	2	2	6
1	2	3	3	0	7
0	1	4	4	1	8
4	3	6	5	6	

(c)

8	0	5	2	10
6	18	10	4	20
10	15	3	20	22
10	1	20	2	28
14	26	18	22	

(d)

∞	∞	0	0	∞	∞	10
∞	∞	∞	∞	0	0	5
0	0	∞	∞	0	∞	14
∞	∞	0	∞	∞	∞	5
4	3	6	7	12	2	

(e)

10	8	10	20	8	20
10	4	15	10	5	30
10	20	4	12	10	40
20	2	25	5	3	50
15	25	35	45	20	

(f)

∞	∞	0	∞	0	0	9
∞	∞	∞	0	∞	∞	10
∞	∞	0	∞	0	0	7
0	∞	∞	∞	∞	∞	12
5	6	8	4	6	9	

(g)

∞	0	∞	∞	0	∞	∞	75
0	∞	0	∞	0	∞	0	103
∞	0	0	0	∞	0	∞	86
∞	∞	0	∞	0	∞	0	200
0	∞	∞	0	∞	0	∞	136
91	43	112	54	120	75	95	

(h)

10	10	1	0	4	8	8	23
3	4	9	15	0	6	5	19
0	5	5	10	5	5	5	18
10	5	10	0	3	3	0	30
8	10	5	4	2	2	2	26
21	14	10	20	12	19	20	

2. Suppose Step 4b is reached in the application of the distribution problem algorithm to a problem for which each k_{ij} is equal to either 0 or ∞. Let R and C be the labeled rows and columns. Explain why $i \in R$ and $j \notin C$ implies that $k_{ij} = 0$.

3. Suppose there is a column j of a distribution problem for which $\sum_{\text{all } i} k_{ij} < b_j$. Then the problem has no feasible solution. Determine sets R' and C' for which the inequality of Theorem 2 applies.

4. Prove that the distribution problem algorithm will terminate after a finite number of steps if applied to a problem with all data rational and not necessarily integral.

5. Consider the following assignment problem. Suppose there are n jobs to be assigned to n individuals, but each individual is capable of doing only some of the available jobs. The problem is to determine if there is an assignment of indivduals to jobs so that all individuals are assigned jobs for which they are qualified.

(a) Formulate this assignment problem as a distribution problem. (*Hint.* Associate individuals to sources, jobs to destinations, and let each of the a_i's and b_j's equal 1.)

(b) Apply the distribution problem algorithm to determine if proper assignments can be made for the following assignment problems. (An "x" in the ijth entry indicates that individual I_i is qualified for job J_j.)

(i)

	J_1	J_2	J_3	J_4	J_5	J_6	J_7	J_8
I_1	x	x					x	
I_2		x		x				x
I_3			x					x
I_4			x			x	x	
I_5	x			x	x			
I_6	x					x		
I_7		x		x	x		x	
I_8	x					x		

(ii)

	J_1	J_2	J_3	J_4	J_5	J_6	J_7	J_8
I_1	x			x		x		
I_2		x	x			x		x
I_3				x		x		
I_4	x		x		x		x	
I_5	x	x		x	x		x	x
I_6	x					x		
I_7		x	x		x			x
I_8	x			x		x		

(c) For any set I of individuals, let $J(I)$ denote the set of all jobs J_j for which at least one individual in I is qualified. Prove that a proper assignment of indviduals and jobs exists if and only if for every set I, the number of elements in $J(I)$ is greater than or equal to the number of elements in I. (*Hint*. To prove that the condition is sufficient; use Corollary 2, Theorem 1, and Problem 2.)

6. Determine a way of using the distribution problem algorithm to resolve the assignment problem of assigning to all indviduals jobs for which they are qualified as described in Problem 5, but under the assumption that there are more jobs than individuals.

SECTION 7.2 THE TRANSPORTATION PROBLEM

The transportation problem is the problem of determining a minimal cost shipping schedule of a commodity between plants or sources or origins and markets or distribution centers or destinations. Specifically, suppose there are m origins and n destinations, and that a supply of a_i units of the commodity is available at the ith origin and a demand of b_j units is to be met at the jth destination. Suppose the cost of shipping a unit from origin i to destination j is c_{ij}. Then a mathematical model of the problem is to

$$\text{Minimize} \quad \sum_{i=1}^{m} \sum_{j=1}^{n} c_{ij} x_{ij}$$

subject to

$$\sum_{j=1}^{n} x_{ij} \leq a_i, \qquad i = 1, \ldots, m$$

$$\sum_{i=1}^{m} x_{ij} = b_j, \qquad j = 1, \ldots, n$$

$$x_{ij} \geq 0, \qquad 1 \leq i \leq m, \qquad 1 \leq j \leq n \tag{1}$$

where x_{ij} denotes the number of units to be shipped from origin i to destination j.

Clearly this problem would have no feasible solutions if the total supply $\sum_i a_i$ were less than the total demand $\sum_j b_j$. In fact, for the rest of this section we will assume that $\sum_i a_i = \sum_j b_j$ and, with this assumption, the above problem is equivalent to the problem of

$$\text{Minimizing} \quad \sum_i \sum_j c_{ij} x_{ij}$$

subject to

$$\sum_j x_{ij} = a_i, \qquad 1 \leq i \leq m$$

$$\sum_i x_{ij} = b_j, \qquad 1 \leq j \leq n$$

$$x_{ij} \geq 0 \tag{2}$$

We lose no generality here by assuming equality of supply and demand; if we encounter a problem with $\sum a_i > \sum b_j$, we can simply create an additional destination, with demand $\sum a_i - \sum b_j$ and shipping costs of 0 from each origin. The formulation of this expanded problem would then have total supply equal to total demand, and any solution to this problem would give a solution to the original problem, with the interpretation that all units scheduled to be sent to this additional destination are surplus units that would remain at their respective origins. We emphasize here in passing that with the assumption that $\sum a_i = \sum b_j$, all our transportation problems have feasible solutions, and the problem is solely to determine a minimal cost feasible solution. This is in contrast to the distribution problem of the last section, where the question of the existence of feasible solutions to the problem was a primary consideration.

The algorithm developed in this section to solve the transportation problem of (2) is due essentially to Ford and Fulkerson ([10] or [11]). It is based on the dual problem to the problem of (2). Recall, in each iteration of the standard simplex algorithm developed in Chapter 3, we moved from feasible solution to feasible solution of the problem, attempting to improve the value of the objective function at each step. In the algorithm that we develop now, in each iteration we will move from feasible solution to feasible solution of the dual problem to (2), improving at each step the dual objective function. We first construct the dual to (2).

In Section 4.1, we saw that the dual to the problem of minimizing $b \cdot Y$ subject to $A^t Y \geq c$, $Y \geq 0$, is to maximize $c \cdot X$ subject to $AX \leq b$, $X \geq 0$. This fact is used to determine the dual to (2).

Now, as long as $\sum_i a_i = \sum_j b_j$, the problem of (2) is equivalent to the problem of

$$\text{Minimizing} \quad \sum_i \sum_j c_{ij} x_{ij}$$

subject to

$$\sum_j x_{ij} \leq a_i, \qquad 1 \leq i \leq m$$

$$\sum_i x_{ij} \geq b_j, \qquad 1 \leq j \leq n$$

$$x_{ij} \geq 0 \tag{3}$$

To show this, suppose $\{x_{ij}\}$ is any feasible solution to (3). Then the first set of inequalities of (3) requires that the total amount shipped $\sum_{i,j} x_{ij} = \sum_i \sum_j x_{ij} \leq \sum_i a_i$. But the second set of inequalities require that $\sum_{i,j} x_{ij} = \sum_j \sum_i x_{ij} \geq \sum_j b_j$. Since $\sum a_i = \sum b_j$, we have that $\sum_i \sum_j x_{ij} = \sum_i a_i$. But, for each i, $\sum_j x_{ij} \leq a_i$. Thus we must have $\sum_j x_{ij} = a_i$ for all i. Similarly, for each j, $\sum_i x_{ij}$ must equal b_j. Hence, $\{x_{ij}\}$ is a feasible solution to (2), and the problems are equivalent.

If we now multiply each inequality in the first set by -1, (3) becomes a minimization problem with $m + n$ ($\geq$) constraints.

$$\text{Minimize} \quad \sum_i \sum_j c_{ij} x_{ij}$$

subject to

$$\sum_j - x_{ij} \geq -a_i, \qquad 1 \leq i \leq m$$

$$\sum_i x_{ij} \geq b_j, \qquad 1 \leq j \leq n$$

$$x_{ij} \geq 0 \tag{4}$$

Thus the dual is a maximization problem with ($\leq$) constraints involving $m + n$ variables. To formulate the problem, introduce dual variables $y_1, \ldots, y_m$, corresponding to the origin constraints of (4), and $z_1, \ldots, z_n$, corresponding to the destination constraints of (4). Then the dual to (4) is the problem of

$$\text{Maximizing} \quad \sum_{j=1}^{n} b_j z_j - \sum_{i=1}^{m} a_i y_i$$

subject to

$$z_j - y_i \leq c_{ij}, \qquad 1 \leq i \leq m,\ 1 \leq j \leq n$$

$$y_i, z_j \geq 0, \qquad 1 \leq i \leq m,\ 1 \leq j \leq n \tag{5}$$

To see this notice first that the coefficient matrix for the constraints of (4) has mn columns corresponding to the mn variables x_{ij} of (4). And each x_{ij} appears in exactly two constraints: with coefficient $+1$ in the jth destination inequality and with coefficient -1 in the ith origin inequality. Hence we have in (5) the mn inequalities $z_j - y_i \leq c_{ij}$. Finally, the coefficients of the objective function in (5) come directly from the constant term column of the constraints of (4).

We illustrate with an example.

Example 1

Consider the transportation problem with two origins, three destinations, and data given by Table 7.17.

Table 7.17

	Destinations 1	2	3	
Origins 1	3	5	7	30 *Origin Supplies*
2	4	7	11	50
	20	25	35	
	Destination Demands			

As in Section 7.1, the entires in the right-hand column of the table represent the supplies and the entries of the bottom row the demands, but now, the *ij*th entry in the main body of the table is the cost c_{ij}. The associated linear programming problem is to

$$\text{Minimize} \quad 3x_{11} + 5x_{12} + 7x_{13} + 4x_{21} + 7x_{22} + 11x_{23}$$

subject to

$$\begin{aligned}
x_{11} + x_{12} + x_{13} \qquad\qquad\qquad\qquad &= 30 \\
x_{21} + x_{22} + x_{23} &= 50 \\
x_{11} \qquad\qquad + x_{21} \qquad\qquad &= 20 \\
x_{12} \qquad\qquad + x_{22} \qquad &= 25 \\
x_{13} \qquad\qquad + x_{23} &= 35 \\
x_{ij} \geq 0 &
\end{aligned}$$

This problem, as we have seen, is equivalent to the problem of

$$\text{Minimizing} \quad 3x_{11} + 5x_{12} + 7x_{13} + 4x_{21} + 7x_{22} + 11x_{23}$$

subject to

$$\begin{aligned}
-x_{11} - x_{12} - x_{13} \qquad\qquad\qquad\qquad &\geq -30 \\
- x_{21} - x_{22} - x_{23} &\geq -50 \\
x_{11} \qquad\qquad + x_{21} \qquad\qquad &\geq \ \ 20 \\
x_{12} \qquad\qquad + x_{22} \qquad &\geq \ \ 25 \\
x_{13} \qquad\qquad + x_{23} &\geq \ \ 35 \\
x_{ij} \geq 0 &
\end{aligned}$$

Thus the dual is to

$$\text{Maximize} \quad -30y_1 - 50y_2 + 20z_1 + 25z_2 + 35z_3$$

subject to

$$\begin{aligned}
&z_1 - y_1 \leq 3,\ z_2 - y_1 \leq 5,\ z_3 - y_1 \leq 7 \\
&z_1 - y_2 \leq 4,\ z_2 - y_2 \leq 7,\ z_3 - y_2 \leq 11 \\
&y_i, z_j \geq 0
\end{aligned}$$

Although the problem of (5) is all we need to develop the solution algorithm to the transportation problem of (2), it is more convenient to work with a linear programming problem equivalent to (5) but with unrestricted variables.

Theorem 1 *The problem of* (5) *is equivalent to the problem of*

$$\textit{Maximizing} \quad \sum_{i=1}^{m} a_i u_i + \sum_{j=1}^{n} b_j v_j$$

subject to

$$u_i + v_j \le c_{ij}, \qquad 1 \le i \le m, 1 \le j \le n$$
$$u_i \text{ and } v_j \text{ unrestricted} \tag{6}$$

Proof First, if $\{y_i, z_j\}$ is any feasible solution to (5), then

$$u_i = -y_i, \qquad 1 \le i \le m$$
$$v_j = z_j, \qquad 1 \le j \le n$$

is a feasible solution to (6), and at these points the corresponding objective functions have the same value.

Conversely, suppose $\{u_i, v_j\}$ is a feasible solution to (6). Then we would like to define the y_i and z_j so that they relate to the u_i and v_j just as above, but now we must modify the above equations so that we are guaranteed nonnegative values for the new variables. In fact, let

$$M = \operatorname*{Max}_{i,j} \{|u_i|, |v_j|\}$$

and define

$$y_j = M - u_i, \qquad 1 \le i \le m$$
$$z_j = M + v_j, \qquad 1 \le j \le n$$

Then, for all i and j, $y_i \ge 0$, $z_j \ge 0$, and $z_j - y_i = v_j + u_i \le c_{ij}$. And the value of the objective function of (5) at this feasible solution is

$$\begin{aligned}\sum_j b_j z_j - \sum_i a_i y_i &= \sum_j b_j(M + v_j) - \sum_i a_i(M - u_i) \\ &= \sum_j b_j v_j + M \sum_j b_j - M \sum_i a_i + \sum_i a_i u_i \\ &= \sum_i a_i u_i + \sum_j b_j v_j\end{aligned}$$

the value of the objective function of (6) at $\{u_i, v_j\}$. # # #

The fundamental step of the solution algorithm to the transportation problem of (2) involves a distribution problem. In the algorithm, we start with a feasible solution $\{u_i, v_j\}$ to (6), the dual of (2). Then a distribution problem between the given origins and destinations is defined as follows: supplies and demands as given in the original transportation problem, and capacities k_{ij} defined to be either infinite or 0, depending on whether $u_i + v_j = c_{ij}$ or $u_i + v_j < c_{ij}$. If this distribution problem has a feasible solution, such a solution is also a minimal cost solution to the transportation problem. If the distribution problem is not feasible, a better solution to the dual problem (6) is constructed by changing the values of some of the u_i's and v_j's. The variables with value changed correspond to the labeled rows and columns of Theorem 1 of the last section. With this new solution to the dual, we begin again, considering the associated distribution problem, and so on. We now outline the steps of the algorithm.

The Transportation Problem Algorithm

Consider a transportation problem as in (2) with m origins, n destinations, supplies a_i, demands b_j, transportation costs c_{ij}, and total supply $\sum a_i$ equal to total demand $\sum b_j$.

1. Construct an initial solution to the dual problem (6) by defining

$$u_i = \operatorname*{Min}_j\{c_{ij}\}, \qquad i = 1,\ldots,m$$

and then

$$v_j = \operatorname*{Min}_i\{c_{ij} - u_i\}, \qquad j = 1,\ldots,n$$

2. Associate with $\{u_i,v_j\}$ a distribution problem: origins and their supplies and destinations and their demands as in the transportation problem, and link capacities

$$k_{ij} = \begin{cases} \infty, & u_i + v_j = c_{ij} \\ 0, & u_i + v_j < c_{ij} \end{cases}$$

3. Attempt to solve this distribution problem.

(a) If there are feasible solutions to the distribution problem, then any such solution is a minimal cost shipping schedule for the original transportation problem.

(b) If the distribution problem has no feasible solutions, determine the labeled rows R and columns C of Theorem 1, Section 7.1.

4. Define a new solution $\{u_i',v_j'\}$ to the dual. Let

$$d = \operatorname*{Min}_{\substack{i \in R \\ j \notin C}}\{c_{ij} - (u_i + v_j)\}$$

and define

$$u'_i = \begin{cases} u_i + d, & i \in R \\ u_i, & i \notin R \end{cases} \qquad \text{and} \qquad v_j' = \begin{cases} v_j - d, & j \in C \\ v_j, & j \notin C \end{cases}$$

Return to Step 2, now using this solution $\{u_i',v_j'\}$ to the dual.

Before discussing convergence and proving the claims of the algorithm, namely, that in Step 3a, a feasible solution to the associated distribution problem is a minimal cost solution to the transportation problem, and that in Step 4, a better feasible solution to the dual is defined, we illustrate the algorithm with two examples. The table that we use for the recording of the steps of the algorithm will be an expansion on the table as suggested in Example 1. To record the present values of the dual variables, we introduce a new left-hand column for the u_i values and a new top row for the v_j values. Since the capacities for the associated distribution problem can be only 0 or ∞, we can denote the existence of a shipping link from, say origin i to destination j, by simply placing a circle next to the cost c_{ij} in the main body of the table. Then the x_{ij} values calculated when attempting to solve the corresponding distribution problem can be placed in these circles.

Example 1 (continued)

The data are presented in Table 7.18.

Table 7.18

3	5	7	30
4	7	11	50
20	25	35	

We first construct an initial solution to the dual, letting $u_1 = 3$, the minimal cost in the first row, and $u_2 = 4$, the minimal cost in the second row. Since we need $u_i + v_j \leq c_{ij}$ for all i and j, the largest value v_1 could be is

$$\text{Min}\{c_{11} - u_1, c_{21} - u_2\} = \text{Min}\{0,0\} = 0$$

and similarly

$$v_2 = \text{Min}\{5 - 3, 7 - 4\} = 2$$
$$v_3 = \text{Min}\{7 - 3, 11 - 4\} = 4$$

See Table 7.19. Now we indicate with circles the usable shipping links for the

Table 7.19

	$v_1 = 0$	$v_2 = 2$	$v_3 = 4$	
$u_1 = 3$	3	5	7	30
$u_2 = 4$	4	7	11	50
	20	25	35	

associated distribution problem, that is, those entries for which $u_i + v_j = c_{ij}$ (Table 7.20). Notice that in attempting to solve this distribution problem, the only

Table 7.20

	0	2	4	
3	3 ○	5 ○	7 ○	30
4	4 ○	7	11	50
	20	25	35	

data from the tableau that are used are the supplies 30 and 50, the demands 20, 25, and 35, and the presence of the circles. Now this particular distribution problem is not feasible, but a partial solution found by using say the algorithm of the last section is indicated in Table 7.21. There is a surplus at the second row, and these

Table 7.21

	0	2	4		
3	3 ○	5 (25)	7 (5)	30	
4	4 (20)	7	11	50	S
	20 / 2	25	35 ⋮ 30		

units can be sent to the first column only (as $k_{22} = k_{23} = 0$, and $x_{21} = 20 < \infty = k_{21}$). But the first column is not receiving units from any other row, and so the flow cannot be increased. Thus row 2 is the labeled row and column 1 is the labeled column; that is, $R = \{2\}$ and $C = \{1\}$.

To construct a better solution to the dual, calculate

$$d = \underset{\substack{i=2 \\ j=2,3}}{\text{Min}} \{c_{ij} - (u_i + v_j)\}$$

$$= \text{Min}\{7 - (2 + 4), 11 - (4 + 4)\} = 1$$

and so increase u_2 by 1 and decrease v_1 by 1, leaving u_1, v_2, and v_3 unchanged. See Table 7.22

Table 7.22

	~~0~~ − 1	2	4	
3	3	5	7	30
5~~4~~	4	7	11	50
	20	25	35	

Now we proceed just as before, circling the existing shipping links and attempting to solve the associated distribution problem (Table 7.23). Again the distribution

Table 7.23

	−1	2	4		
3	3	5 ◯	7 (30)	30	
5	4 (20)	7 (25)	11	50	S
	20 2	25 2	35 ⋮ 5		

problem is not feasible, and we are led to the sets $R = \{2\}$, corresponding to the row with surplus, and $C = \{1,2\}$, corresponding to the columns from which units could be reassigned. Note that in the application of the transportation problem algorithm, when Step 3b is encountered, not only do we determine that the associated distribution problem is not feasible, but we must also determine the sets of labeled rows R and columns C to which Theorem 1 of the last section applies.

Now here

$$d = \underset{\substack{i=2 \\ j=3}}{\text{Min}}\{c_{ij} - (u_i + v_j)\}$$

$$= \text{Min}\{11 - (4 + 5)\} = 2$$

and so we increase u_2 by 2, decrease v_1 and v_2 by 2, and leave u_1 and v_3 unchanged (Table 7.24).

Table 7.24

	−~~1~~ − 3	~~2~~0	4	
3	3	5	7	30
7~~5~~	4	7	11	50
	20	25	35	

Circling and considering the associated distribution problem, we have Table 7.25.

Table 7.25

	−3	0	4	
3	3	5	7 (30)	30
7	4 (20)	7 (25)	11 (5)	50
	20	25	35	

But now the distribution problem is feasible, and so the indicated solution, $x_{11} = x_{12} = 0, x_{13} = 30, x_{21} = 20, x_{22} = 25, x_{23} = 5$, is a minimal cost shipping schedule for the original transportation problem. The actual shipping cost would be $520 = 7 \cdot 30 + 4 \cdot 20 + 7 \cdot 25 + 11 \cdot 5$.

Example 2

Consider the transportation problem with four origins, six destinations, and data given by Table 7.26.

Table 7.26

7	6	5	8	7	8	16
2	5	6	7	4	6	12
2	2	1	3	3	1	10
1	3	4	3	2	5	18
5	11	3	13	7	17	

Constructing the initial solution to the dual and circling to indicate the associated distribution problem, we have Table 7.27. Attempting to solve the distribution

Table 7.27

	0	1	0	2	1	0	
5	7	6 ○	5 ○	8	7	8	16
2	2 ○	5	6	7	4	6	12
1	2	2 ○	1 ○	3 ○	3	1 ○	10
1	1 ○	3	4	3 ○	2 ○	5	18
	5	11	3	13	7	17	

problem leads to Table 7.28. From that table, we have $R = \{1,2\}$, $C = \{1,2,3\}$, and $d = 1$. We record the new solution to the dual and the associated distribution problem in Table 7.29. Notice that while we have gained three new shipping links (row 1 to columns 4 and 5 and row 2 to column 5) we have lost three from the previous distribution problem (row 3 to columns 2 and 3 and row 4 to column 1). However, none of these lost links were being used in the final attempted solution

Table 7.28

	0	1	0	2	1	0		
5	7	6 (11)	5 (3)	8	7	8	16	*S*
2	2 (5)	5	6	7	4	6	12	*S*
1	2	2 ○	1 ○	3 ○	3	1 (10)	10	
1	1 ○	3	4	3 (13)	2 (5)	5	18	
	5 / 2	11 / 1	3 / 1	13	7 ⋮ 2	17 ⋮ 7		

Table 7.29

	~~0~~ − 1	~~1~~0	~~0~~ − 1	2	1	0	
6~~5~~	7	6 ○	5 ○	8 ○	7 ○	8	16
3~~2~~	2 ○	5	6	7	4 ○	6	12
1	2	2	1	3 ○	3	1 ○	10
1	1	3	4	3 ○	2 ○	5	18
	5	11	3	13	7	17	

Table 7.30

	−1	0	−1	2	1	0		
6	7	6 (11)	5 (3)	8 (2)	7 ○	8	16	4
3	2 (5)	5	6	7	4 (7)	6	12	5
1	2	2	1	3 ○	3	1 (10)	10	
1	1	3	4	3 (11)	2 ○	5	18	S
	5	11	3	13	7	17 ⋮ 7		
	2	1	1	4	4			

Table 7.31

	−~~1~~ − 3	~~0~~ − 2	−~~1~~ − 3	2~~0~~	~~1~~ − 1	0	
8~~6~~	7	6 (11)	5 (3)	8 (2)	7 ○	8 ○	16
5~~3~~	2 (5)	5	6	7	4 (7)	6	12
1	2	2	1	3	3	1 (10)	10
3~~1~~	1	3	4	3 (11)	2 ○	5	18
	5	11	3	13	7	17	

to the first distribution problem, (and this is always the case—see Problem 5), and so in attempting to solve this new distribution problem we need not start with Step 1 of the distribution problem algorithm of the last section but, instead, can use the flow already constructed from the first iteration as the initial flow here. Building on this flow, we are led to the flow in Table 7.30. Here $R = \{1,2,4\}$, $C = \{1,2,3,4,5\}$, and $d = 2$. Table 7.31 records the new solution to the dual along with the last flow. Using the surplus of the fourth row to increase the flow, we are led to Table 7.32. Here $R = \{2,4\}$, $C = \{1,4,5\}$, and $d = 1$, leading to Table 7.33. The last flow has been retained. But now the flow can be increased and, in fact, the given distribution problem has a feasible solution. Table 7.34, the final table, provides a minimal cost flow.

Table 7.32

	−3	−2	−3	0	−1	0		
8	7	6 (11)	5 (3)	8 ○	7 ○	8 (2)	16	
5	2 (5)	5	6	7	4 (7)	6	12	5
1	2	2	1	3	3	1 (10)	10	
3	1	3	4	3 (13)	2 ○	5	18	S
	5 2	11	3	13 4	7 4	17 ⋮ 5		

Table 7.33

	−3̸ − 4	−2	−3	0̸ − 1	−1̸ − 2	0		
8	7	6 (11)	5 (3)	8	7	8 (2)	16	
6 5̸	2 (5)	5	6	7	4 (7)	6 ○	12	5
1	2	2	1	3	3	1 (10)	10	
4 3̸	1	3	4	3 (13)	2 ○	5	18	S
	5 2	11	3	13 4	7 4	17 ⋮ 5 2		

Table 7.34

	-4	-2	-3	-1	-2	0	
8	7	6 (11)	5 (3)	8	7	8 (2)	16
6	2 (5)	5	6	7	4 (2)	6 (5)	12
1	2	2	1	3	3	1 (10)	10
4	1	3	4	3 (13)	2 (5)	5	18
	5	11	3	13	7	17	

We now consider the theoretical questions raised after the formulation of the transportation problem algorithm.

Theorem 1 *When Step 4 of the transportation problem algorithm is implemented, the* $\{u_i',v_j'\}$ *constructed is a new feasible solution to the dual problem* (6), *and the value of the objective function of* (6) *at this point is strictly greater than the value of the function at the* $\{u_i,v_j\}$ *solution point.*

Proof First, from Problem 2 of Section 7.1, for all $i \in R$ and $j \notin C$, $c_{ij} > u_i + v_j$, and so $d > 0$. Thus the point $\{u_i',v_j'\}$ is distinct from the point $\{u_i,v_j\}$.

Next, $u_i' + v_j' \leq c_{ij}$ for all i and j.

If $i \in R, j \in C$,

$$u_i' + v_j' = (u_i + d) + (v_j - d) = u_i + v_j \leq c_{ij}$$

If $i \notin R, j \notin C$,

$$u_i' + v_j' = u_i + v_j \leq c_{ij}$$

If $i \notin R, j \in C$,

$$u_i' + v_j' = u_i + (v_j - d) < u_i + v_j \leq c_{ij}$$

If $i \in R, j \notin C$,

$$u_i' + v_j' = u_i + d + v_j \leq c_{ij} \text{ by the choice of } d$$

Thus $\{u_i',v_j'\}$ is a feasible solution to (6).

Finally,

$$\sum_{i=1}^{m} a_i u_i' + \sum_{j=1}^{n} b_j v_j' = \sum_{i \in R} a_i u_i' + \sum_{i \notin R} a_i u_i' + \sum_{j \in C} b_j v_j' + \sum_{j \notin C} b_j v_j'$$

$$= \sum_{i \in R} a_i(u_i + d) + \sum_{i \notin R} a_i u_i + \sum_{j \in C} b_j(v_j - \mathrm{d}) + \sum_{j \notin C} b_j v_j$$

$$= \sum_{i=1}^{m} a_i u_i + \sum_{j-1}^{n} b_j v_j + d\left(\sum_{i \in R} a_i - \sum_{j \in C} b_j\right)$$

$$> \sum_{i=1}^{m} a_i u_i + \sum_{j=1}^{n} b_j v_j \qquad \text{as} \qquad \sum_{i \in R} a_i - \sum_{j \in C} b_j > 0 \text{ (Theorem 1 of Section 7.1)}$$

###

Theorem 2 *Suppose in the implementation of the transportation problem algorithm, the distribution problem associated with a solution $\{u_i, v_j\}$ of* (6) *has a feasible flow solution $\{x_{ij}\}$. Then this flow $\{x_{ij}\}$ is an optimal solution to* (2).

Proof Since the problems of (2) and (6) are dual, from Corollary 1 of Section 4.3, all we need show is that the value of the objective function of (2) at $\{x_{ij}\}$ equals the value of the objective function of (6) at $\{u_i, v_j\}$. And, using the fact that x_{ij} can be non-zero only if $c_{ij} = u_i + v_j$, we have

$$\sum_{i=1}^{m} \sum_{j=1}^{n} c_{ij} x_{ij} = \sum_{i=1}^{m} \sum_{j=1}^{n} (u_i + v_j) x_{ij}$$

$$= \sum_{i=1}^{m} \sum_{j=1}^{n} u_i x_{ij} + \sum_{i=1}^{m} \sum_{j=1}^{n} v_j x_{ij}$$

$$= \sum_{i=1}^{m} u_i\left(\sum_{j=1}^{n} x_{ij}\right) + \sum_{j=1}^{n} v_j\left(\sum_{j=1}^{m} x_{ij}\right)$$

But the $\{x_{ij}\}$ is a feasible solution to the distribution problem, and so

$$\sum_{i=1}^{m} x_{ij} = b_j \text{ for each } j \text{ and, since } \sum_i a_i = \sum_j b_j,\ \sum_{j=1}^{n} x_{ij} = a_i \text{ for each } i.$$

Thus

$$\sum_{i=1}^{m} \sum_{j=1}^{n} c_{ij} x_{ij} = \sum_{i=1}^{m} a_i u_i + \sum_{j=1}^{n} b_j v_j. \qquad ###$$

Theorem 3 *Suppose the data of the transportation problem of* (2) *are all integral. Then the transportation problem algorithm applied to* (2) *will terminate after a finite number of steps to a minimal cost flow with all integral values.*

Proof The objective function of (2), $\sum_{i,j} c_{ij}x_{ij}$, is bounded below by $c \sum_i a_i$, where $c = \text{Min}_{i,j}\{c_{ij}\}$, and so (2) and therefore its dual (6) have finite optimal solutions (Duality Theorem of Section 4.3). Now, in each iteration of the transportation problem algorithm, the dual objective function is increased by the quantity $d(\sum_{i \in R} a_i - \sum_{j \in C} b_j)$ (Proof of Theorem 1), and this quantity must be at least 1 if the data are integral. Since the dual objective function is bounded the algorithm must therefore eventually terminate at Step 3a, at a distribution problem with a feasible solution. And this distribution problem will have integral solutions $\{x_{ij}\}$, from Corollary 2 of Section 7.1. ###

PROBLEM SET 7.2

1. Solve the transportation problems with the following data tables.

(a)

5	6	7	5	12
9	10	9	6	14
3	4	4	2	10
9	8	11	8	

(b)

4	3	5	2	2	15
6	7	5	5	4	15
5	5	7	3	4	10
8	4	6	10	12	

(c)

8	10	9	8	7	8
9	5	4	6	9	10
10	8	5	8	8	12
5	6	4	8	7	

(d)

12	18	21	8	17	50
10	13	20	8	16	60
13	16	25	12	20	75
15	15	19	13	18	40
25	35	45	55	65	

(e)

4	12	6	5	2	10	20
9	13	7	7	3	10	26
2	10	1	3	1	9	28
5	7	5	6	2	5	22
13	17	15	19	21	11	

2. (a) Solve the transportation problem with the following data table by first creating a fourth destination with a demand equal to the surplus of 3 units and transportation costs of 0.

6	4	5	9
8	3	2	16
8	8	6	

(b) Show that if you initially neglected to establish the fourth destination for the surplus and instead applied the transportation problem algorithm directly to the above problem, you would be led to a distribution problem with the property that some of its feasible solutions are not optimal solutions to the transportation problem.

3. Solve the transportation problems defined by the following tables. (A $c_{ij} = \infty$ indicates the impossibility of shipping between the corresponding origin and destination, and a $c_{ij} < 0$ indicates the effects of government subsidies.)

(a)

13	∞	18	-3	9
17	12	∞	-2	11
19	14	23	-4	13
6	7	9	7	

(b)

5	-2	-7	6	4	20
14	1	-6	16	∞	10
∞	-1	0	∞	13	25
12	-2	-9	14	12	20
10	9	15	20	11	

4. Prove that under the assumption that $\sum a_i = \sum b_j$, any feasible solution to the problem of (1) of this section is also a feasible solution to (2); that is, prove that the problems of (1) and (2) are equivalent.

5. Prove that when going from Step 4 to Step 2 in the transportation problem algorithm, any shipping link that was being used (i.e., the $x_{ij} > 0$) in the attempted solution to the first distribution problem will remain open in the distribution problem of the next iteration (i.e., the $u_i' + v_j' = c_{ij}$). (*Hint.* Use Part b of Theorem 1 of Section 7.1.)

6. Verify for each of the two examples of this section that at the final iteration when a distribution problem with feasible solutions is constructed, the value of the cost function $\sum_{i,j} c_{ij}x_{ij}$ at the feasible flow $\{x_{ij}\}$ equals the value of the dual objective function at the corresponding solution point $\{u_i,v_j\}$ to the dual.

7. Prove Theorem 2 of this section using the Complementary Slackness Theorem, Problem 7 of Section 4.3.

8. Given a transportation problem, let $\{u_i,v_j\}$ be any optimal feasible solution to its dual, the problem of (6). Show that then if $\{x_{ij}\}$ is any minimal cost solution to the transportation problem, $x_{ij} > 0$ implies that $u_i + v_j = c_{ij}$. In other words, suppose the transportation problem algorithm is used to resolve a transportation problem. Then any minimal flow solution to the problem, no matter how it is found, can use only the circled links from the final table of the algorithm resolution to the problem. (*Hint.* Use the Complementary Slackness Theorem, page 00.)

9. Prove that the transportation problem algorithm will lead to an optimal solution after a finite number of steps if the algorithm is applied to a problem with all data rational.

10. Suppose a minimal cost shipping flow $\{x_{ij}\}$ is determined for a standard transportation problem with total supply equal to total demand. Does this solution remain optimal if:
(a) All the costs c_{ij} from an origin i are altered by a fixed amount q?
(b) All the costs c_{ij} to a destination j are altered by a fixed amount q?
Suppose in the initial transportation problem $\sum a_i > \sum b_j$. Would your answers to the above be the same?

Formulate each of the following as a transportation problem, and then solve using the algorithm of this section.

11. (a) Consider the transportation problem with data given by

5	4	1	3	4	5
4	7	2	5	6	8
6	6	2	4	5	7
6	6	5	4	3	

Notice that the total supply of 20 units is less than the total demand of 24 units. The supplier desires to distribute at minimal shipping costs the entire 20 units among the five destinations so that no demand is exceeded.

(b) As above, but now the supplier also incurs a cost of q_j per unit of unmet demand at the jth destination, with $q_1 = 4$, $q_2 = q_4 = 5$, $q_3 = 1$, and $q_5 = 7$.

12. Consider the data for Problem 11a, but assume that the a_i's represent the available supplies, the b_j's represent the number of units that can be sold at the jth destination, and the c_{ij}'s represent the net profit (gross profit less overhead and transportation costs) from the sale of 1 unit shipped from the ith origin and sold at the jth destination. How should the supply be distributed so that profit is maximized?

13. Reconsider Problem 12, using the data from Problem 1c.

14. A firm with two plants must supply three outlets over the next three time periods. The supplies and demands over the three periods and the shipping costs for any period are given in the following tables.

	Plant Supplies 1	*Plant Supplies* 2	*Outlet Demands* 1	*Outlet Demands* 2	*Outlet Demands* 3
Periods 1	15	25	10	5	20
Periods 2	15	25	10	10	10
Periods 3	10	20	10	15	10

Shipping Costs	*Outlets* 1	*Outlets* 2	*Outlets* 3
Plants 1	7	9	12
Plants 2	10	11	16

The period demands at each outlet must be met exactly. Any units produced but undelivered at a plant may be stored at the plant for later delivery with a storage cost of 3/unit/period at Plant 1, and 2/unit/period at Plant 2, or such units may be sold at a profit of 5/unit at Plant 1 and 8/unit at Plant 2. How should the units be distributed so that net expenses are minimized?

15. Reconsider Problem 14 using the following data.

	Plant Supplies 1	Plant Supplies 2	Outlet Demands 1	Outlet Demands 2	Outlet Demands 3
Periods 1	15	15	5	6	2
Periods 2	15	10	8	10	4
Periods 3	10	5	10	7	6

Shipping Costs		Outlets 1	Outlets 2	Outlets 3
Plants	1	2	1	3
Plants	2	6	4	5

	Plant 1	Plant 2
Storage cost/unit/period	3	1
Profit/unit sold	10	8

16. Reconsider the transportation problem of Problem 1b, with the restriction that the maximum number of units that can be shipped from the second source to the third destination is 4.

17. *The Caterer Problem* (*Jacobs* [19]). A caterer must supply 110 napkins on Monday, 90 on Tuesday, 130 on Wednesday, and 170 on Thursday. The caterer initially has no napkins on hand. He can buy napkins new at 7 cents/napkin, and can have used napkins laundered for use the next day at 4 cents/napkin, or laundered for use in 2 days or more at 2 cents/napkin. At the end of the week all used napkins have no value. How can the caterer meet these demands at minimal costs. (*Hint*. Consider this as a transportation problem with four sources—the new napkin outlet, and Monday's, Tuesday's, and Wednesday's collections of used napkins.)

18. Reconsider Problem 17, with demands of 70 for Monday, 60 for Tuesday, 80 for Wednesday, 100 for Thursday, and 90 for Friday, and new napkins costs of 12 cents, next-day laundry service at 5 cents, 2-day laundry service at 3 cents, and 3 day or more service at 2 cents. Furthermore, assume that all used napkins are worth 1 cent at the end of the week.

19. Problem 8 of Section 2.6.

SECTION 7.3 THE ASSIGNMENT PROBLEM

As a direct application of the algorithm of the last section we consider the assignment problem. Suppose a plant manager has eight different jobs to be performed for the next month and eight different machines to do the jobs. Suppose that for each different machine and job, there is a known cost to be incurred if the given machine is assigned to perform the given job. This cost factor could include setup time expenses, production expenses, and so on. Obviously the manager seeks an assignment of machines to jobs that would minimize the total monthly costs. One way to solve this problem would be simply to list all the possible assignments, compute the costs associated with each, and choose an assignment that delivers the minimal cost. But even for this simple problem such an approach would not be all that efficient, because there are $8! = 40{,}320$ ways of assigning the eight machines to the eight jobs. As we will show in this section, this problem can be formulated as a transportation problem, so the algorithm of the previous section could be used as an effective tool for determining the minimal cost assignment.

For the general assignment problem, suppose there are m individuals or machines or whatever, $I_1, I_2, \ldots, I_m$, to be assigned to n jobs, $J_1 J_2, \ldots, J_n$, and that for each I_i and J_j, there is an associated cost c_{ij} to be incurred if I_i is assigned J_j. We will assume that $m = n$; if this is not the case, we can add either additional individuals or jobs to the problem so that the numbers are equal, associating a cost factor of zero to each new individual and all jobs, or to each new job and all individuals. Any optimal solution to this modified problem would translate directly into a solution to the original. With this assumption, the problem is to determine an assignment of all n individuals to the n jobs so that the associated total cost is minimized.

This problem can be easily formulated as an integer programming problem. Introduce n^2 variables x_{ij}, with the interpretation that

$$x_{ij} = \begin{cases} 1, & I_i \text{ is assigned } J_j \\ 0, & I_i \text{ is not assigned } J_j \end{cases}$$

Then the assignment problem is equivalent to the problem of

$$\text{Minimizing} \quad \sum_{i=1}^{n} \sum_{j=1}^{n} c_{ij} x_{ij}$$

subject to

$$\sum_{i=1}^{n} x_{ij} = 1, \qquad 1 \le j \le n$$

$$\sum_{j=1}^{n} x_{ij} = 1, \qquad 1 \le i \le n$$

$$0 \le x_{ij} \le 1 \text{ and integral, } 1 \le i, j \le n \tag{1}$$

To see this, note that the bounds and integral restrictions on the x_{ij} limit their values to be either 0 or 1. Thus, the first set of constraints in (1) demand that for each j, there is exactly one i for which $x_{ij} = 1$; that is, for each job there is exactly one individual assigned the job and, similarly, the second set of constraints in (1) demand that for each i, there is exactly one j for which $x_{ij} = 1$; that is, each individual is assigned exactly one job. Thus the desired interpretation of the x_{ij} variables is accomplished. Note finally that the objective function of (1) properly measures the cost associated with an assignment $\{x_{ij}\}$.

Now the problem of (1) closely resembles the general transportation problem, the problem of (2) of Section 7.2. However, the problems differ in that the x_{ij} variables of the assignment problem (1) are restricted to be both integral and not greater than 1. This later restriction, that the $x_{ij} \leq 1$, can be discarded immediately, however; the equations of (1) and the nonnegativity of the x_{ij}'s make this upper bound constraint redundant. Let us now assume that the c_{ij} are all integral. We lose no generality here, because if the c_{ij} are rational, there exists a constant $c \neq 0$ such that cc_{ij} is integral for all i and j, and the assignment problem with costs cc_{ij} would be equivalent to the original; if some of the c_{ij}'s are irrational, they can be approximated by rationals. With this assumption the problem of (1) is equivalent to the following.

$$\text{Minimize} \quad \sum_{i=1}^{n} \sum_{j=1}^{n} c_{ij} x_{ij}$$

subject to

$$\sum_{i=1}^{n} x_{ij} = 1, \qquad 1 \leq j \leq n$$

$$\sum_{j=1}^{n} x_{ij} = 1, \qquad 1 \leq i \leq n$$

$$x_{ij} \geq 0, \qquad 1 \leq i, j \leq n \tag{2}$$

This follows directly from Theorem 3 of the last section; that theorem guarantees that the problem of (2) has optimal solutions with all x_{ij} integral.

To summarize, if all the c_{ij} are integral, the assignment problem, the integer programming problem of (1), is equivalent to the linear programming problem of (2). But (2) can be considered as a transportation problem, with $m = n$ and all a_i and b_j equal to 1. Thus the transportation problem algorithm can be used to

Table 7.35

	J_1	J_2	J_3	J_4	J_5	J_6
I_1	14	13	12	17	15	10
I_2	8	12	9	11	10	12
I_3	3	4	2	6	5	7
I_4	7	8	6	9	9	6
I_5	11	17	14	16	15	∞

determine a solution to (2), and that is equivalent to determining an optimal assignment solution to (1).

Example

Consider the assignment problem involving five individuals, six jobs, and rating matrix given by the Table 7.35. We seek the assignment that leaves one job unassigned and minimizes total costs. Note that $c_{56} = \infty$ indicates that I_5 cannot be

Table 7.36

14	13	12	17	15	10	1
8	12	9	11	10	12	1
3	4	2	6	5	7	1
7	8	6	9	9	6	1
11	17	14	16	15	∞	1
0	0	0	0	0	0	1
1	1	1	1	1	1	

Table 7.37

	~~0~~ − 2	0	~~0~~ − 2	0	0	~~0~~ − 2	
12 ~~10~~	14	13	12	17	15	10 ①	1
10 ~~8~~	8 ○	12	9	11	10 ①	12	1
4 ~~2~~	3	4 ○	2 ①	6	5	7	1
8 ~~6~~	7	8 ①	6 ○	9	9	6 ○	1
13 ~~11~~	11 ①	17	14	16	15	∞	1
0	0	0 ○	0	0 ①	0 ○	0	1
	1	1	1	1	1	1	

assigned J_6. Adding an extra row, the initial table for the corresponding transportation problem is given by Table 7.36. Applying the transportation problem algorithm through two iterations leads to Table 7.37. Thus one optimal assignment is I_1 to J_6, I_2 to J_5, I_3 to J_3, I_4 to J_2, I_5 to J_1, and J_4 unassigned. The minimal cost of any optimal assignment is $41 = 10 + 10 + 2 + 8 + 11 + 0$.

PROBLEM SET 7.3

1. Is the optimal assignment listed in the example unique? In any optimal assignment from the final table, must J_4 go unassigned? Why must the value of any optimal assignment be 41?

2. Determine optimal assignments and total minimal costs for the assignment problems defined by the following rating matrices.

(a)

17	27	20
23	25	19
11	18	13

(b)

6	6	7	8	3
2	4	2	1	3
6	9	8	9	7
1	3	2	3	0

(c)

4	8	7	2	5
7	9	10	3	7
5	10	11	6	6
2	6	9	1	8
9	8	7	5	9
3	8	6	4	8

(d)

12	16	14	10	5	12	18	13
8	6	7	9	8	11	10	12
13	18	16	14	9	11	14	17
20	18	17	19	12	13	15	14
13	15	12	10	6	18	13	13
6	5	8	9	8	7	4	7
1	4	7	6	3	3	2	5
11	9	10	12	7	5	7	11

3. Solve the following assignment problems but, for these problems, assume that the c_{ij}'s represent profit and so the objective function $\sum_{i,j} c_{ij}x_{ij}$ is to be maximized.

(a)

7	8	3	10
6	5	0	11
5	6	2	12
6	5	1	9

(b)

6	9	5	7	4
3	8	4	5	7
2	7	3	4	6
5	6	5	7	5

4. Each of six individuals are to be assigned to one of six different jobs, and all six jobs must be completed. The individuals have ranked the jobs in order of preference, giving a 1 to the most desirable job, and so on. These rankings are given in the following table.

	J_1	J_2	J_3	J_4	J_5	J_6
I_1	3	2	1	6	4	4
I_2	1	2	2	4	6	5
I_3	3	1	2	∞	∞	4
I_4	2	1	4	3	5	6
I_5	4	4	1	2	6	3
I_6	1	3	2	6	4	5

Note that the rankings include some ties, and that I_3 is not qualified for two jobs. How should the assignments be made?

5. Two jobs, the first requiring three workers and the second two workers, must be completed. There are seven workers that are qualified for the jobs. The wages a worker would receive depend on both the worker and the job assigned, and are as follows.

	W_1	W_2	W_3	W_4	W_5	W_6	W_7
J_1	105	90	85	95	80	65	80
J_2	125	105	135	115	100	105	95

Furthermore, each of the first four workers must still be paid 50 units if unassigned, whereas each of the last three workers receive only 25 units if unassigned. Determine a minimal cost assignment.

6. (a) The shop of Problem 5 is given an order to be completed in 2 weeks. The order requires 9 man-weeks of labor for a J_1 type job, and 4 man-weeks of labor for a J_2 type job. Assuming that the wages listed in Problem 5 represent wages per week for the workers, how should the seven workers of the shop be assigned over the 2-week period so that the order is completed and the costs are minimized?
(b) As in part a, but assume now that the shop has 3 weeks to complete the order.

7. A machine shop is given six orders. However, the shop has only five machines and therefore cannot fulfill one order. The profits realized from assigning a machine to an order, and the penalty costs incurred if an order is not completed, are as follows.

		Orders					
		1	2	3	4	5	6
	1	75	80	85	70	70	x
	2	56	55	65	50	48	48
Machines (profit)	3	20	35	30	x	24	x
	4	50	46	42	38	42	32
	5	65	62	48	55	52	45
Penalty Cost		20	25	5	15	24	18

An "x" in the table indicates that a machine is not suited for the associated job. Determine an assignment that optimizes net profit.

8. True or false: In the final tableau corresponding to the resolution of an assignment problem, the sum of the dual variables, $\sum_i u_i + \sum_j v_j$, equals the total cost of an optimal assignment?

9. Prove that in any optimal assignment, at least one individual is assigned a job for which her cost factor is minimal; that is, at least one individual is assigned her best job.

10. We have seen that the assignment problem of (1) of this section is equivalent to a transportation problem. Assume that the data for the transportation problem of (2) of Section 7.2 are integral and prove the converse; that is, prove that the problem of (2) of Section 7.2 is equivalent to an assignment problem.

11. Suppose a rating matrix of an assignment problem is altered by the addition of a fixed constant to all the entries of either a row or a column of the matrix. Show that an assignment is optimal for the original problem if and only if it is optimal for the altered problem.

12. Given an assignment problem with an $n \times n$ rating matrix (c_{ij}), show that there are n numbers $r_1, \ldots, r_n$ such that the assignment problem with rating matrix $(c_{ij} + r_i)$ [i.e., r_i is added to the ith row of (c_{ij}), for each i] has the property that the cost of an optimal assignment is $\sum_{j=1}^{n} \text{Min}_i \{c_{ij} + r_i\}$ (i.e., the sum of the minimal entries from each column). (*Hint.* Make use of the u_i's from the final table of a resolution to the original problem.)

13. Given an assignment problem with rating matrix (c_{ij}), show that there are numbers r_i and s_j such that the assignment problem with ranking matrix (c'_{ij}), where $c'_{ij} = c_{ij} + r_i + s_j$ [i.e., r_i is added to the ith row of (c_{ij}) and s_j to the jth column] has the property that all the c'_{ij}'s are nonnegative and the cost of an optimal assignment is 0.

14. Suppose the transportation problem algorithm is used to solve an assignment problem. Show that for any optimal assignment, no matter how it is found, if I_i is assigned J_j, $u_i + v_j = c_{ij}$, where u_i and v_j are the dual variables of the final table corresponding to the original resolution of the problem. (*Hint.* Use Problems 11–13, or else Problem 8 of Section 7.2.)

SECTION 7.4 AN EXAMPLE INVOLVING UNCERTAINTY

In Chapter 5 we developed techniques for measuring the effects changes in the constants of a linear programming problem had on an optimal solution to the problem. In this section we consider a way of working with uncertainty in the values of these constant terms. And, as a secondary result, in the resolution of the example of this section, a special technique that can be used to solve linear programming problems with upper bounds on many of its variables is demonstrated.

For many applications, it is more realistic to assume that the terms of the problem are not fixed but, instead, can range over sets of values subject to estimated probability distributions. For example, consider the problem of a manufacturer of goods to be sold on the market. In general he seeks to determine a production schedule that maximizes profits and/or minimizes costs. But first he is faced with the problem of estimating the demand on the market for his product. It may be that he cannot assume that everything he produces will be sold, but that unpredictable conditions such as weather, strength of the competition, whims of the public, and so on, influence the salability of his product. However, it could be that from past records a probability distribution for the number of units sold can be estimated with some degree of accuracy. In this section we will, by means of an example, demonstrate one possible technique by which information such as this can be incorporated into a linear programming problem. The basic idea behind the approach is set forth in a paper by A. Ferguson and G. Dantzig [9] (see also Chapter 28 of Dantzig's book [6]), in which the realistic problem of the allocation of commercial aircraft to meet uncertain demands is discussed.

Example

Consider the problem of the boat manufacturer described in Example 1 of Section 2.3 (see also Problem 2 of Section 4.2 and Example 2 of Section 5.1). The manufacturer produces two types of small boats, a rowboat and a canoe, with the total number produced restricted by the availability of aluminum, machine time, and finishing labor. Realizing a profit of \$50 on the sale of a rowboat and \$60 on the sale of a canoe, the specific linear programming problem is concerned with optimizing profits, and is to

$$\text{Maximize} \quad 50R + 60C$$

subject to

$$\begin{aligned} 50R + 30C &\leq 2000 \\ 6R + 5C &\leq 300 \\ 3R + 5C &\leq 200 \end{aligned} \tag{1}$$

where $R \geq 0$ and $C \geq 0$ are the number of rowboats and canoes produced, respectively. Of course, in this formulation of the problem we are assuming that all boats produced are sold. With this assumption an optimal production schedule can be easily determined geometrically. The graph of the set of feasible solutions is sketched in Figure 7.1. The value of the objective function $50R + 60C$ can be computed at the four vertices of this set, and the maximum value, \$2750, attained at the point (25,25), easily determined. Thus, if all boats produced can be sold, the manufacturer realizes a maximal profit of \$2750 by making 25 boats of each type.

Let us suppose now that the market for the boats is not fixed, but is contingent on various factors, the primary one being the weather conditions in early summer. Also assume that the manufacturer is attempting to meet these summer demands by a spring production and thus must estimate from past experience the demands for boats. It is known that the family rowboat market is more variable than the sports canoe market. More specifically, assume it is equiprobable that there will be either 20 potential rowboat buyers or 50 potential rowboat buyers, and that there will be either 20, 30, or 40 people wanting canoes, with probabilities $\frac{1}{4}, \frac{1}{2}, \frac{1}{4}$, respectively. Our problem is to develop a model that takes this new information into account.

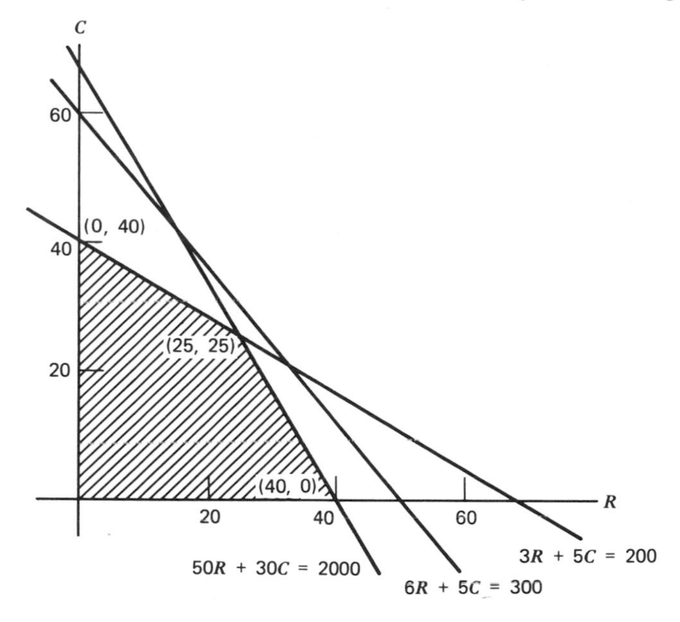

Figure 7.1

One somewhat simplistic approach is to restrict the number of rowboats and canoes produced to the expected number of potential buyers. The expected number of rowboat buyers is $\frac{20}{2} + \frac{50}{2} = 35$, and the expected number of canoe buyers is $\frac{20}{4} + \frac{30}{2} + \frac{40}{4} = 30$. Thus, using this approach, the manufacturer would simply add two new constraints to the set in (1), that $R \leq 35$ and $C \leq 30$. Since the optimal value of the objective function of (1) is attained at a point that also satisfies these new constraints, it follows that the point (25,25) provides the solution to this modified problem also. However, the profit expected from this production schedule is not the \$2750 of the original problem, but is less because it is no longer certain that all boats produced will be sold. In fact, we are certain that only the first 20 of each type of boat will be sold. Since the probability that 50 people will desire row boats is $\frac{1}{2}$, the probability that the last 5 rowboats produced will be sold is $\frac{1}{2}$. Similarly, the probability that the last 5 canoes produced will be sold is $\frac{3}{4}$ (the probability that there will be *at least* 30 buyers is $\frac{1}{2} + \frac{1}{4}$). Thus the expected profit is

$$\$50 \cdot 20 + \$50 \cdot 5 \cdot \tfrac{1}{2} + \$60 \cdot 20 + \$60 \cdot 5 \cdot \tfrac{3}{4} = \$2550$$

This computation of the expected profit suggests a way of refining our model that may lead to another production schedule with a higher expected profit. Our model should take into consideration the differences in expected profit associated with the sale of the first 20 rowboats and the next 30, and the differences associated with the sale of the first 20 canoes, the next 10, and the next 10. We can do this as follows. Consider the total number R of rowboats produced as divided into two increments, the number between 0 and 20, denoted by R_1, and the number between 20 and 50, denoted by R_2. Note that no more than 50 such boats can be sold, and so we restrict our attention to only these profitable increments. Similarly, consider C to be divided into 3 increments C_1, C_2, and C_3 where C_1 denotes the number between 0 and 20, C_2 the number between 20 and 30, and C_3 the number between 30 and 40. For example, corresponding to the above (25,25) production schedule, we have $R_1 = 20$, $R_2 = 5$, $C_1 = 20$, $C_2 = 5$, and $C_3 = 0$. For these five new variables we have the following constraints.

$$R_1 + R_2 = R$$
$$0 \leq R_1 \leq 20$$
$$0 \leq R_2 \leq 30$$
$$C_1 + C_2 + C_3 = C$$
$$0 \leq C_1 \leq 20$$
$$0 \leq C_2 \leq 10$$
$$0 \leq C_3 \leq 10$$

R and C must still satisfy the original three production constraints given in (1). However, the expected profit function can now be expressed precisely using the new variables. The expected profit associated with the sale of $R = R_1 + R_2$ rowboats is $\$50 \cdot R_1 + \$50 \cdot R_2 \cdot \frac{1}{2}$, since the probability of selling the first 20 of the boats is 1 and the probability of selling the next 30 is $\frac{1}{2}$. Similarly, the sale of $C = C_1 + C_2 + C_3$ canoes, with the C_i's restricted as above, will realize an expected profit of $\$60 \cdot C_1 + \$60 \cdot C_2 \cdot \frac{3}{4} + \$60 \cdot C_3 \cdot \frac{1}{4}$, since the probability of selling the first 20 canoes is 1, the next 10 is $\frac{3}{4}$, and the last 10 is $\frac{1}{4}$. Combining all this, we have the following linear programming problem.

$$\text{Maximize} \quad 50R_1 + 25R_2 + 60C_1 + 45C_2 + 15C_3$$

subject to

$$\begin{aligned}
50R + 30C &\leq 2000 \\
6R + 5C &\leq 300 \\
3R + 5C &\leq 200 \\
R_1 + R_2 &= R \\
C_1 + C_2 + C_3 &= C \\
R_1 &\leq 20 \\
R_2 &\leq 30 \\
C_1 &\leq 20 \\
C_2 &\leq 10 \\
C_3 &\leq 10 \\
R,C,R_1,R_2,C_1,C_2,C_3 &\geq 0 \qquad (2)
\end{aligned}$$

One problem that may have occurred to the reader is that if our interpretation of the variables R_1 and R_2 and of C_1, C_2, and C_3 is to be valid, we want to consider only those solutions to the above problem for which: $R_2 = 0$ whenever $R_1 < 20$; $C_2 = C_3 = 0$ whenever $C_1 < 20$; and $C_3 = 0$ whenever $C_1 = 20$ and $C_2 < 10$. However, by the nature of the objective function of (2), any optimal solution must have this property; the coefficient of R_1 in the objective function is greater than the coefficient of R_2, and a similar relationship holds for the coefficients of C_1, C_2, and C_3.

We now proceed to solve the problem of (2). Although the simplex method could be applied to the problem as it stands, frequently problems such as this with upper bounds on many of the variables can be resolved without the introduction of a full set of slack variables. We will solve this problem in such a manner.

First note that from the graph in Figure 7.1 of the solution set to the constraints of (1), it is obvious that the inequality $6R + 5C \leq 300$ is satisfied by any point satisfying the other two constraints of (1), and so this inequality can be dropped from (2). Introducing slack variables X_1 and X_2 in the first and third inequalities of (2), we have

$$\begin{aligned} 5R + 3C + X_1 \qquad &= 200 \\ 3R + 5C \qquad + X_2 &= 200 \end{aligned} \tag{3}$$

Pivoting at the R term of the first equation and then at the C term of the second, (3) is equivalent to

$$\begin{aligned} R + \tfrac{5}{16}X_1 - \tfrac{3}{16}X_2 &= 25 \\ C - \tfrac{3}{16}X_1 + \tfrac{5}{16}X_2 &= 25 \end{aligned}$$

Using these two equations and eliminating the variables R and C from the constraints in (2), we have that the problem of (2) is equivalent to the problem of

$$\text{Maximizing} \quad 50R_1 + 25R_2 + 60C_1 + 45C_2 + 15C_3 = z$$

subject to

$$\begin{aligned} R_1 + R_2 \qquad\qquad\qquad\quad + \tfrac{5}{16}X_1 - \tfrac{3}{16}X_2 &= 25 \\ C_1 + C_2 + C_3 - \tfrac{3}{16}X_1 + \tfrac{5}{16}X_2 &= 25 \end{aligned}$$

$$\begin{aligned} &0 \leq R_1 \leq 20 \qquad 0 \leq C_1 \leq 20 \\ &0 \leq R_2 \leq 30 \qquad 0 \leq C_2 \leq 10 \\ &\quad\;\; X_1 \geq 0 \qquad 0 \leq C_3 \leq 10 \\ &\quad\;\; X_2 \geq 0 \end{aligned} \tag{4}$$

We know from our previous work that $R_1 = 20$, $R_2 = 5$, $C_1 = 20$, $C_2 = 5$, $C_3 = X_1 = X_2 = 0$ is a solution to the constraints of (4) with $z = 2550$. Notice that at this solution the values of the variables R_2 and C_2 are strictly between their upper and lower bounds. Thus we use the two equations of (4) to eliminate R_2

and C_2 from the equation in (4) defining z. Subtracting 25 times the first equation plus 45 times the second, we have

$$25R_1 + 15C_1 - 30C_3 + \tfrac{5}{8}X_1 - \tfrac{75}{8}X_2 = -1750 + z \tag{5}$$

In (5), the coefficients of R_1, C_1, and X_1 are positive and the coefficients of C_3 and X_3 are negative. This suggests that the value of z can be increased by moving to another solution of the constraints in (4) for which either R_1, C_1, or X_1 is larger, or C_3 or X_2 is smaller. But at the $R_1 = 20$, $R_2 = 5$, $C_1 = 20$, $C_2 = 5$, $C_3 = X_1 = = X_2 = 0$ solution, the values of R_1 and C_1 are at their maximums and the values of C_3 and X_2 are at their minimums. Consider, however, X_1. Letting $R_1 = 20$, $C_1 = 20$, $C_3 = X_2 = 0$, the equations of (4) become

$$\begin{aligned} 20 + R_2 + \tfrac{5}{16}X_1 &= 25 \\ 20 + C_2 - \tfrac{3}{16}X_2 &= 25 \end{aligned} \qquad \text{or} \qquad \begin{aligned} R_2 &= 5 - \tfrac{5}{16}X_1 \\ C_2 &= 5 + \tfrac{3}{16}X_2 \end{aligned} \tag{6}$$

Since $R_2 \geq 0$, the first equality implies that $X_1 \leq 16$. Since $C_2 \leq 10$, the second equality implies that $X_1 \leq \frac{80}{3}$. Thus, under these conditions the largest possible value for X_1 is 16. Using (6) to solve for R_2 and C_2, the corresponding solution to the constraints of (4) is

$$R_1 = 20, R_2 = 0, C_1 = 20, C_2 = 8, C_3 = 0, X_1 = 16, X_2 = 0$$

with $z = 2560$ at this point. Moreover, at this solution only the values of the variables C_2 and X_1 are not equal to one of the limits of their bounds. Thus we eliminate these two variables from the expression for the objective function. Using the first constraining equation in (4) and the equation of (5), we have

$$23R_1 - 2R_2 + 15C_1 - 30C_3 - 9X_2 = -1800 + z \tag{7}$$

From (7), we see that the maximum value of z will be attained when R_1 and C_1 assume their maximum values and R_2, C_3, and X_2 assume their minimum values. But this is precisely the situation in the above solution.

Thus the maximum expected profit for the boat manufacturer is \$2560 and is realized by producing 20 rowboats and 28 canoes. Note that this approach has led us to a solution with an expected profit \$10 greater than the expected profit corresponding to the solution that used only expected values for the demands.

PROBLEM SET 7.4

1. In the example of this section, show that if the probabilities of there being 20 or 50 rowboat buyers are changed to $\frac{2}{5}$ and $\frac{3}{5}$, respectively, the maximum expected profit is attained at the $R = C = 25$ solution.

2. Suppose in our example the probabilities of potential boat buyers are given by the following table.

Rowboat Buyers	*Probability*	*Canoe Buyers*	*Probability*
20	$\frac{1}{6}$	15	$\frac{4}{5}$
30	$\frac{1}{3}$	30	$\frac{1}{10}$
40	$\frac{1}{2}$	50	$\frac{1}{10}$

(a) Compute the optimal production schedule using only expected values for the demands.
(b) Compute the optimal production schedule using the approach of this section.
(c) Do the two above answers agree?

3. Consider the situation of the dealer of home heating oil described in Example 1 of Section 2.5. Reformulate the problem using the approach of this section so as to incorporate the following information on the probability distributions for the demands of oil for the three time periods.

	Gallons of Oil that can be Sold	*Probability*
First Month	4000	$\frac{1}{3}$
	6000	$\frac{1}{3}$
	8000	$\frac{1}{3}$
Second Month	6000	$\frac{1}{2}$
	8000	$\frac{1}{2}$
Third Month	8000	1

4. Consider the linear programming problem of

$$\text{Maximizing} \quad z = 3x_2 - 5x_3 - 2x_4 + 8x_5$$

subject to

$$x_1 + x_2 - x_3 + 3x_5 = 16$$
$$2x_1 + 2x_3 + x_4 - x_5 = 24$$
$$0 \le x_1 \le 10, 0 \le x_2 \le 10, 0 \le x_3 \le 5, 0 \le x_4 \le 5, 0 \le x_5$$

(a) Show that $x_1 = 10, x_3 = 0, x_5 = 0, x_2 = 6, x_4 = 4$ is a feasible solution.

(b) In this solution, only the variables x_2 and x_4 are strictly between their respective bounds. Use the two equations to express the objective function in terms of the remaining variables x_1, x_3, and x_5. Note that this resulting expression suggests that we try to increase the value of either x_1 or x_3, or decrease the value of x_5. Thus we work with x_3. Why?

(c) For $x_1 = 10$ and $x_5 = 0$, show that the two equations and the bounds force $x_3 \leq 2$, leading to the solution $x_1 = 10$, $x_4 = 0$, $x_5 = 0$, $x_2 = 8$, $x_3 = 2$.

(d) In this solution x_2 and x_3 are strictly between their respective bounds. Put the system of two equations into canonical form with these as basic variables and use these equations to express z in terms of x_1, x_4, and x_5.

(e) Note that the resulting expression for z suggests that x_1 enter the basis. With $x_4 = x_5 = 0$, determine the lower bound value for x_1 and the corresponding solution point.

(f) At this solution point, what variables are strictly between their bounds? Expressing z in terms of the remaining variables, show that the optimal solution point has been attained and that the maximum value of z is 15.

5. Using the solution technique outlined in this section (and in the above problem), solve the following linear programming problems.

(a) Maximize $4x_1 + x_2 + x_3$

subject to

$$3x_1 + x_2 - x_3 = 14$$
$$2x_1 + 2x_3 + x_4 = 10$$
$$0 \leq x_1, x_2 \leq 4; 0 \leq x_3, x_4 \leq 6$$

Start with the solution (4,2,0,2).

(b) Minimize $x_1 + x_2 + 3x_3 - 2x_4 + 7x_5$

subject to

$$x_1 + 2x_3 - x_4 + 8x_5 = 10$$
$$x_2 - x_3 + 2x_4 - 3x_5 = 20$$
$$0 \leq x_1; 0 \leq x_2, x_3 \leq 8; 0 \leq x_4, x_5 \leq 12$$

Start with the solution (6,4,8,12,0).

8 Two-Person, Zero-Sum Games

SECTION 8.1 INTRODUCTION TO GAME THEORY

In this chapter and the next, some topics from the theory of games are discussed. In this section the subject is introduced primarily by means of examples. The remainder of the chapter is devoted to a special class of games, the two-person, zero-sum games. For these games, a complete theory can be presented. Starting with the definition of a two-person, zero-sum game, we will formulate some general principles on which to base our notion of a "solution" to the game. Then we will show, using the Duality Theorem of Chapter 4, that such games always have solutions, and we will develop techniques for finding these solutions. Thus, for these games a mathematical model can be formulated precisely and analyzed completely. Although the applicability of this theory is limited due to the restrictive assumptions that we make, the theory does serve as the springboard for the study of the more general classes of games, some of which are introduced in the next chapter.

By a game we mean roughly a situation of conflict between two or more people, in which each contestant, player, or participant has some, but not total, control over the outcome of the conflict. We assume that all players have complete knowledge of all actions, moves, or choices available to themselves and their opponents, and knowledge of the results of the conflict associated with any given selection of actions. Assuming that each player acts rationally to maximize his gain, our basic problem is to develop a theory that will help us to understand and predict human behavior or economic phenomena. As we will see, however, the translation of the phrase "each player acts rationally to maximize his gain" into mathematical terms is not always straightforward, but can lead to various interpretations, approaches, and solutions of a game. This is in contrast to the situation in linear programming in which the optimization problems lead to a well-defined and generally accepted theory.

Parlor games such as poker, tic-tac-toe, and chess are of the type of game described above. On the other hand games, such as roulette or craps are not, since one simply plays against certain odds. And even though some of our examples and problems are of the parlor game variety, the broader scope of the theory must be recognized and appreciated. Indeed, the book that established game theory as a mathematical theory of its own and abundant with potential applications is the work of von Neumann and Morgenstern entitled *Theory of Games and Economic Behavior*, written in 1944. Since then much work on the theory has been done by both mathematicians and social scientists, but its full range of applications is still in a state of exploration.

We now give some examples of some elementary games.

Example 1a

Two players each have two cards, a one (or ace) and a two. Each player selects one of her cards, with her choice unknown to her opponent. They then compare the two selected cards. If the sum of the face values of the two selected cards is even, Player 1, denoted by P_1, wins that sum from Player 2, denoted by P_2. If the sum is odd, P_2 wins that amount from P_1.

This is an example of a *two-person, zero-sum* game: two-person, because there are exactly two participants, and zero-sum, because the sum of the payments made to the players at the end of the game (with a negative payment indicating a loss) is equal to zero. In a two-person game, zero-sum means simply that what one player wins, the other player loses.

Each player has two possible courses of action or *strategies*, in this case, to play either a one or a two. In general, by the term *strategy* we mean a rule of action or set of instructions that tells a player what to do under all possible contingencies.

The information critical to this game can be recorded very easily using matrix notation. If we let s_1 and s_2 denote the two strategies for P_1, with s_i meaning that P_1 plays card i and, similarly, let t_1 and t_2 denote P_2's two strategies, then all the possible outcomes for the game from P_1's point of view can be recorded by the following tableau.

	t_1	t_2
s_1	2	−3
s_2	−3	4

In this tableau, the entry in the ith row and jth column represents the amount that P_2 pays P_1 if P_1 plays s_i and P_2 plays t_j, with a negative sign indicating that P_2 wins the associated amount from P_1. Note that since the game is zero sum, we need only record the outcomes for one player, and we fix that player to be P_1. The 2×2 matrix

$$\begin{bmatrix} 2 & -3 \\ -3 & 4 \end{bmatrix}$$

is called the *payoff matrix* of the above game.

Example 1b

Here the rules are just as in the game of Example 1a, with the only difference being that P_1 is given initially one additional card, a three. P_1 has now three strategies, to play either the one, two, or three. Denoting these strategies by s_1, s_2, and s_3, respectively, the tableau associated with this game is the following.

	t_1	t_2
s_1	2	−3
s_2	−3	4
s_3	4	−5

Example 1c

The rules are as in Example 1a, with each player receiving two cards, but now, after the players have selected their card but before the cards are compared, P_1 declares "even" or "odd," with P_1 winning if and only if the sum of the face values of the selected cards is of the parity she has declared. P_1 has now four strategies, to declare even and play the one or play the two, or to declare odd and play the one or play the two. Denote these strategies by (even; 1), (even; 2), (odd; 1), and (odd; 2). P_2 still has only the two strategies t_1 and t_2.

	t_1	t_2
(even; 1)	2	−3
(even; 2)	−3	4
(odd; 1)	−2	3
(odd; 2)	3	−4

Example 1d

The rules are as in Example 1c, except that now P_1 must declare "even" or "odd" before P_2 selects her card. P_1's set of strategies remain as in Example 1c, but P_2 has now the opportunity to react to P_1's declaration of parity. Since a strategy for P_2 is a set of instructions telling P_2 what to do in all possible circumstances, here some possible strategies for P_2 are to "play the one if P_1 declares even and the two if P_1 declares odd," or to "always play the one," etc. P_2 has, in fact, a total of four strategies, and they can be denoted by the set of ordered pairs (i,j), $1 \leq i, j \leq 2$, where strategy (i,j) is defined to mean to play card i if P_1 declares even and card j if P_1 declares odd. The payoff tableau is as follows.

	(1,1)	(1,2)	(2,1)	(2,2)
(even; 1)	2	2	−3	−3
(even; 2)	−3	−3	4	4
(odd; 1)	−2	3	−2	3
(odd; 2)	3	−4	3	−4

Example 2

Two players have two cards. P_1 has a red 5 and a black 4; P_2 has a red 7 and a black 8. Each player selects one of his cards, with his choice unknown to his opponent, and then the players compare the selected cards. If the selected cards are of the same color, P_1 wins the difference in face values from P_2; if the selected cards are of different colors, P_2 wins the difference in face values from P_1. Denoting P_1's two strategies by $R5$ and $B4$, and P_2's by $R7$ and $B8$, the game tableau is:

	$R7$	$B8$
$R5$	2	−3
$B4$	−3	4

This payoff matrix is identical to the payoff matrix for the game of Example 1a, and so the two games are essentially equivalent. Thus, to study both games, we simply need to consider the two-person game having two strategies for each player and payoff matrix

$$\begin{bmatrix} 2 & -3 \\ -3 & 4 \end{bmatrix}$$

Example 3

Two major automobile manufacturers compete for a fixed market of new car buyers. The buyers in this group are attracted to a particular automobile for two reasons: the styling, features, and quality of the automobile; and the intensity of the manufacturer's advertising campaign. Each of the manufacturers have fixed amounts of money, say M_1 and M_2 dollars, to divide between their Research and Development Division and their Product Promotion Division. Moreover, suppose the number of these new car buyers attracted (or lost) to Manufacturer 1 because of product development is given the function $f(x,y)$, where x and y are the amounts of money spent by Manufacturers 1 and 2, respectively, on research and development. Similarly, suppose the function $g(x,y)$ measures the number of buyers attracted (or lost) to Manufacturer 1 because of advertising. Assuming that both manufacturers have the above information, how should they allocate their resources?

This is an example of a two-person infinite game. Each player, (i.e., each manufacturer) must decide how to divide his resources between the two critical divisions, and each player has essentially an infinite number of choices available. To determine the payoff for this game, suppose Manufacturer 1 allots x dollars to his Research and Development Division, $0 \leq x \leq M_1$, and the remainder to his Product Promotion Division. Similarly, define y for Manufacturer 2, $0 \leq y \leq M_2$. Then the total number of car buyers attracted (or lost) to Manufacturer 1 is given by the function

$$A(x,y) = f(x,y) + g(M_1 - x, M_2 - y).$$

Games of this type have been studied, and a complete theory can be developed as long as the payoff function $A(x,y)$ is reasonably well-behaved (e.g., when A is continuous). However, we will say no more about them in this text. For more information, see the books on game theory listed in the Bibliography, especially the books by Dresher [8] and McKinsey [24].

Example 4

A town puts up for closed bidding its annual trash collection contract. Three firms compete for the job by submitting sealed bids, with the job awarded to that firm submitting the lowest bid less than or equal to \$50,000, since the town itself will manage the collection operation if all bids received exceed that figure. Because of the differences in costs of labor and efficiency of equipment, the actual cost of providing the service to the town varies from firm to firm, and these overhead costs are known to all the parties. Specifically, the cost to Firm 1 to provide the service would be \$38,000, to Firm 2, \$40,000, and to Firm 3, \$44,000. Economic necessity demands that the bids submitted by each firm exceed their individual overhead costs. How should each of the firms bid?

This is an example of a three-person game. Each of the firms must submit a bid determined by their overhead costs and the profit they wish to realize. The payoff (i.e., the profit actually realized by each of the firms as a result of the three bids) can then be easily determined. For example, if Firms 1, 2, and 3 bid \$44,000, \$43,000, and \$48,000, respectively, Firm 2 would be awarded the contract and realize a profit of \$3000, while the other two firms earn nothing.

The distinguishing factor between two-person game theory and n-person theory ($n \geq 3$) is the existence of the potential for the players to form coalitions in n-person games. In fact, n-person games are usually described in terms of their "characteristic function," a function that measures the strength of each of the possible coalitions of players. Suppose, for example, that it is possible for any subgroup of the three firms in the above example to agree beforehand on the bids to be set by the members of that subgroup. We can then define a function v that measures the maximum profit that the subgroup can guarantee itself if all members in the group cooperate. For example, if Firms 1 and 2 decide to form a coalition, Firm 1 can bid \$44,000, and Firm 2 can bid anything greater. Since it is known that the bid of Firm 3 must exceed \$44,000, Firm 1 would be awarded the contract and realize a profit of \$6000. Thus the coalition of Firms 1 and 2 is worth \$6000, or $v(\{1,2\}) =$ \$6000. Now verify that

$$v(\{1\}) = \$2000$$
$$v(\{1,2,3\}) = \$12{,}000$$
$$v(\{2\}) = v(\{3\}) = v(\{2,3\}) = \$0$$

We have now seen examples of the major types of games. Examples 1 and 2 are examples of finite, two-person, zero-sum games, the class of games that will be studied in this chapter. The games described in Section 1.3 are two-person, non-zero-sum games, the class of games that will be considered in the next chapter. Example 3 of this section is an example of an infinite game, and Example 4 an example of an n-person game. The theory of n-person games is still very much in the development stage; for more information refer again to the game theory books in the Bibliography, especially those by Luce and Raiffa [23] and by Owen [28].

PROBLEM SET 8.1

1. Determine the payoff matrices for the following two-person, zero-sum games.
   (a) P_1 has two cards, a red 1 and a black 7. P_2 has three cards, a red 3, a red 4, and a black 7. Each selects a card, with his choice unknown to his opponent. The selected cards are then compared: if they are of the same color, P_1 wins the sum in face values of the selected cards from P_2; otherwise, P_2 wins that amount from P_1.
   (b) Just as in part a, except that if the colors of the selected cards are not the same, P_2 wins an amount x from P_1.
   (c) *Colonel Blotto.* The Colonel has three divisions available to attempt to capture a town accessible by two different roads. The defender has four divisions to divide between the two routes to defend the town. If the Colonel's forces outnumber the defender's along a route, the Colonel wins the number of enemy divisions on that road plus the town, equivalent to two enemy divisions. If the defender's forces outnumber the attacker's, he wins the attacker's divisions. If the numbers are equal, it is a draw. The total payoff is the sum of the results along the two routes.
   (d) *Morra.* Two players simultaneously show one or two fingers and shout out a number.

If the number announced by a player is the same as the total number of fingers shown by both players, then she wins that amount from her opponent. If both players guess correctly, the game is a draw.

2. What would happen in the following situation? A stranger offers to possibly give either \$3 or \$5 to a group of three friends. If two members of the group can decide on how to divide the money between only themselves, the stranger will give \$5 to these two. If all three can agree on a way of sharing the money, the stranger will give the three \$3. If there is no agreement between any two or all three members of the group, the stranger keeps the money.

3. A submarine has one attempt to intercept and destroy an enemy ship transporting strategic cargo while the ship passes through a small unprotected area. The submarine is sailing directly at the enemy, and initially the two are 2000 yd apart. The submarine can fire a round of torpedoes at any time, but the probability of a hit increases as the ships come closer because of the increase in the accuracy of the weapon and the decrease in the available time for the enemy to evade the torpedoes. Suppose, in fact, that the probability of a hit is given by the function $p(x) = (2000 - x)/2000, 0 \leq x \leq 2000$, where x is the distance in yards between the ships. However, the cargo ship has one antisubmarine missile at its disposal, and the ship's captain is aware of the approaching submarine. If the missile can be landed within 50 yd of the submarine, its effect would so disrupt the operation of the submarine that its torpedoes would not be able to be launched and safe passage through the unprotected waters would be guaranteed the cargo ship. However, the accuracy of this missile also increases as the target comes closer. Suppose it is given by the function

$$q(y) = \begin{cases} \dfrac{50}{y}, & y \geq 50 \\ 1, & y \leq 50 \end{cases}$$

where y is again the distance in yards between the ships.

Assume that cargo ship's captain knows at all times the distance between the two ships and also whether or not the submarine has released its torpedoes. Similarly assume that the submarine commander knows at all times whether or not the missile has been fired from the ship, even if the missile does not land near the submarine.

The captain and the commander must decide at what points to fire their weapons. The goal of the submarine commander is to prevent the passage of the cargo ship; the goal of the captain is to deliver the strategic cargo. Determine a payoff function, a function of the points at which the two weapons are fired, that measures the probability of the success of the submarine commander.

SECTION 8.2 SOME PRINCIPLES OF DECISION MAKING IN GAME THEORY

All the problems discussed in this book fall into the general category of problems in decision making. By the term "decision making" we mean in general a situation in which, from a given set of possible courses of action, a specific course of action must be selected that is in some way preferred over the alternatives. This field of decision making is extremely broad, as one can realize from the not too restrictive definition given above. However, the field can be simplified somewhat by a partitioning according to whether a decision is made under conditions of certainty, risk, or uncertainty.

A decision is made under conditions of certainty if all the available courses of action lead to specific, fixed outcomes. Linear programming is an example of such. For example, in the standard diet problem, the cost and nutritional content of any given diet was known, and the decision problem reduced to a problem of optimizing a linear function on a domain restricted by a system of linear constraints. The optimization problems of freshman calculus are other examples of problems in decision making under conditions of certainty.

A decision is made under conditions of risk if a given course of action can lead not to a unique outcome, but to a set of possible outcomes, each outcome in the set occurring with a specified probability known to the decision maker. For example, when playing roulette, one must decide on what color or number or combination of numbers to bet on. The possible outcomes for these actions, either the winning of an amount determined by the type and amount of the original bet or the losing of the original bet, occur with known, definite probabilities depending (presumably) only on the type of bet.

A decision is made under conditions of uncertainty if a given course of action can lead again to a set of possible outcomes, but where the probabilities of these outcomes occurring is unknown. In a general sense game theory falls into this category, because the players in a game are ignorant of their opponents' moves. However, as we will see shortly, we strive to reduce this element of uncertainty in game theory by attempting to predict our opponents' courses of action, based on some reasonable principles. In fact, the phrase "decision making under uncertainty" is usually reserved for a definite theory that excludes game theory but includes a study of the experimentation and statistical analysis that can be used to reduce the element of uncertainty.

Our primary concern for the remainder of this chapter will be with two-person, zero-sum games, in which each player has only a finite number of possible courses of actions (i.e., strategies). Let us agree to call these games *maxtrix games* since, as we have seen in the first section, such a game can be represented by its payoff matrix. Recall, if $\{s_1,s_2,\ldots,s_m\}$ is P_1's strategy set and $\{t_1,t_2,\ldots,t_n\}$ is P_2's strategy set, the payoff matrix is that $m \times n$ matrix $A = (a_{ij})$, in which, for $1 \le i \le m$ and $1 \le j \le n$, the entry a_{ij} is the outcome of the game for P_1 if P_1 uses strategy s_i and P_2 uses strategy t_j. In the examples we have seen so far the numerical values of the a_{ij}'s translated immediately into dollars or some other monetary unit and, in general, in game theory we assume that the desirability of an outcome for a player can be measured by a real number. That this is always possible is not immediately obvious. Consider, for example, a game-type situation between two conglomerates where the outcome is either the successful elimination of the competition or an antitrust suit. However, the problem of translating the desirability of various alternatives into numerical values has been studied, and the resulting theory is called *utility theory*. This topic is discussed in the next chapter; for the time being we will assume that the outcomes of all of our games can be expressed using numerical values.

To illustrate the conditions of uncertainty present in these simple games, consider the game represented by the following payoff matrix.

	t_1	t_2
s_1	0	2
s_2	8	-8
s_3	-5	5

Here P_1 has three strategies and P_2 has two. Suppose now that we were in the position of P_1, and about to play the game. What strategy should we choose in order to maximize our winnings? We have three courses of action but, clearly, the resulting outcomes of these actions is uncertain, being contingent on the play of P_2. For example, playing strategy s_2 will either win or lose for us 8 units, depending on whether P_2 plays t_1 or t_2. If we anticipate P_2 to play t_1, then s_2 would be our obvious choice, but if we expect P_2 to play t_2, then playing s_3 would realize our maximum gain. On the other hand, by playing s_1, we are assured of not losing, and possibly of winning 2. But if we do play this strategy, P_2 can simply play t_1, with no one winning and our seemingly advantageous position nullified. But anticipating P_2 to play t_1 suggests the response s_2, as mentioned above. We could go on and on with these circular arguments, leading us to no definite position. What we need are some reasonable and acceptable principles of play that, if followed, will enable us to in some way predict our opponent's play and suggest our own play. Only with some such precise statement of the goals of the players can we hope to develop a mathematical model and, depending on the applicability of the principles, a theory that can assist in understanding and predicting human behavior.

A very simple rule for play that we could consider is that each player choose a strategy that has as a possible outcome the most favorable outcome for the game for that player. Thus P_1 should choose strategy s_i if the ith row of A contains the largest entry for A. For example, for the game with payoff matrix

	t_1	t_2
s_1	100	-50
s_2	6	5

P_1 would play s_1 if he is using this rule to guide his choice, because the largest entry of the payoff matrix, 100, is in the first row. Is this in any way a reasonable choice? Certainly for this game the answer must be "no"! P_2 is going to play t_2 and, by playing s_1, P_1 would lose 50 units, whereas by playing s_2, P_1 would win 5 units. Thus this proposed rule of play is unacceptable.

Another suggestion somewhat similar to the first is that each player choose that strategy for which the sum of the entries of the payoff matrix in the associated row or column are the most favorable. However, in the above example the sum of the entries in the first row of the payoff matrix is 50 and the sum in the second row is 11. Thus the use of this rule as a guide places P_1 in the same untenable position as before, so this rule too must be rejected.

The above suggested principles proved unacceptable because they represented a rather naive approach to the game. A player following one of these principles, while attempting to maximize his gain, would be completely ignoring his opponent's potential moves, an opponent whose interests are strictly opposed to his own as the game is zero sum—what one player wins, the other loses. A more reasonable approach, especially for this type of game in which interests are in direct conflict, would be to play the strategy that can guarantee the largest gain regardless of what the opponent does. In other words, a player determines the least amount that he can gain from playing each of his strategies and then chooses that strategy corresponding to the maximum of these least amounts. For example, in the above 2×2 game, P_1 could lose 50 by playing s_1, but is certain of winning at least 5 by playing s_2. Thus, using this principle as a guide, P_1 should play s_2, because this play guarantees the larger gain independent of P_2's action. Similarly, P_2 could lose up to 100 units by playing t_1, and only 5 by playing t_2, and so his choice would be t_2.

This principle represents a much more conservative approach to the game. By following it each player is giving complete recognition to his opponent's capabilities and then is acting to maximize his guaranteed gain, or better, his security level. The word "gain" indicates winnings, whereas we mean either maximized winnings or minimized losses, whichever the case may be. In a game with high stakes representing some economic, military, or social situation it seems realistic to expect the players to play conservatively, not take risky chances, and give full recognition to the capabilities of their opponent. Thus we establish this principle as a rule of action governing the play of the participants of a game, and we refer to it as *Principle I*.

Principle I

Each player acts to maximize his security level.

Example 1

Consider the game given by the tableau

	t_1	t_2	t_3	t_4
s_1	-2	0	-1	0
s_2	1	2	0	1
s_3	-1	-3	0	2

By playing s_1, P_1 could lose up to 2 units, by playing s_2, P_1 would at least break even, and by playing s_3, P_1 could lose up to 3 units. Guided by Principle I then, P_1 would choose strategy s_2. Similarly, P_2 could lose up to 1, 2, and 2 units by playing t_1, t_2, and t_4, respectively, but by playing t_3 he would at least break even. Thus P_2 is lead to strategy t_3.

There is one obvious complication to using only this principle as a guide to the play of a game. Since we are assuming that both players are intelligent, the acceptance of this principle implies that each player can anticipate the opponent's move. Thus each player would certainly consider the possibility of improving his outcome for the game by changing his strategy in anticipation of the expected move of the opponent. Consider, for example, the game of Example 1. P_1 realizes that P_2 maximizes his security level by playing t_3, and he thus asks if playing s_2 is the best response to strategy t_3. In this case the answer is "yes," since he has nothing to gain by playing s_3 and could lose 1 by playing s_1. Similarly, P_2 would expect P_1 to play s_2 but, by considering all the entries in the second row of the payoff matrix, P_2 would still choose strategy t_3. However, this resulting stability with the strategy choices of s_2 and t_3 need not always be present.

Example 2

Consider the game given by the tableau

	t_1	t_2
s_1	-2	4
s_2	1	-3

Principle I leads P_1 to choose s_1 since, by playing s_1, P_1 could lose 2 but by playing s_2, he could lose 3. Similarly, P_2 is lead to strategy t_1, where his losses are at most 1. However, if P_1 suspects P_2 to play t_1, he should alter his strategy and play s_2, since he would then win 1 instead of losing 2. But now we can go on. If P_2 anticipates P_1 to play s_2, he should change his strategy from t_1 to t_2. And so forth.

What the above example indicates is that Principle I alone does not adequately reflect the behavior of rational players for all games. In fact, what is suggested is that strategies are sought that are stable or *in equilibrium*, that is (for two-person games), a pair of prescribed strategies such that if one player stays with his prescribed strategy the other player has nothing to gain by deviating from his. In Example 1, the pair of strategies s_2 and t_3 is in equilibrium; in Example 2, the pair s_1 and t_1 is not. In that any reasonable description of the play of rational players should include this element of potential reaction of a player to the anticipated move of his opponent, we consider the need for stability in establishing *Principle II*.

Principle II

The players tend to strategy pairs that are in equilibrium.

In the next few sections these two principles will be translated into precise mathematical statements. Then, once we broaden our concept of strategy to include what will be called "mixed strategies," we will show that for any matrix game there is a pair of strategies that satisfies these two principles, and the outcome of the game played with these strategies would measure the value or worth of the game to the two players.

In that we can find such a strategy pair, we say that matrix games are "solved." But it must be emphasized that these games are completely determined and the play of the participants accurately predicted only upon the acceptance of the two principles. Our model will provide a norm for human behavior and a measure of the value of a game only if the players use the two principles as a guide to play. In any mathematical model the applicability of the theory is limited to those examples for which the underlying axioms of the theory are valid.

PROBLEM SET 8.2

1. Find strategy pairs that satisfy Principles I and II for the games with the following payoff matrices.

   (a) $\begin{bmatrix} 3 & 1 & 2 \\ 1 & 0 & 5 \end{bmatrix}$ (b) $\begin{bmatrix} 7 & 1 & 5 & 9 \\ 1 & 0 & 3 & 2 \\ 6 & 3 & 6 & 4 \end{bmatrix}$

2. Consider the game with payoff matrix

$$\begin{bmatrix} 1 & -1 \\ -1 & 1 \end{bmatrix}$$

   How would you play it? What do you expect is meant by the term "mixed strategy?" Does your suggested definition apply to possible strategies for the game

$$\begin{bmatrix} 3 & -1 \\ -1 & 1 \end{bmatrix}$$

3. We all have an intuitive idea of what it means to say a game is "fair." Would you accept the following definition for matrix games: A matrix game is fair if the sum of all the a_{ij}'s is zero?

4. On a cloudy morning a baseball fan must decide whether to travel 50 miles that day to see her favorite team play. Let us say that the fan feels that witnessing her team play is worth 20 units (of satisfaction?), but that traveling to the stadium only to have the game postponed on account of rain is worth -10 units. On the other hand, by staying home and

working around the house, she can accumulate up to 3 units. Is the fan making a decision under conditions of certainty, risk, or uncertainty? Is she playing a game with Nature as the opponent? If so, should she accept our two principles as a guide to her action?

SECTION 8.3 SADDLE POINTS

Principle I, as developed in the last section, states that each player seeks to maximize his security level. Guided by this principle, Player 1, as we have seen, should determine the minimal entry of each row of the payoff matrix and then consider playing that strategy corresponding to a row that attains the maximum of these minimum values. In symbols, the minimum entry of row i is simply $\text{Min}_{1 \leq j \leq n} a_{ij}$, and so P_1 seeks that row, say row h, such that

$$\underset{1 \leq j \leq n}{\text{Min}}\, a_{hj} = \underset{1 \leq i \leq m}{\text{Max}}\ \underset{1 \leq j \leq n}{\text{Min}}\, a_{ij}$$

Define $u_1 = \text{Max}_i\, \text{Min}_j\, a_{ij}$, that is, the maximum of the row minimums.

Similarly, Player 2 should determine the maximal entry in each column and then consider playing that strategy corresponding to a column that attains the minimum of these maximum values. In symbols, P_2 seeks that column, say column k, such that

$$\underset{1 \leq i \leq m}{\text{Max}}\, a_{ik} = \underset{1 \leq j \leq n}{\text{Min}}\ \underset{1 \leq i \leq m}{\text{Max}}\, a_{ij}$$

Define $u_2 = \text{Min}_j\, \text{Max}_i\, a_{ij}$, that is, the minimum of the column maximums.

Example 1

In the following, the row minimums are written to the right of the payoff matrix, and the circled numbers are the maximum of these minimums. Similarly, the column maximums are written below the matrix, and the smallest are circled.

(a)
$$\begin{bmatrix} 10 & 5 & 5 & 20 & 3 \\ 10 & 15 & 10 & 17 & 25 \\ 7 & 12 & 8 & 9 & 8 \\ 5 & 13 & 9 & 10 & 5 \end{bmatrix} \begin{matrix} 3 \\ \textcircled{10} \\ 7 \\ 5 \end{matrix}$$
$$\begin{matrix} \textcircled{10} & 15 & \textcircled{10} & 20 & 25 \end{matrix}$$

$u_1 = 10, h = 2; u_2 = 10, k = 1$ *or* 3.

(b)
$$\begin{bmatrix} 1 & 3 \\ 4 & 2 \end{bmatrix} \begin{matrix} 1 \\ \textcircled{2} \end{matrix}$$
$$\begin{matrix} 4 & \textcircled{3} \end{matrix}$$

$u_1 = 2, h = 2; u_2 = 3, k = 2.$

Now, the question we must consider, given Principle II, is "When is the strategy pair (s_h, t_k) in equilibrium?" In Example 1a, it can be seen that either pair (s_2, t_1) or (s_2, t_3) is in equilibrium; if P_2 anticipates P_1 to play s_2, his best response is either t_1 or t_3, and if P_1 anticipates P_2 to play either t_1 or t_3, he can gain nothing by deviating from strategy s_2. However, in Example 1b, the pair (s_2, t_2) is not in equilibrium, since P_1 can benefit by deviating from s_2, provided P_2 plays t_2. Thus, only in Example 1a is it reasonable to say that we have a "solution" to the game. Notice that in the first game $u_1 = 10 = u_2$, but that in the second, $u_1 = 2 \neq 3 = u_2$. In fact, equality of the u_1 and u_2 is the simple condition we seek for determining the existence of stability, as the corollary to the following theorem will imply.

Theorem 1 $u_1 \leq u_2$.

Proof By our definition of h and k, the maximum of the row minimums occurs in row h and the minimum of the column maximums occurs in column k. Suppose entry a_{hj} is the minimum of row h, and a_{ik} is the maximum of column k. Since entry a_{hk} is in row h and column k, we have $a_{hj} \leq a_{hk}$ and $a_{hk} \leq a_{ik}$. Thus

$$u_1 = a_{hj} \leq a_{hk} \leq a_{ik} = u_2$$

Note that the possibility of $j = k$ or $i = h$ has not been excluded.

A simple way to visualize this proof is to consider the payoff matrix

$$\begin{bmatrix} a_{hj} \leq a_{hk} \\ \quad\;\; \mathrel{\rotatebox{90}{$\geq$}} \\ \quad\;\; a_{ik} \end{bmatrix} a_{hj} = u_1$$

$$a_{ik} = u_2$$

###

Corollary

If $u_1 = u_2$, $a_{hj} = a_{hk} = a_{ik}$, and entry a_{hk} is both a minimum of row h and a maximum of column k.

Thus $u_1 = u_2$ implies that entry a_{hk} is both a row minimum and a column maximum. It follows immediately that the strategy pair s_h, t_k is in equilibrium, and so these strategies satisfy both our principles. Thus in this case we say that the strategies s_h, t_k are a *solution* to the game, and that the *value* of the game is this common value $u_1 = u_2$.

We single out this property of the entry a_{hk} by a special definition. It turns out that the existence of such an entry is not only a necessary but also a sufficient condition for equality of u_1 and u_2.

Definition An entry a_{hk} of a payoff matrix A is a *saddle point* if $\text{Min}_{1 \leq j \leq n} a_{hj} = a_{hk} = \text{Max}_{1 \leq i \leq m} a_{ik}$, that is, if a_{hk} is the minimum of row h and the maximum of column k.

Theorem 2 $u_1 = u_2$ *if and only if the payoff matrix A has a saddle point.*

Proof The above corollary proves that $u_1 = u_2$ implies the existence of a saddle point. Suppose now that A has a saddle point, say the entry a_{hk}. Since a_{hk} is a column maximum, all the other entries in the kth column must be less than or equal to a_{hk}. Thus the minimum values of all the rows other than row h must be less than or equal to a_{hk}, since the kth entry in each row has this property. But since a_{hk} is also the minimum of row h, a_{hk} equals the maximum of the row minimums; that is, $a_{hk} = u_1$. Similarly, $a_{hk} = u_2$. Hence $u_1 = u_2$. # # #

In summary, we have shown that the play of those matrix games for which $u_1 = u_2$ is completely determined by our two principles. Player 1 should play any row in which the maximum of the row minimums is attained, and Player 2 should play any column in which the minimum of the column maximums is attained. The value of such a game is $u_1 = u_2$, since this is the expected outcome. And a necessary and sufficient condition for u_1 to equal u_2 is that the payoff matrix A has a saddle point. In fact, if a_{hk} is a saddle point, P_1 should play row h, P_2 should play column k, and the value of the game is $a_{hk} = u_1 = u_2$.

PROBLEM SET 8.3

1. Do the following payoff matrices have saddle points? If they do, what is a solution and value of the corresponding game?

(a) $\begin{bmatrix} 9 & 7 & 8 & 10 \\ 6 & 5 & 12 & 8 \\ 8 & 10 & 5 & 9 \end{bmatrix}$

(b) $\begin{bmatrix} 2 & 6 & 1 & 2 \\ 3 & 5 & 4 & 3 \\ 1 & 0 & 2 & 4 \end{bmatrix}$

(c) $\begin{bmatrix} 1 & 1 & 1 & 2 \\ 1 & 2 & 0 & 0 \\ 1 & 2 & 1 & 1 \end{bmatrix}$

2. Suppose a matrix has two saddle points. Prove that they have the same numerical value. (*Hint.* For a very short proof, use the result in the proof of Theorem 2.) Thus the value of any game with a saddle point is unique.

3. Why is the word "saddle" used to describe a point of a matrix that is both a row minimum and a column maximum?

4. Suppose entries a_{ij} and a_{hk} are saddle points of a matrix. What can you say about the entries a_{ik} and a_{hj}?

5. Show that if the strategy pair s_h, t_k is in equilibrium, then the entry a_{hk} of the payoff matrix is a saddle point.

SECTION 8.4 MIXED STRATEGIES

As we have seen, games with saddle points are completely determined by the two principles set out in Section 8.2. For other games, however, we can use Principle I to lead us to suggested strategies for each player, but these strategies do not turn out to be in equilibrium. Thus, for these games we are still faced with the problem of determining if some sort of strategies exist for the two players that satisfy Principles I and II and, if so, how to find them.

Although the reader may feel as though we have placed ourselves in an impossible situation, because certainly no strategies as we known them could be stable for say the game with payoff matrix

$$\begin{bmatrix} 1 & -1 \\ -1 & 1 \end{bmatrix}$$

the solution to our dilemma is both realistic and elementary, and consists of simply broadening our concept of possible strategies. Consider, for example, the above game. Even though each player has only two choices, it is intuitively clear that what the players would do would be to choose either of their strategies with equal probability, and in a manner that would not give their opponent an opportunity to predict their move. Thus Player 1, each time he plays the game, would randomly select either s_1 or s_2, choosing s_1 with probability $\frac{1}{2}$ and s_2 with probability $\frac{1}{2}$. This extension of the concept of strategy is called a *mixed strategy*.

Definition A *mixed strategy* for P_1 is a vector $X = (x_1, x_2, \ldots, x_m)$ of nonnegative real numbers satisfying the condition $x_1 + x_2 + \cdots + x_m = 1$, with the interpretation that P_1 plays strategy s_i with probability x_i, $1 \leq i \leq m$.

Similarly, a mixed strategy $Y = (y_1, y_2, \ldots, y_n)$ for P_2 is defined. The set of all mixed strategies for P_1 will be denoted by S, and for P_2 by T. A strategy calling for the play of only one row or column is called a *pure strategy* and can be considered as a special case in the set of mixed strategies. For example, the m-tuple $(1,0,\ldots,0)$ is a mixed strategy corresponding to the pure strategy of P_1 always playing s_1.

The interpretation of a mixed strategy is important. For example, a mixed strategy of $(\frac{2}{3},\frac{1}{3})$ for P_1 does not mean that P_1 should play s_1 twice and then s_2 once, but that at each play of the game P_1 should play s_1 with probability $\frac{2}{3}$ and s_2 with probability $\frac{1}{3}$. One way to implement this would be for P_1 to roll a die at each play and use s_1 if the 1, 2, 3, or 4 comes up, and s_2 if the 5 or 6 comes up.

Since the notion of a mixed strategy does embody an idea that a player may realistically use in a game, we now need a way of evaluating the outcome of games for which mixed strategies are employed. For this we use the concept of "expected value" from probability theory. The expected value of an event is defined simply to be the sum of the values of each of the possible outcomes of the event times the probability that the outcome occurs. For example, if a fair die were to be rolled and you were to win an amount in dollars equal to the number rolled if that number were even, and otherwise you were to lose \$3, the expected value to you would be $\$2(\frac{1}{6}) + \$4(\frac{1}{6}) + \$6(\frac{1}{6}) - \$3(\frac{1}{2}) = \$0.50$. And for a game with payoff matrix $A = (a_{ij})$, if P_1 uses strategy $X = (x_1, x_2, \ldots, x_m)$ and P_2 uses strategy $Y = (y_1, y_2, \ldots, y_n)$, the outcome a_{ij} will occur with probability $x_i y_j$, since this is the probability of both P_1 playing s_i and P_2 playing t_j. Thus the *expected payoff* for the game is the sum of all the products $x_i a_{ij} y_j$, that is, the sum

$$\sum_i \sum_j x_i a_{ij} y_j$$

It can be easily shown that this sum is simply the product XAY^t by a direct computation of XAY^t. (Note that since mixed strategies are expressed as row vectors, the transpose of the vector Y gives us the appropriate column vector for multiplication on the right of A.)

Definition The *expected payoff* for a game with payoff matrix $A = (a_{ij})$, $1 \leq i \leq m$, $1 \leq j \leq n$, in which P_1 uses strategy $X = (x_1, x_2, \ldots, x_m)$ and P_2 uses strategy $Y = (y_1, y_2, \ldots, y_n)$, is

$$XAY^t = \sum_{1 \leq i \leq m} \sum_{1 \leq j \leq n} x_i a_{ij} y_j$$

Example 1

Consider the game with payoff matrix

$$A = \begin{bmatrix} 1 & 3 \\ 4 & 0 \end{bmatrix}$$

We have that $u_1 = 1$, and so P_1 can secure for himself 1 unit by always playing s_1. Consider, however, the effect for P_1 of using the mixed strategy $X = (\frac{1}{2}, \frac{1}{2})$. If P_2 responds with t_1, the expected payoff is $1(\frac{1}{2}) + 4(\frac{1}{2}) = \frac{5}{2}$, and if P_2 responds with t_2, the expected payoff is $\frac{3}{2}$. In fact, if P_2 uses strategy $Y = (y_1, y_2)$, the expected payoff is

$$\begin{aligned}(\tfrac{1}{2} \ \ \tfrac{1}{2})\begin{bmatrix}1 & 3\\ 4 & 0\end{bmatrix}\begin{bmatrix}y_1\\ y_2\end{bmatrix} &= (\tfrac{5}{2} \ \ \tfrac{3}{2})\begin{bmatrix}y_1\\ y_2\end{bmatrix}\\ &= \tfrac{5}{2}y_1 + \tfrac{3}{2}y_2\\ &\geq \tfrac{3}{2}y_1 + \tfrac{3}{2}y_2\\ &= \tfrac{3}{2}(y_1 + y_2)\\ &= \tfrac{3}{2}\end{aligned}$$

Thus, by using the strategy $(\frac{1}{2},\frac{1}{2})$, P_1 can secure for himself an expected payoff of $\frac{3}{2}$ because, no matter what P_2 plays, we have $(\frac{1}{2},\frac{1}{2})AY^t \geq \frac{3}{2}$. This is not to say that on any one play P_1 would win no less than $\frac{3}{2}$, but it means that by adhering to this particular mixed strategy, P_1 would win on the average at least $\frac{3}{2}$ per game. Thus, by using this strategy, P_1 has increased his security level from 1 to $\frac{3}{2}$. The obvious question that must now be considered, permitting the use of mixed strategies, is: what is P_1's maximum security level? For example, verify that the strategy $X = (\frac{2}{3},\frac{1}{3})$ provides P_1 with a security level even higher than $\frac{3}{2}$.

As suggested by this example, we must now reconsider the application of the two basic principles in light of the fact that each player has at his disposal an infinite set of mixed strategies. Our development initially will follow the same steps as the development in Section 8.3, with the difference being that the sets of pure strategies are replaced with their generalization, the sets of mixed strategies.

Consider a game with payoff matrix A from, say, Player 1's point of view. Principle I dictates that he is to attempt to maximize his security level. Thus, for each mixed strategy X, P_1 must determine the worst possible outcome of the game if he were to use this strategy, allowing that P_2 also has available mixed strategies. This would be then

$$\operatorname*{Min}_{Y \in T} XAY^t$$

This minimum represents P_1's guaranteed minimal gain if he were to use strategy X. Since he wants to maximize his security level, P_1 should seek a strategy that attains the maximum of these minimums, that is, a strategy X_0 such that

$$\operatorname*{Min}_{Y \in T} X_0AY^t = \operatorname*{Max}_{X \in S} \operatorname*{Min}_{Y \in T} XAY^t$$

These are the terms we will now be dealing with, and we make them precise with the following definitions.

Definition For a game with payoff matrix A, define P_1's and P_2's *optimal security levels*, denoted by v_1 and v_2, respectively, by

$$v_1 = \operatorname*{Max}_{X \in S} \operatorname*{Min}_{Y \in T} XAY^t \qquad \text{and} \qquad v_2 = \operatorname*{Min}_{Y \in T} \operatorname*{Max}_{X \in S} XAY^t$$

Definition A (mixed) strategy X_0 is an *optimal strategy* for P_1 if

$$\operatorname*{Min}_{Y \in T} X_0 A Y^t = v_1$$

A (mixed) strategy Y_0 is an *optimal strategy* for P_2 if

$$\operatorname*{Max}_{X \in S} X A Y_0{}^t = v_2$$

Notice how these terms are simply a generalization of the terms of the last section, with the X_0 and Y_0 comparing with the s_h and t_k, and the v_1 and v_2 comparing with the u_1 and u_2. However, as we will see, we can now develop a complete theory for all games based on Principles I and II.

By the above definitions we have that P_1's optimal security level v_1 is the largest expected payoff P_1 can be assured of, and that by using an optimal strategy X_0, P_1 is, in fact, guaranteed an expected payoff of at least v_1. We note in passing that the strategy sets S and T are infinite, and so, for the time being, to be precise we should replace the words "maximum" and "minimum" with "least upper bound" and "greatest lower bound," and question the boundedness of the set of possible values of XAY^t and the existence of optimal strategies. However, in the next section we will show that the use of the terms "maximum" and "minimum" is fully justified, and that optimal strategies always exist. Thus, for the sake of simplicity and understanding, we use these terms now, and for the remainder of this section we assume that optimal strategies always exist.

In the last section the first major result was that $u_1 \le u_2$. We have this same relationship between the v_1 and v_2.

Theorem 1 $v_1 \le v_2$.

Proof Let X_0 and Y_0 be optimal strategies for P_1 and P_2, respectively. Then

$$v_1 = \operatorname*{Min}_{Y \in T} X_0 A Y^t \le X_0 A Y_0{}^t \le \operatorname*{Max}_{X \in S} X A Y_0{}^t = v_2 \qquad \# \# \#$$

By playing their optimal strategies X_0 and Y_0, we see that both players P_1 and P_2 are optimizing their security levels of v_1 and v_2, respectively, and thus are satisfying Principle I. The next question is whether or not the strategy pair (X_0, Y_0) is in equilibrium; that is, is it stable? For example, by playing X_0, P_1 is assured of an expected outcome of at least v_1. However, P_1 anticipates P_2 to play Y_0, so P_1 should seek that $X \in S$ that attains $\operatorname{Max}_{X \in S} X A Y_0{}^t = v_2$. If this maximum were attained at X_0 and if Y_0 had the same corresponding property, the pair (X_0, Y_0) would be in equilibrium and thus provide a solution to the game based on the two principles. In fact, we have from the above theorem that $v_1 \le X_0 A Y_0{}^t \le v_2$. We will show in the next section that for all games optimal strategies always exist and that $v_1 = X_0 A Y_0^t = v_2$. Thus it will follow that a pair (X_0, Y_0) of optimal strategies will satisfy both basic principles and therefore provide a solution to the game.

In preparation for the proof of the main theorem of the next section and in order to provide practice in working with mixed strategies, we end this section with a theorem that simplifies the definitions of v_1 and v_2 and allows us actually to compute optimal strategies for simple games without saddle points.

First we need some notation. For a matrix A, let $A_{(i)}$ denote the ith row of A and $A^{(j)}$ the jth column (as in Section 5.2). We have that, for a game with payoff matrix A, if P_1 uses pure strategy s_i and P_2 uses stragegy Y, the expected payoff is $A_{(i)} Y^t$ and, similarly, if P_1 uses strategy X and P_2 uses pure strategy t_j, the payoff is $XA^{(j)}$. Using this notation, the product XAY^t can be decomposed two ways, because we have

$$\begin{aligned} XAY^t = (XA)Y^t &= \sum_{1 \le j \le n} XA^{(j)} y_j \\ = X(AY^t) &= \sum_{1 \le i \le m} x_i A_{(i)} Y^t \end{aligned}$$

where $X = (x_1, x_2, \ldots, x_m)$ and $Y = (y_1, y_2, \ldots, y_n)$.

Theorem 2 *For a fixed strategy X for P_1,*

$$\min_{Y \in T} XAY^t = \min_{1 \le j \le n} XA^{(j)}$$

Similarly, for a fixed strategy Y for P_2,

$$\max_{X \in S} XAY^t = \max_{1 \le i \le m} A_{(i)} Y^t$$

Proof Take a fixed strategy X for P_1. Then, for each j, $XA^{(j)}$ is a real number. Let

$$w = \min_{1 \le j \le n} XA^{(j)}$$

Then, for any $Y \in T$,

$$\begin{aligned} XAY^t &= \sum_{1 \le j \le n} XA^{(j)} y_j \\ &\ge \sum_{1 \le j \le n} w y_j \\ &= w \sum_{1 \le j \le n} y_j \\ &= w \cdot 1 = w \end{aligned}$$

But clearly the outcomes $XA^{(j)}$ are contained in the set $\{XAY^t \mid Y \in T\}$, as they correspond to the outcomes when P_2 uses his pure strategies.

Thus

$$\underset{Y \in T}{\text{Min}}\ XAY^t = \underset{1 \le j \le n}{\text{Min}}\ XA^{(j)}$$

Similarly,

$$\underset{X \in S}{\text{Max}}\ XAY^t = \underset{1 \le i \le m}{\text{Max}}\ A_{(i)}Y^t$$

for a fixed Y in T. # # #

As a result, we can simplify the definition of the optimal security levels v_1 and v_2. We have now

$$v_1 = \underset{X \in S}{\text{Max}}\ \underset{Y \in T}{\text{Min}}\ XAY^t = \underset{X \in S}{\text{Max}}\ \underset{1 \le j \le n}{\text{Min}}\ XA^{(j)} \quad \text{and}$$

$$v_2 = \underset{Y \in T}{\text{Min}}\ \underset{X \in S}{\text{Max}}\ XAY^t = \underset{Y \in T}{\text{Min}}\ \underset{1 \le i \le m}{\text{Max}}\ A_{(i)}Y^t$$

Thus, for example, if P_1 wishes to determine his security level for a strategy X, he need not consider the set of all possible outcomes XAY^t for $Y \in T$, but only those outcomes $XA^{(j)}$ corresponding to the use of pure strategies by P_2.

Example 2

Consider the game with payoff matrix

$$A = \begin{bmatrix} 1 & 3 \\ 4 & 0 \end{bmatrix}$$

as discussed in Example 1.

We have

$$\begin{aligned} v_1 &= \underset{X=(x_1,x_2) \in S}{\text{Max}}\ \underset{1 \le j \le 2}{\text{Min}}\ XA^{(j)} \\ &= \underset{(x_1,x_2) \in S}{\text{Max}}\ \text{Min}\{x_1 + 4x_2, 3x_1\} \end{aligned}$$

Now, for any $(x_1,x_2) \in S$, $x_1 + x_2 = 1$ and $0 \le x_1,x_2 \le 1$. Hence $S = \{(x_1,x_2) | 0 \le x_1 \le 1 \text{ and } x_2 = 1 - x_1\}$. Therefore

$$\begin{aligned} \underset{(x_1,x_2) \in S}{\text{Max}}\ \text{Min}\{x_1 + 4x_2, 3x_1\} &= \underset{0 \le x_1 \le 1}{\text{Max}}\ \text{Min}\{x_1 + 4(1 - x_1), 3x_1\} \\ &= \underset{0 \le x_1 \le 1}{\text{Max}}\ \text{Min}\{4 - 3x_1, 3x_1\} \end{aligned}$$

Consider the graph in Figure 8.1. The heavy line segments represent $\text{Min}\{4 - 3x_1, 3x_1\}$. The maximum of this function occurs at the point $x_1 = \frac{2}{3}$ where the two lines

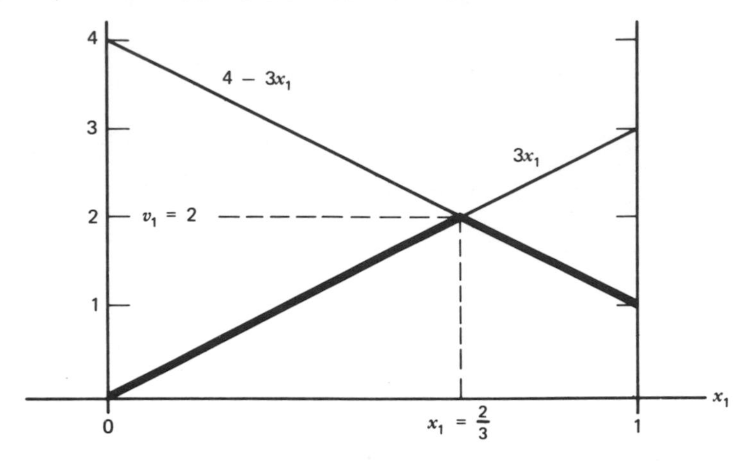

Figure 8.1

intersect, and the value of the function at this point is 2. Thus $v_1 = 2$, and an optimal strategy for P_1 is the mixed strategy $(\frac{2}{3},\frac{1}{3})$. Note that for any $Y = (y_1,y_2) \in T$,

$$(\tfrac{2}{3},\tfrac{1}{3})AY^t = 2y_1 + 2y_2 = 2(y_1 + y_2) = 2$$

PROBLEM SET 8.4

1. Continuation of Example 2.

(a) Calculate graphically

$$v_2 = \min_{Y=(y_1,y_2)\in T} \max_{1\le i\le 2} A_{(i)}Y^t$$

Determine also an optimal strategy Y_0 for P_2.

(b) Let $X_0 = (\frac{2}{3},\frac{1}{3})$. Show that the pair (X_0,Y_0) is in equilibrium, and so this game is "solved."

2. Using the technique introduced in Example 2, compute v_1, v_2, and optimal strategies for the game with matrix

$$\begin{bmatrix} -1 & 4 \\ 3 & -6 \end{bmatrix}$$

Is $v_1 = v_2$? Which player, if any, has the advantage?

3. Consider the game with payoff matrix

$$A = \begin{bmatrix} 0 & 0 & 3 & 1 \\ 2 & 1 & 4 & 2 \\ 1 & 3 & 0 & 0 \end{bmatrix}$$

(a) By considering the security level of the mixed strategy for P_1 of $X_0 = (0,\frac{3}{4},\frac{1}{4})$, show that $v_1 \geq \frac{3}{2}$.
(b) Similarly, by considering P_2's mixed strategy of $Y_0 = (0,\frac{1}{2},0,\frac{1}{2})$, show that $v_2 \leq \frac{3}{2}$.
(c) Using Theorem 1, what can you conclude about v_1 and v_2? Conclude that X_0 and Y_0 are optimal strategies.
(d) Using these results, show that the pair (X_0,Y_0) is in equilibrium.

4. Intuitively the game with payoff matrix

$$\begin{bmatrix} 1 & -1 \\ -1 & 1 \end{bmatrix}$$

should be fair, with optimal strategies of $X_0 = (\frac{1}{2},\frac{1}{2})$ and $Y_0 = (\frac{1}{2},\frac{1}{2})$. Using Problem 3 as a model, prove that this is the case.

5. Suppose that the payoff matrix A of a game has a saddle point a_{hk}.
(a) Prove that $v_1 \geq a_{hk}$.
(b) Prove that $v_2 \leq a_{hk}$.
(c) Conclude that $v_1 = a_{hk} = v_2$.
(d) Prove that the pure strategies s_h and t_k are optimal.

6. Prove that for any game $u_1 \leq v_1$ and $v_2 \leq u_2$.

SECTION 8.5 THE FUNDAMENTAL THEOREM

In this section we prove the fundamental theorem of two-person, zero-sum game theory. The theorem states simply that the optimal security levels of the two players are equal and, as we have seen in the last section, this will imply the stability of optimal strategies and therefore a solution of these games determined by the two basic principles formulated in Section 8.2. The proof that we give is based on the Duality Theorem of linear programming, and is due to G. Dantzig. One advantage with this proof is that it is constructive; that is, the proof provides a practical technique for actually computing optimal strategies and the value $v_1 = v_2$ for any game. The theorem was first proved by J. von Neumann [32] in 1928 using the fixed-point theorem of Brouwer. Since then, many other proofs, both topological and algebraic, have been developed.

To assist the reader we first repeat here the essential definitions from the previous section, making use of Theorem 2 of that section. Recall that S and T denote the sets of mixed strategies for P_1 and P_2, respectively, and $A_{(i)}$ and $A^{(j)}$ the ith row and jth column, respectively, of a matrix A.

Definition For a game with payoff matrix A, the *optimal security levels* v_1 and v_2 of P_2 and P_2 are defined by

$$v_1 = \underset{X \in S}{\text{Max}} \underset{Y \in T}{\text{Min}} XAY^t = \underset{X \in S}{\text{Max}} \underset{1 \leq j \leq n}{\text{Min}} XA^{(j)}$$

$$v_2 = \underset{Y \in T}{\text{Min}} \underset{X \in S}{\text{Max}} XAY^t = \underset{Y \in T}{\text{Min}} \underset{1 \leq i \leq m}{\text{Max}} A_{(i)} Y^t$$

A strategy X_0 is an *optimal strategy* for P_1 if

$$v_1 = \underset{Y \in T}{\text{Min}} X_0 AY^t = \underset{1 \leq j \leq n}{\text{Min}} X_0 A^{(j)}$$

A strategy Y_0 is an *optimal strategy* for P_2 if

$$v_2 = \underset{X \in S}{\text{Max}} XAY_0{}^t = \underset{1 \leq i \leq m}{\text{Max}} A_{(i)} Y_0{}^t$$

Theorem 1 (The Fundamental Theorem) *For any matrix game, optimal strategies for both players exist, and moreover, the optimal security levels v_1 and v_2 are equal.*

Proof Suppose the game has payoff matrix $A = (a_{ij})$, $1 \leq i \leq m$, $1 \leq j \leq n$. In order to be able to apply linear programming techniques to the problem of determining v_1, v_2, and optimal strategies, we need to be able to assume that v_1 and v_2 are positive. This will certainly be the case if all the entries a_{ij} are positive. Thus we divide our proof into two cases, considering first the case in which all $a_{ij} > 0$, and then the general case. The heart of the proof is contained in the first case since, as we will see, in the second case a minor modification of an arbitrary matrix A will allow us to apply the results from the first part.

Case 1 All $a_{ij} > 0$.
Consider the game from P_1's point of view. His optimal security level is

$$v_1 = \underset{X \in S}{\text{Max}} \underset{1 \leq j \leq n}{\text{Min}} XA^{(j)}$$

In order to determine v_1, for any strategy $X \in S$ the minimum of the $XA^{(j)}$ must first be determined, that is, the minimum of the n quantities

$$\begin{aligned}
&a_{11}x_1 + a_{21}x_2 + \cdots + a_{m1}x_m \\
&a_{12}x_1 + a_{22}x_2 + \cdots + a_{m2}x_m \\
&\vdots \\
&a_{1n}x_1 + a_{2n}x_2 + \cdots + a_{mn}x_m
\end{aligned}$$

Now the minimum of these n quantities is less than or equal to each of the n quantities, and is in fact equal to at least one of them, the smallest one. Thus the minimum is the largest real number w satisfying the n inequalities

$$\begin{aligned}
&a_{11}x_1 + a_{21}x_2 + \cdots + a_{m1}x_m \geq w \\
&a_{12}x_1 + a_{22}x_2 + \cdots + a_{m2}x_m \geq w \\
&\vdots \\
&a_{1n}x_1 + a_{2n}x_2 + \cdots + a_{mn}x_m \geq w
\end{aligned} \tag{1}$$

Hence for each $X \in S$ the maximum w satisfying (1) must first be determined. Next, consider v_1, which is the maximum over all the $X \in S$ of these w's. Since

$$S = \{(x_1,x_2,\ldots,x_m) | x_1 + x_2 + \cdots + x_m = 1, x_i \geq 0, 1 \leq i \leq m\}$$

it follows that v_1 is equal to the maximum w satisfying

$$\begin{aligned}
&a_{11}x_1 + \cdots + a_{m1}x_m \geq w \\
&\vdots \\
&a_{1n}x_1 + \cdots + a_{mn}x_m \geq w \\
&\quad x_1 + \cdots + x_m = 1 \\
&\quad x_i \geq 0, 1 \leq i \leq m
\end{aligned} \tag{2}$$

Moreover, a point $X = (x_1,\ldots,x_m)$ at which this maximum is attained is an optimal strategy for P_1.

Now, since all the $a_{ij} > 0$, we know that v_1 must be positive, and so we can restrict our attention to only those $w > 0$ that satisfy (2). Dividing the equation and the inequalities in (2) by w, we have the problem of

Maximizing w

subject to

$$\begin{aligned}
&a_{11}\frac{x_1}{w} + \cdots + a_{m1}\frac{x_m}{w} \geq 1 \\
&\vdots \\
&a_{1n}\frac{x_1}{w} + \cdots + a_{mn}\frac{x_m}{w} \geq 1 \\
&\quad \frac{x_1}{w} + \cdots + \frac{x_m}{w} = \frac{1}{w} \\
&\frac{x_i}{w} \geq 0, 1 \leq i \leq m
\end{aligned} \tag{3}$$

Let $x_i' = (x_i/w)$, $1 \leq i \leq m$. Since the problem of maximizing w is equivalent to the problem of minimizing $1/w$, the optimization problem of (3) is equivalent to the problem of

$$\begin{aligned}
&\text{Minimizing} \quad x_1' + x_2' + \cdots + x_m' \\
&\text{subject to} \\
&a_{11}x_1' + \cdots + a_{m1}x_m' \geq 1 \\
&\vdots \\
&a_{1n}x_1' + \cdots + a_{mn}x_m' \geq 1 \\
&x_i' \geq 0,\ 1 \leq i \leq m
\end{aligned} \tag{4}$$

In order to express this concisely in vector notation, let $X' = (x_1',x_2',\ldots,x_m')^t$, $b = (1,1,\ldots,1)^t$, and $c = (1,1,\ldots,1)^t$, where b is an m-tuple and c an n-tuple. Then (4) is simply the problem of

$$\begin{aligned}
&\text{Minimizing} \quad b \cdot X' \\
&\text{subject to} \\
&A^tX' \geq c,\ X' \geq 0
\end{aligned}$$

In summary, the reciprocal of the minimal value of the objective function $b \cdot X'$ of the linear programming problem of (4) is equal to v_1. Moreover, since $w(x_1', \ldots, x_m') = (x_1, \ldots, x_m)$, multiplication of the coordinates of a point X' at which this minimum is attained by v_1 gives an optimal strategy for P_1.

We will now show, by a completely parallel development, that the problem of determining the security level and an optimal strategy for P_2 leads us to the dual problem of the problem of (4). We have

$$v_2 = \underset{Y \in T}{\text{Min}} \ \underset{1 \leq i \leq m}{\text{Max}} \ A_{(i)} Y^t$$

Now, for a fixed $Y = (y_1, y_2, \ldots, y_n) \in T$, the maximum of the m quantities $A_{(i)} Y^t$ is the smallest real number z satisfying

$$\begin{aligned}
&a_{11}y_1 + a_{12}y_2 + \cdots + a_{1n}y_n \leq z \\
&a_{21}y_1 + a_{22}y_2 + \cdots + a_{2n}y_n \leq z \\
&\vdots \\
&a_{m1}y_1 + a_{m2}y_2 + \cdots + a_{mn}y_n \leq z
\end{aligned}$$

Since v_2 is the minimum over all the $Y \in T$ of these z's, it follows that v_2 is equal to the minimum z satisfying

$$\begin{aligned}
&a_{11}y_1 + \cdots + a_{1n}y_n \leq z \\
&\vdots \\
&a_{m1}y_1 + \cdots + a_{mn}y_n \leq z \\
&y_1 + y_2 + \cdots + y_n = 1 \\
&y_j \geq 0,\ 1 \leq j \leq n
\end{aligned} \tag{5}$$

and a point at which this minimal value is attained is an optimal strategy for P_2. Again, we know that $v_2 > 0$, and therefore all z satisfying the constraints of (5) must be positive. Dividing the inequalities and equation of (5) by z gives the problem of

$$\begin{aligned}
&\text{Minimizing} \quad z\\
&\text{subject to}\\
&a_{11}\frac{y_1}{z} + \cdots + a_{1n}\frac{y_n}{z} \le 1\\
&\vdots\\
&a_{m1}\frac{y_1}{z} + \cdots + a_{mn}\frac{y_n}{z} \le 1\\
&\qquad \frac{y_1}{z} + \cdots + \frac{y_n}{z} = \frac{1}{z}\\
&\frac{y_j}{z} \ge 0,\ 1 \le j \le n
\end{aligned} \tag{6}$$

Let $y'_j = (y_j/z)$, $1 \le j \le n$. Then (6) is equivalent to the problem of

$$\begin{aligned}
&\text{Maximizing} \quad y'_1 + y'_2 + \cdots + y'_n\\
&\text{subject to}\\
&a_{11}y'_1 + \cdots + a_{1n}y'_n \le 1\\
&\vdots\\
&a_{m1}y'_1 + \cdots + a_{mn}y'_n \le 1\\
&y'_j \ge 0,\ 1 \le j \le n
\end{aligned} \tag{7}$$

In vector notation, this problem is to

$$\begin{aligned}
&\text{Maximize} \quad c \cdot Y'\\
&\text{subject to}\\
&AY' \le b,\ Y' \ge 0
\end{aligned}$$

And, as before, the reciprocal of the maximum value of the objective function $c \cdot Y'$ of (7) is equal to v_2, and multiplication of the coordinates of a point Y' at which this maximum is attained by v_2 gives an optimal strategy for P_2.

But now the problems of (4) and (7) are dual linear problems with, in the terminology of Section 4.1, the problem of maximizing $c \cdot Y'$ subject to $AY' \le b$, $Y' \ge 0$ as the primal problem and the problem of minimizing $b \cdot X'$ subject to $A^tX' \ge c$, $X' \ge 0$ as the dual problem. Moreover, the dual problem must have a finite optimal solution, because the objective function $b \cdot X' = x'_1 + \cdots + x'_m$ is bounded below by zero and, since all the entries of A are positive, there exist feasible solutions to the system of constraints $A^tX' \ge c$. Thus it follows from the Duality

Theorem of Section 4.3 that both problems have finite solutions attaining the same optimal value. Hence there exist optimal strategies for both players P_1 and P_2, and the security levels v_1 and v_2 are equal.

Case 2 (General Case)

Suppose some entries a_{ij} are nonpositive. Choose any constant r with the property that $a_{ij} + r > 0$ for all i and j (e.g., r could equal $1 - \text{Min}_{i,j}\, a_{ij}$). Let E be the $m \times n$ matrix with all entries equal to 1, and consider the game with payoff matrix $A + rE$. The expected payoff for any pair of strategies (X,Y) is

$$X(A + rE)Y^t = XAY^t + rXEY^t$$

But, since X and Y are strategies, XE equals the n-vector $(1,1,\ldots,1)$, and $(1,1,\ldots,1)Y^t = 1$. The expected payoff is $XAY^t + r$, which is the expected payoff for the game with matrix A plus the constant r. Since these expected payoffs XAY^t and $X(A + rE)\,Y^t$ differ only by the constant r, it follows that the games with payoff matrices A and $A + rE$ will have the same optimal strategies, and that the players' security levels will differ only by the constant r. But all the entries of the matrix $A + rE$ are positive, and so the results of Case 1 can be applied to the corresponding game. Thus, for the game with payoff matrix A, optimal strategies exist and the players' security levels are equal. # # #

Example

As an example of the technique developed in this proof, consider the game discussed in the last section with payoff matrix

$$\begin{bmatrix} 1 & 3 \\ 4 & 0 \end{bmatrix}$$

Since both entries in the first row are positive, v_1 is positive, and the two linear programming problems associated with this game are simply to

$$\begin{array}{lll} \text{Minimize } x'_1 + x'_2 & \text{and} & \text{Maximize } y'_1 + y'_2 \\ \text{subject to} & & \text{subject to} \\ x'_1 + 4x'_2 \geq 1 & & y'_1 + 3y'_2 \leq 1 \\ 3x'_1 \qquad\quad \geq 1 & & 4y'_1 \qquad\quad \leq 1 \\ x'_1, x'_2 \geq 0 & & y'_1, y'_2 \geq 0 \end{array}$$

Suppose we wish to solve these problems using the simplex method. The maximization problem associated with P_2's determination of his security level and an optimal strategy can be handled without using artificial variables. Introducing two slack variables y'_3 and y'_4, the simplex method leads to the following tableaux of Table 8.1.

Table 8.1

	y_1'	y_2'	y_3'	y_4'	
y_3'	1	(3)	1	0	1
y_4'	4	0	0	1	1
	-1	-1	0	0	0
y_2'	$\frac{1}{3}$	1	$\frac{1}{3}$	0	$\frac{1}{3}$
y_4'	(4)	0	0	1	1
	$-\frac{2}{3}$	0	$\frac{1}{3}$	0	$\frac{1}{3}$
y_2'	0	1	$\frac{1}{3}$	$-\frac{1}{12}$	$\frac{1}{4}$
y_1'	1	0	0	$\frac{1}{4}$	$\frac{1}{4}$
	0	0	$\frac{1}{3}$	$\frac{1}{6}$	$\frac{1}{2}$

Thus the maximum of $y_1' + y_2'$ is $\frac{1}{2}$, and this value is attained at the point $(\frac{1}{4}, \frac{1}{4})$. Therefore $v_2 = 2$, and an optimal strategy for P_2 is $2(\frac{1}{4}, \frac{1}{4}) = (\frac{1}{2}, \frac{1}{2})$. Furthermore, we know also that $v_1 = 2$ and, using the entries in the bottom row of the slack variable columns, an optimal strategy for P_1 is $2(\frac{1}{3}, \frac{1}{6}) = (\frac{2}{3}, \frac{1}{3})$.

Although the above problem was solved in the last section using other methods, it should be clear that this simplex method can be applied to any matrix game. More will be said about this in the next section, in which computational techniques are discussed. For the remainder of this section we will consider some of the theoretical implications of the Fundamental Theorem.

In the last section it was suggested that a pair (X_0, Y_0) of optimal strategies would be stable if $v_1 = v_2$. We now prove that this is the case. For any pair of optimal strategies (X_0, Y_0), the proof of Theorem 1 of the last section shows that

$$v_1 = \min_{Y \in T} X_0 A Y^t \leq X_0 A Y_0{}^t \leq \max_{X \in S} X A Y_0{}^t = v_2$$

Since we know now that $v_1 = v_2$, the above two inequalities must be equalities. Now for any $X \in S$ and $Y \in T$,

$$X A Y_0{}^t \leq \max_{X \in S} X A Y_0{}^t \qquad \text{and} \qquad X_0 A Y^t \geq \min_{Y \in T} X_0 A Y^t$$

Putting these two facts together gives the following corollary.

Corollary

Let X_0 and Y_0 be optimal strategies. Then, for any $X \in S$ and $Y \in T$,

$$XAY_0^t \leq X_0AY_0^t \leq X_0AY^t$$

But this is precisely what is meant by saying that the pair of strategies (X_0, Y_0) is in equilibrium; P_1 has nothing to gain by deviating from the strategy X_0 if P_2 plays Y_0 and, similarly, P_2 gains nothing by deviating from Y_0 if P_1 adheres to X_0. Thus, any pair of optimal strategies (X_0, Y_0) satisfies the two principles set out in Section 8.2; Principle I since, from the definition of an optimal strategy, by using these strategies the players are maximizing their security levels; and Principle II from what has just been proven. Thus matrix games are completely determined by the two principles, and this determination is called the *solution* to the game.

Definition Any pair (X_0, Y_0) of optimal strategies is said to be a *solution* to a matrix game. If the payoff matrix for the game is A, the common value $v = v_1 = X_0AY_0^t = v_2$ is the *value* of the game. If $v = 0$, the game is *fair*.

A solution is therefore a suggested course of play for both players given that their play is to be determined by Principles I and II. The value of the game is the optimal security level for both players and the expected outcome of the game if the players use the suggested strategies. Thus, for matrix games we have been able to develop a complete mathematical model based on the two principles. However, this model can be applied to a game-theoretic situation only if these principles are representative of the approach of the players to the situation. An excellent discussion of some of the limitations of the applicability of the theory from the viewpoint of a social scientist is contained in the book *Games and Decisions* by Luce and Raiffa [23].

It may have occurred to the reader that a solution and value of a game with a saddle point was defined back in Section 8.3. It is therefore contingent on us to show that this definition agrees with the above definition for any game in this particular case. Although this can be done directly (see Problem 5 of Section 8.4), it will also follow from the following theorem, a converse to the above corollary.

Theorem 2 *Let a matrix game have payoff matrix A. Suppose the strategy pair (X_1, Y_1) is in equilibrium; that is, for any $X \in S$ and $Y \in T$,*

$$XAY_1^t \leq X_1AY_1^t \leq X_1AY^t$$

Then X_1 and Y_1 are optimal strategies, and $X_1AY_1^t$ is the value of the game.

Proof Let v denote the value of the game. Then

$$v = v_1 = \max_{X \in S} \min_{Y \in T} XAY^t$$
$$\geq \min_{Y \in T} X_1AY^t = X_1AY_1{}^t$$

and

$$v = v_2 = \min_{Y \in T} \max_{X \in S} XAY^t$$
$$\leq \max_{X \in S} XAY_1{}^t = X_1AY_1{}^t$$

Therefore $v = X_1AY_1{}^t$ and $\text{Min}_{Y \in T} X_1AY^t = v$, and so, by definition, X_1 is an optimal strategy for P_1. Similarly, $\text{Max}_{X \in S} XAY_1{}^t = v$ implies that Y_1 is an optimal strategy for P_2. ###

Corollary

Suppose a_{hk} is a saddle point of the payoff matrix A. Let X_1 be the pure strategy s_h and Y_1 the pure strategy t_k. Then X_1 and Y_1 are optimal, and the value of the game is a_{hk}.

Proof Note that $X_1AY_1{}^t = a_{hk}$. Now for any $X = (x_1,x_2, \ldots ,x_m) \in S$, since a_{hk} is a column maximum,

$$XAY_1{}^t = \sum_{1 \leq i \leq m} x_i A_{(i)} Y_1{}^t$$
$$= \sum x_i a_{ik}$$
$$\leq \sum x_i a_{hk}$$
$$= a_{hk} \sum x_i = a_{hk} = X_1AY_1{}^t$$

Similarly, for any $Y \in T$,

$$X_1AY^t \geq X_1AY_1{}^t$$

Therefore, from the above theorem, X_1 and Y_1 are optimal and $X_1AY_1{}^t = a_{hk}$ is the value of the game. ###

PROBLEM SET 8.5

1. The following refer to the proof of Theorem 1.
   (a) In Case 1, it is claimed that feasible solutions exist for the linear programming problem of minimizing $b \cdot X'$ subject to $A^tX' \geq c$. Prove that this is true.
   (b) In Case 2, it is claimed that the security levels of the players differ only by the constant r for the two games with payoff matrices A and $A + rE$. Prove this using the definition of a security level.

2. True or false: From Theorem 1 of Section 8.4, $v_1 \leq v_2$ for any matrix game. From Theorem 2 of this section, $X_1 A Y_1{}^t \leq v_1$ and $v_2 \leq X_1 A Y_1{}^t$. Therefore $v_1 = v_2$, and so Theorem 2 provides an alternate proof of the Fundamental Theorem?

3. Suppose that X_0 and Y_0 are optimal strategies for a game with payoff matrix A and value v. Prove that for any i and j, $X_0 A^{(j)} \geq v$ and $A_{(i)} Y_0{}^t \leq v$.

4. Suppose that X_0 and Y_0 are optimal strategies and v the value of a game with matrix A. Suppose that $X_0 A^{(k)} > v$. Prove that the kth coordinate of Y_0 is zero.

5. True or false: If v is the value of a game and Y_0 an optimal strategy for P_2, then $X_1 A Y_0{}^t = v$ implies that X_1 is an optimal strategy for P_1?

$$\left(\textit{Hint}.\ \text{Let } A = \begin{bmatrix} 1 & 0 \\ 1 & 2 \end{bmatrix}.\right)$$

6. True or false: Suppose A is the payoff matrix of a game with value v. Suppose X_1 is a strategy for P_1 such that $X_1 A^{(j)} \geq v$ for all j. Then X_1 is an optimal strategy?

7. (a) You, as P_1, pay your opponent P_2 \$2, and then play once and only once the game with payoff matrix

$$\begin{bmatrix} 1 & 3 \\ 4 & 0 \end{bmatrix}$$

What would be your strategy?

(b) As above, but this time you play the game 25 times, giving P_2 \$2 before each play.

8. True or false: A game has a pair of optimal strategies (X_0, Y_0) that are both pure strategies if and only if the game has a saddle point?

9. True or false: Suppose X_0 is an optimal strategy for P_1 for the matrix game A. Then, for any $X \in S$ and $Y \in T$, $XAY^t \leq X_0 A Y^t$? (In other words, an optimal strategy is the best response to any strategy of the opponent?)

10. True or false: One player has an optimal pure strategy if and only if the game has a saddle point?

11. A matrix game A is said to be symmetric if $A^t = -A$. Prove that the value of a symmetric game is 0.

SECTION 8.6 COMPUTATIONAL TECHNIQUES

The solution of games with saddle points is straightforward, and the existence of saddle points can be easily determined by computing the u_1 and u_2 defined in Section 8.3. and using the corollary to Theorem 1 in that section. For other games we list the following techniques.

Linear Programming

As seen in the last section, the problem of determining optimal strategies and the value of a matrix game is equivalent to two dual linear programming problems that can be solved using the simplex method. However, this method is not directly applicable if the value of the game is not positive. For an arbitrary game a constant must first be chosen such that when this constant is added to each entry of the original payoff matrix, the game corresponding to this new matrix has a positive value. Then the simplex method can be applied to this new game, with the value of the original game equal to the value of the new game less the constant. Note that it may not be necessary to make all the entries in the modified payoff matrix positive; for example, if the matrix has at least one row with all positive entries, the value of the corresponding game is positive (the possibility of P_1 using the pure strategy of playing that particular row shows that his security level is positive).

Example 1

P_1 and P_2 each extend either one, two, or three fingers, and the difference in the amounts put forth is computed. If this difference is 0, the payoff is 0; if the difference is 1, the player putting forth the smaller amount wins 1; and if the difference is 2, the player putting forth the larger amount wins 2.

Each player has three pure strategies. Let s_i denote P_1's pure strategy of extending i fingers, $1 \leq i \leq 3$, and similarly define t_j, $1 \leq j \leq 3$, for P_2. The payoff tableau is then

	t_1	t_2	t_3
s_1	0	1	−2
s_2	−1	0	1
s_3	2	−1	0

By symmetry it is reasonable to expect the value of this game to be 0. To verify this and compute optimal strategies, we first add 2 to each entry of the above matrix, giving the following matrix, which corresponds to a game with value at least 1 as all the entries in the last two rows are greater than or equal to 1.

$$\begin{bmatrix} 2 & 3 & 0 \\ 1 & 2 & 3 \\ 4 & 1 & 2 \end{bmatrix}$$

The associated linear programming problem corresponding to P_2's determination of an optimal strategy and security level is to

$$\text{Maximize} \quad y_1' + y_2' + y_3'$$

subject to

$$\begin{aligned} 2y_1' + 3y_2' \qquad &\leq 1 \\ y_1' + 2y_2' + 3y_3' &\leq 1 \\ 4y_1' + y_2' + 2y_3' &\leq 1 \\ y_1', y_2', y_3' &\geq 0 \end{aligned}$$

Table 8.2

	y_1'	y_2'	y_3'	y_4'	y_5'	y_6'	
y_4'	2	3	0	1	0	0	1
y_5'	1	2	(3)	0	1	0	1
y_6'	4	1	2	0	0	1	1
	-1	-1	-1	0	0	0	0
y_4'	2	(3)	0	1	0	0	1
y_3'	$\frac{1}{3}$	$\frac{2}{3}$	1	0	$\frac{1}{3}$	0	$\frac{1}{3}$
y_6'	$\frac{10}{3}$	$-\frac{1}{3}$	0	0	$-\frac{2}{3}$	1	$\frac{1}{3}$
	$-\frac{2}{3}$	$-\frac{1}{3}$	0	0	$\frac{1}{3}$	0	$\frac{1}{3}$
y_2'	$\frac{2}{3}$	1	0	$\frac{1}{3}$	0	0	$\frac{1}{3}$
y_3'	$-\frac{1}{9}$	0	1	$-\frac{2}{9}$	$\frac{1}{3}$	0	$\frac{1}{9}$
y_6'	($\frac{32}{9}$)	0	0	$\frac{1}{9}$	$-\frac{2}{3}$	1	$\frac{4}{9}$
	$-\frac{4}{9}$	0	0	$\frac{1}{9}$	$\frac{1}{3}$	0	$\frac{4}{9}$
y_2'	0	1	0	$\frac{5}{16}$	$\frac{1}{8}$	$-\frac{3}{16}$	$\frac{1}{4}$
y_3'	0	0	1	$-\frac{7}{32}$	$\frac{5}{16}$	$\frac{1}{32}$	$\frac{1}{8}$
y_1'	1	0	0	$\frac{1}{32}$	$-\frac{3}{16}$	$\frac{9}{32}$	$\frac{1}{8}$
	0	0	0	$\frac{1}{8}$	$\frac{1}{4}$	$\frac{1}{8}$	$\frac{1}{2}$

Adding three slack variables and solving leads to the tableaux of Table 8.2.

The value of the modified game is 2, and so the value of the original game is 0, as suggested. Since the optimal value of the above problem is attained at $(y'_1, y'_2, y'_3) = (\frac{1}{8}, \frac{1}{4}, \frac{1}{8})$, an optimal strategy for P_2 is $2(\frac{1}{8}, \frac{1}{4}, \frac{1}{8}) = (\frac{1}{4}, \frac{1}{2}, \frac{1}{4})$. Similarly, the solution to the dual problem, found in the bottom row in the slack variable columns, is $(x_1, x_2, x_3) = (\frac{1}{8}, \frac{1}{4}, \frac{1}{8})$, and so an optimal strategy for P_1 is also $2(\frac{1}{8}, \frac{1}{4}, \frac{1}{8}) = (\frac{1}{4}, \frac{1}{2}, \frac{1}{4})$.

2 × 2 Games

The solution of games with a 2×2 payoff matrix can be given by simple formulas. Before we state these, we state a theorem that enables one to immediately determine if a 2×2 game has a saddle point. The proof of this first theorem is outlined in Problem 3.

Consider the game with payoff matrix

$$A = \begin{bmatrix} a & b \\ c & d \end{bmatrix}$$

Theorem 1 *A has no saddle points if and only if a and d are both larger than or both less than b and c, that is, if and only if either* $a > b$, $a > c$, $d > b$, $d > c$ *or* $a < b$, $a < c$, $d < b$, $d < c$.

Theorem 2 *Suppose A has no saddle points. Let* $r = a + d - b - c$. *Then the value* v *of the game is*

$$v = \frac{ad - bc}{r}$$

and optimal strategies X_0 *and* Y_0 *for* P_1 *and* P_2 *are*

$$X_0 = \left(\frac{d - c}{r}, \frac{a - b}{r}\right)$$

$$Y_0 = \left(\frac{d - b}{r}, \frac{a - c}{r}\right)$$

Proof Note that by Theorem 1, $r \neq 0$ and X_0 and Y_0 are strategies. By direct calculation, we have

$$X_0 A = \left(\frac{ad - bc}{r}, \frac{ad - bc}{r}\right)$$

and

$$AY_0^t = \left(\frac{ad - bc}{r}, \frac{ad - bc}{r}\right)^t$$

Thus for any strategies $X = (x_1,x_2) \in S$ and $Y = (y_1,y_2) \in T$,

$$X_0AY^t = \frac{ad - bc}{r}(y_1 + y_2) = \frac{ad - bc}{r}$$

and

$$XAY_0^{\,t} = (x_1 + x_2)\frac{ad - bc}{r} = \frac{ad - bc}{r}$$

From Theorem 2 of Section 8.5, it follows that X_0 and Y_0 are optimal strategies, and the value of the game is

$$X_0AY_0^{\,t} = \frac{ad - bc}{r}$$

###

These formulas are easy to remember. Once r is determined, the value of the game is the determinant of A divided by r. The numerators in the optimal strategy for P_1 are the differences between the entries in the rows of A, with the difference of the entries in the second row going into the first component. Moreover, since the components of a strategy are nonnegative, all that is critical is the magnitude of these quantities, and X_0 can be remembered as

$$\left(\frac{|d - c|}{|r|}, \frac{|a - b|}{|r|}\right)$$

In a similar manner, P_2's optimal strategy Y_0 can be interpreted, simply replacing the word "row" with the word "column." The fact that the sum of the components of a strategy is 1 provides a partial check of one's calculations.

Example 2

$$A = \begin{bmatrix} 1 & 3 \\ 4 & 0 \end{bmatrix}$$

This matrix has no saddle point, because 1 and 0 are both less than 3 and 4. Thus we can apply Theorem 2. We have $r = -6$, and $v = -12/-6 = 2$,

$$X_0 = \left(\frac{-4}{-6}, \frac{1 - 3}{-6}\right) = \left(\frac{2}{3}, \frac{1}{3}\right)$$

$$Y_0 = \left(\frac{-3}{-6}, \frac{1 - 4}{-6}\right) = \left(\frac{1}{2}, \frac{1}{2}\right)$$

Example 3

$$A = \begin{bmatrix} 10 & -8 \\ -7 & 5 \end{bmatrix}$$

Theorem 2 again applies, and

$$\begin{aligned} r &= 30 \\ v &= -\tfrac{6}{30} = -\tfrac{1}{5} \\ X_0 &= (\tfrac{12}{30},\tfrac{18}{30}) = (\tfrac{2}{5},\tfrac{3}{5}) \\ Y_0 &= (\tfrac{13}{30},\tfrac{17}{30}) \end{aligned}$$

Dominance

Consider the game with payoff matrix

$$A = \begin{bmatrix} 0 & -2 & 0 & 3 \\ 6 & 3 & -1 & -4 \\ 8 & -1 & 7 & 3 \end{bmatrix}$$

Notice that the four entries in the third row are all greater than or equal to the corresponding entries in the first row, and that the three entries in the third column are less than or equal to the corresponding entries in the first column. It follows intuitively that P_1 would never use strategy s_1, since he can do at least as well with s_3 and, similarly, that P_2 would never use strategy t_1. In fact, row 1, which is said to be *dominated* by row 3, and column 1, which is dominated by column 3, can be deleted from the payoff matrix and optimal strategies and a value for the original game computed by using the resulting 2×3 matrix. We make this precise with the following definition.

Definition Consider a game with an $m \times n$ payoff matrix $A = (a_{ij})$. Then row h is *dominated* by row i if $a_{hj} \leq a_{ij}$ for all j, $1 \leq j \leq n$; and column k is *dominated* by column j if $a_{ik} \geq a_{ij}$ for all i, $1 \leq i \leq m$.

Theorem 3 *Suppose row h of a payoff matrix A is dominated. Then there is an optimal strategy $X_0 = (x_1,x_2,\ldots,x_m)$ for P_1 with $x_h = 0$. An optimal strategy for P_1 for the game with payoff matrix A but with the* hth *row removed is an optimal strategy for the original game (after the addition of an* hth *component equal to zero). Similarly for a dominated column.*

The proof of Theorem 3 is left to the reader (see Problem 4). Certainly if our definition of optimal strategy is to be at all reasonable, Theorem 3 must hold.

Example 4

Consider the game with payoff matrix

$$A = \begin{bmatrix} 0 & -2 & -1 & 0 \\ 3 & 5 & 6 & -1 \\ 5 & -1 & -3 & -2 \end{bmatrix}$$

Column 1 is dominated by column 4, and thus finding a solution reduces to consideration of the game

$$\begin{bmatrix} 0 & -2 & -1 & 0 \\ 3 & 5 & 6 & -1 \\ 5 & -1 & -3 & -2 \end{bmatrix}$$

In this 3×3 game, row 3 is dominated by row 2, and so we consider

$$\begin{bmatrix} 0 & -2 & -1 & 0 \\ 3 & 5 & 6 & -1 \\ 5 & -1 & -3 & -2 \end{bmatrix}$$

Now column 3 is dominated by column 2, leaving just a 2×2 matrix

$$\begin{bmatrix} 0 & -2 & -1 & 0 \\ 3 & 5 & 6 & -1 \\ 5 & -1 & -3 & -2 \end{bmatrix}$$

Using the formulas of Theorem 2, we find that the value of the game with this 2×2 payoff matrix is $-\frac{1}{4}$, with $(\frac{3}{4},\frac{1}{4})$ and $(\frac{1}{8},\frac{7}{8})$ optimal strategies for P_1 and P_2, respectively. Thus the original game has value $-\frac{1}{4}$, and optimal strategies

$$X_0 = (\tfrac{3}{4},\tfrac{1}{4},0)$$
$$Y_0 = (0,\tfrac{1}{8},0,\tfrac{7}{8})$$

$2 \times n$ and $m \times 2$ Games

Games in which one player has only two pure strategies have several methods of solution. One method that uses a graph to determine the optimum of a set of linear functions was described in Example 2 of Section 8.4. To solve a game using this method, follow that example, using the definition of

$$v = v_1 = \operatorname*{Max}_{X \in S} \operatorname*{Min}_{1 \le j \le n} XA^{(j)}$$

if it is P_1 that has only the two pure strategies, and the definition of

$$v = v_2 = \operatorname*{Min}_{Y \in T} \operatorname*{Max}_{1 \le i \le m} A_{(i)}Y^t$$

if it is P_2.

Another method makes use of the ease of determining solutions of 2×2 games. From the corollary to the theorem of Section 3.8, it follows that for any game with an $m \times n$ payoff matrix, there are optimal strategies for both players with at most the minimum of m and n non-zero components. Thus, for games in which one player has only two pure strategies, there is an optimal strategy for the other player that has at most two non-zero components, and so the original game reduces to a 2×2 game. Moreover, that 2×2 game would correspond to the 2×2 submatrix of the original $m \times 2$ or $2 \times n$ matrix that attains the most advantageous value for the player that has more than two pure strategies. In other words, in an $m \times 2$ game, P_1 should compute the values of all the 2×2 subgames [there are $\binom{m}{2} = m(m-1)/2$ of them] and play the game that has the largest value. Similarly, in a $2 \times n$ game, P_2 would select the 2×2 subgame with the smallest value.

Example 5

Consider the game with payoff matrix

$$A = \begin{bmatrix} 1 & -5 \\ -4 & 4 \\ -2 & 3 \\ 0 & -5 \end{bmatrix}$$

It can be easily seen that there is no saddle point, but that the last row is dominated by the first row. Thus, to find the solution, we need only consider the first three rows.

Using the first technique described for $m \times 2$ games, we have

$$\begin{aligned} v = v_2 &= \underset{Y \in T}{\text{Min}} \ \underset{1 \le i \le 3}{\text{Max}} \ A_{(i)} Y^t \\ &= \underset{Y \in T}{\text{Min}} \ \text{Max}\{y_1 - 5y_2, -4y_1 + 4y_2, -2y_1 + 3y_2\} \\ &= \underset{0 \le y_1 \le 1}{\text{Min}} \ \text{Max}\{6y_1 - 5, -8y_1 + 4, -5y_1 + 3\} \end{aligned}$$

Consider the graph in Figure 8.2. The heavy line represents the maximum of $\{6y_1 - 5, -8y_1 + 4, -5y_1 + 3\}$, and the minimum occurs when $y_1 = \frac{8}{11}$ at the

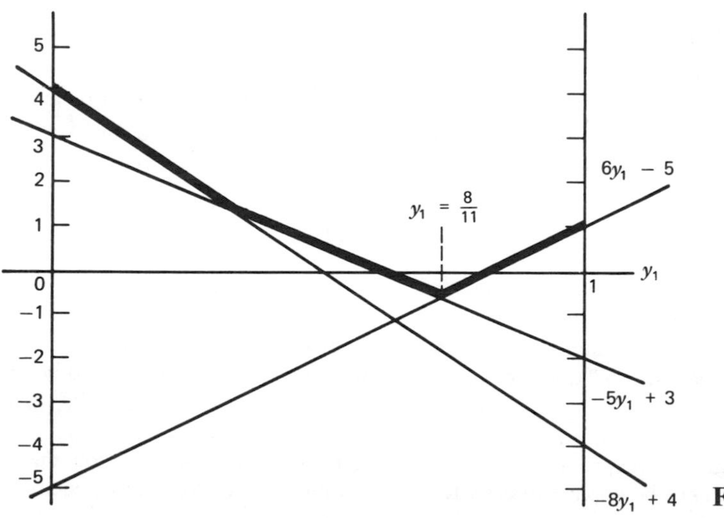

Figure 8.2

intersection of the lines determined by $6y_1 - 5$ and $-5y_1 + 3$. The common value of $6y_1 - 5$ and $-5y_1 + 3$ at $y_1 = \frac{8}{11}$ is $-\frac{7}{11}$. Thus the value of the game is $-\frac{7}{11}$, and an optimal strategy for P_2 is $(\frac{8}{11},\frac{3}{11})$. The two intersecting lines determined by $6y_1 - 5$ and $-5y_1 + 3$ correspond to the first and third rows of the matrix. An optimal strategy for P_1 for the associated 2×2 subgame with matrix

$$\begin{bmatrix} 1 & -5 \\ -2 & 3 \end{bmatrix}$$

is $(\frac{5}{11},\frac{6}{11})$, and so an optimal strategy for P_1 for the original game is $(\frac{5}{11},0,\frac{6}{11},0)$.

Using the second method described for $m \times 2$ games, we consider the three possible 2×2 subgame and compute their values.

$$\begin{bmatrix} 1 & -5 \\ -4 & 4 \end{bmatrix}, \qquad \text{value} = \frac{4-20}{14} = -\frac{8}{7}$$

$$\begin{bmatrix} 1 & -5 \\ -2 & 3 \end{bmatrix}, \qquad \text{value} = \frac{3-10}{11} = -\frac{7}{11}$$

$$\begin{bmatrix} -4 & 4 \\ -2 & 3 \end{bmatrix}, \qquad \text{value} = -2$$

The second game corresponding to the first and third rows of A gives P_1 the largest value. It has value $-\frac{7}{11}$, and optimal strategies $(\frac{5}{11},\frac{6}{11})$ and $(\frac{8}{11},\frac{3}{11})$ for P_1 and P_2, respectively. Thus the value of the original game is $-\frac{7}{11}$, with optimal strategies

$$X_0 = (\tfrac{5}{11},0,\tfrac{6}{11},0) \qquad \text{and} \qquad Y_0 = (\tfrac{8}{11},\tfrac{3}{11})$$

PROBLEM SET 8.6

1. The only technique discussed in this section that cannot be applied to matrix games with saddle points is the use of the formulas given in Theorem 2 for 2×2 games. Consider the game with payoff matrix

$$A = \begin{bmatrix} 2 & 3 \\ 1 & 4 \end{bmatrix}$$

(a) Solve by using linear programming.
(b) Solve by using only dominance.
(c) Show that the formulas of Theorem 2 do not give the solution to this game. However, in this case the "r" does not equal zero. Where does the proof of Theorem 2 break down?

2. As suggested by the proof of the Fundamental Theorem in the last section, the simplex method cannot be applied directly to a game with a value less than or equal to zero.

Convince yourself of this by trying to use the simplex method without altering the original matrix to the games with payoff matrices

$$\begin{bmatrix} 2 & -1 \\ -3 & 1 \end{bmatrix} \quad \text{and} \quad \begin{bmatrix} 2 & -1 \\ -4 & 2 \end{bmatrix}$$

3. Proof of Theorem 1. Consider the game with payoff matrix

$$A = \begin{bmatrix} a & b \\ c & d \end{bmatrix}$$

(a) Prove that if $a < b$, $a < c$, $d < b$, $d < c$, then A has no saddle point.
(b) Prove that if the two entries in any one row or column are equal, then A has a saddle point.
(c) Prove that if A has no saddle point and $a < b$, then $a < c$, $d < c$, $d < b$.
(d) Complete the proof of Theorem 1.

4. Prove Theorem 3. (*Hint.* We know that for any game, optimal strategies exist. Show that if one such optimal strategy for P_1 does not have $x_h = 0$, it can be modified to a strategy with the hth component equal to zero that also satisfies the definition of optimality.)

5. (a) Solve the games with the following payoff matrices using both methods of solution for $2 \times n$ and $m \times 2$ games.

(i) $\begin{bmatrix} 2 & 0 & 1 & 4 \\ 1 & 3 & 2 & 0 \end{bmatrix}$

(ii) $\begin{bmatrix} -1 & 7 \\ 4 & -1 \\ 2 & 1 \end{bmatrix}$

(b) Show that for the game in (i), P_2 does not have a unique optimal strategy.
(c) Using the graph associated with the game of (ii), show that no matter what strategy P_2 employs, P_1 is always better off using a strategy with the third component equal to zero.

6. Solve the games with the following payoff matrices.

(a) $\begin{bmatrix} -12 & 9 & 5 & 0 \\ 4 & 7 & 2 & 1 \\ 0 & -5 & 10 & -5 \end{bmatrix}$

(b) $\begin{bmatrix} 8 & -1 & -3 & 6 \\ -3 & 10 & 8 & -4 \\ 3 & -4 & -5 & 4 \end{bmatrix}$

(c) $\begin{bmatrix} -1 & 0 & 1 & 1 \\ 2 & 1 & -2 & -1 \\ 1 & 3 & -1 & 0 \end{bmatrix}$

7. *Analyze.* P_1 selects a number from {1,2}, and P_2 selects a number from {1,2,3}, with each choice made in ignorance of the opponent's choice. If the sum of the two choices is even, P_1 wins that amount from P_2; if the sum is odd, P_2 wins the sum from P_1.

8. *Analyze.* P_1 and P_2 both select a number from {1,2,3}, each in ignorance of the opponent's selection. If both select the same number, P_2 wins an amount equal to the common selection from P_1. If not, P_1 wins an amount from P_2 equal to the difference of the two selections.

9. True or false: Suppose A is an $m \times m$ matrix. Then the two games, one with payoff matrix A and the other with payoff matrix A^t, have the same value, and an optimal strategy for P_i for one game is an optimal strategy for P_{3-i} for the other game, $i = 1,2$?

10. Solve the games of Problem 1 of Section 8.1. (For the game of part b, first determine x so that the game is fair.)

11. True or false: A 2×2 game has a saddle point if and only if the game can be resolved using dominance?

SECTION 8.7 GAMES PEOPLE PLAY

In this section we will consider several games modeled after the common card game of poker. Poker is an example of a zero-sum game in which the outcome of the game is determined both by chance and the skill of the players—the element of chance in determining the deal, and the skill of the players in their ability to successfully balance conservative play with bluffing. In fact, in the first book on game theory by von Neumann and Morgenstern [33] a section was devoted to a study of a form of poker; since then, various other articles on the game have appeared.

Certainly the standard game of poker is beyond our analysis, even after restricting the game to two players, because of the many different hands a player can be dealt. In the examples we develop the range of possible draws will be much more limited—in fact, one player will receive only one card, and that will be either a "high" of "low" card. However, our examples will preserve the opportunity for the players to bluff or to play conservatively, and so will preserve the element of poker that makes the game interesting.

Note, too, that these games are the first that we have encountered in which outcomes are partially determined by chance. For such games the entries in the payoff matrix are the expected value or weighted average for Player 1 of all the possible outcomes.

Game 1

Rules. Players 1 and 2 ante an amount $a > 0$ into the pot. One card is dealt to P_1 from a deck containing an equal number of "high" and "low" cards. After looking at his card, P_1 can either "pass" or "raise." If P_1 passes, P_1 wins the pot if he had been dealt a high card and loses the pot if he had been dealt a low card. If P_1 raises, he adds an amount $b > 0$ to the pot, and then Player 2 has two options. P_2 can either "fold" or "call." If P_2 folds, P_1 wins the pot (without revealing his hand). If P_2 calls, he also adds the amount b to the pot, and then P_1 wins or loses the pot if he has been dealt a high or low card, respectively.

Thus if P_1 is dealt a high card, he is guaranteed of winning at least the amount a from P_2, and he can win $a + b$ if he raises and P_2 elects to call. On the other hand, if P_1 is dealt a low card, he can pass and lose the amount a, or he can raise (i.e., bluff), with the hope that P_2 will assume that P_1 has a high card and elect to fold. If P_2 folds, P_1 wins a even though he has a low card. However, if P_2 suspects that P_1 does not have a high card but is bluffiing, P_2 can call, and now P_1 loses not a but $a + b$.

In order to apply our theory to this game, we first must list all the possible strategies for the two players. Recall that a strategy is a rule that tells a player what to do in any possible situation that he may find himself in the game. Consider P_1's situation. He is dealt either a high or low card, and so a strategy for him must tell him what to do in either case. Thus a possible strategy for P_1 would be to raise if he has a high card and pass if he has a low card. We will denote this strategy by (R,P), where the first component of the ordered pair directs P_1 if he has a high card, and the second component if he has a low card, with the letter "R" corresponding to raising and "P" to passing. It can be seen that P_1 has four possible strategies, denoted by (R,R), (R,P), (P,R), and (P,P). Thus (P,P) directs P_1 to pass no matter what he has been dealt, and (P,R) directs him to pass on a high card and raise on a low card. The strategy set for P_2 is simpler as he is unaware of the card dealt P_1. If P_1 elects to pass, P_2 has no options and the game is terminated. If P_1 elects to raise, P_2 can either call or fold, and so a strategy for P_2 is a rule that directs P_2's response to a raise by P_1. Denote these two possible strategies by call and fold.

Thus the payoff matrix associated with this game is 4×2. Since the outcome of the game is contingent not only on the strategies employed by the players but also on the card dealt P_1, the entries of the matrix are found by computing the expected value of the outcome to P_1 for the eight possible strategy pairs. For example, suppose P_1 uses strategy (P,R), and P_2 uses strategy call. Then if P_1 is dealt a high card, he passes and wins a, and if he is dealt a low card, he raises and P_2 calls, and so P_1 loses $a + b$. Since we are assuming that the deck contains an equal number of high and low cards and that the card dealt P_1 is randomly chosen, the probabilities of both a high card hand and a low card hand are $\frac{1}{2}$. Thus the expected value of the outcome corresponding to this strategy pair is $a/2 + (-a - b)/2 = -b/2$. Similarly, if P_1 uses strategy (R,P) and P_2 uses strategy fold, on a high card deal P_1 raises, P_2 folds, and P_1 wins a, and on a low card deal, P_1 passes and loses a. The expected outcome is therefore $a/2 + (-a)/2 = 0$.

The other six entries in the payoff matrix can be similarly computed. The result is:

	Call	*Fold*
(R,R)	0	a
(R,P)	$b/2$	0
(P,R)	$-b/2$	a
(P,P)	0	0

With this the translation of the game into a matrix game is complete, and we can apply the theory as developed in this chapter. Of course, we must assume that the decisions of the players are governed by the two principles set out in Section 8.2 in order for our notion of a solution to the game to be meaningful and applicable. Is this reasonable?

An initial inspection of the above game matrix indicates that the last two rows are dominated by the first row and so can be deleted without affecting the solution. Notice that the two associated strategies correspond to P_1 passing if dealt a high card. Actually, we could have reasoned to the ineffectiveness of such an action by P_1 by just considering the rules of the game. If P_1 is dealt a high card, he is guaranteed of winning at least the amount a regardless of P_2's play, and could win $a + b$ if he raises and P_2 elects to call. Thus P_1, if dealt a high card, has nothing to lose and something to gain by raising, and so should always raise in this case.

The game matrix is thus reduced to the 2×2 matrix.

	Call	*Fold*
(R,R)	0	a
(R,P)	$b/2$	0

The first row corresponds to P_1 raising with a low card (i.e., bluffing) and the second row to passing with a low card (i.e., conservative play). Similarly, the first column corresponds to P_2 challenging a raise and the second to P_2 playing conservatively. Since a and b are both positive, the solution to this game can be computed using the formulas of Theorem 2 of Section 8.6. The value v of the game and optimal strategies X_0 and Y_0 are given by

$$v = \frac{ab}{2a + b}$$

$$X_0 = \frac{1}{2a + b}(b, 2a)$$

$$Y_0 = \frac{1}{2a + b}(2a, b)$$

Hence, no matter what a and b are, the game is advantageous to P_1. Both players should mix bluffing with conservative play, with the amounts of each in reverse order. For example, if $a = 1$ and $b = 1$, the value of the game is $\frac{1}{3}$, and to realize this value P_1 must bluff $\frac{1}{3}$ of the time and P_2 must call $\frac{2}{3}$ of the time. If $a = 1$ and $b = 2$, the value of the game is $\frac{1}{2}$ and both players should play conservatively $\frac{1}{2}$ of the time.

Game 2

Rules. The ante, the deal, and the options for P_1 are just as in Game 1, and if P_1 elects to raise, P_2's options are also as in Game 1. However, if P_1 elects to pass, the play is not terminated, but P_2 can either pass to raise. If P_2 elects to pass, play is terminated and the pot distributed as before. If P_2 elects to raise, both he and P_1 add an amount b to the pot, and then the pot is distributed as before.

Thus, the only difference between this game and the first game is that if P_1 chooses to pass, P_2 now has the opportunity to increase the pot to $a + b$, and P_1 cannot withdraw if P_2 so wishes to increase the pot. P_1's strategy set remains the same, but now a strategy for P_2 must direct his response to both possible plays of P_1. Thus a possible strategy for P_2 would be to fold if P_1 raises and raise if P_1 passes. We will denote this strategy by (F,R), where the first component of the ordered pair is P_2's response to a raise by P_1, and the second component his response to a pass, with the obvious abbreviations for fold, call, pass, and raise. P_2 has four pure strategies, denoted by (C,R), (C,P), (F,R), and (F,P). For example, the strategy (C,P) directs P_2 to call if P_1 raises and pass if P_1 passes.

For this game the payoff matrix is 4×4. The entries again are expected values and are computed just as before. The following tableau results.

	(C,R)	(C,P)	(F,R)	(F,P)
(R,R)	0	0	a	a
(R,P)	0	$b/2$	$-b/2$	0
(P,R)	0	$-b/2$	$(2a + b)/2$	a
(P,P)	0	0	0	0

While seeming to offer more variety, this game is not as interesting as the first game; the payoff matrix has saddle points, the first and fourth entries of the first column. The game has value zero, and optimal strategies for P_1 are either to raise all the time or pass all the time. The optimal strategy for P_2 is to call if P_1 raises and raise if P_1 passes. Notice that this strategy for P_2 insures that the game is always played for the amount $a + b$ and that the winner is determined by the deal, and so P_1's threat of bluffiing is effectively nullified.

Game 3

Rules. This game is played just as Game 2, with one minor difference. If P_1 elects to pass and P_2 elects to raise, both players must add the amount $2b$ to the pot before it is distributed in the usual manner.

Thus, if P_1 raises and P_2 calls, the stakes are $a + b$; but if P_1 passes and P_2 raises, the stakes are $a + 2b$. It is not at all obvious how P_1 should proceed if he is dealt a high card. If he raises he wins either a or $a + b$, depending on P_2's actions; if he passes, he wins at least a and if P_2 raises he wins $a + 2b$. But if P_2 anticipates P_1 to always pass if he has a high card, then a raise by P_1 indicates that he has a low card and in such circumstances a raise by P_2 will net P_2 the amount $a + 2b$. Thus, in order to still be able to successfully bluff with a low card by raising on a low card hand, it seems that P_1 must occasionally raise on a high card hand.

In order to develop a more precise analysis of this game, we need to consider the payoff matrix. The strategy sets for both players are identical to those in Game 2, and the resulting 4×4 payoff matrix this time is given by:

	(C,R)	(C,P)	(F,R)	(F,P)
(R,R)	0	0	a	a
(R,P)	$-b/2$	$b/2$	$-b$	0
(P,R)	$b/2$	$-b/2$	$a + b$	a
(P,P)	0	0	0	0

As can be seen, there are no saddle points, but the fourth row is dominated by the first. However, no other simplifications are possible unless the ratio of a to b is known (and then, the only case of domination occurs between the first and last columns if $b/2 \leq a$). Thus we have a game with essentially a 3×4 payoff matrix to evaluate. However, an analysis is not difficult in this case. Since all the entries in the first row are either 0 or a, P_1's security level v_1 and therefore the value of the game is at least zero. Suppose now P_2 restricts his choice of mixed strategy to only those strategies involving the pure strategies (C,R) and (C,P), that is, the two strategies corresponding to the first two columns of the payoff matrix. P_2 would then be forcing P_1 to play the game with the 3×2 payoff matrix,

$$\begin{bmatrix} 0 & 0 \\ -\frac{b}{2} & \frac{b}{2} \\ \frac{b}{2} & -\frac{b}{2} \end{bmatrix}$$

But as can be easily seen the value of this game is zero, and so P_2's security level v_2 and therefore the value of the original game is at most zero. Hence the game has in fact value zero, and optimal strategies can be determined by considering the above 3×2 game. For example, an optimal strategy for P_2 would be $(\frac{1}{2},\frac{1}{2},0,0)$, and optimal strategies for P_1 would be either (1,0,0,0) or $(0,\frac{1}{2},\frac{1}{2},0)$ regardless of the value of b.

Game 4

Rules. This game is played as Game 2, with the only difference being in the amounts bet. If P_1 elects to raise, he must add in this game $2b$ to the pot, and if P_2 elects to call, he adds only b to the pot. Similarly, if P_1 elects to pass and P_2 elects to raise, P_2 adds here $2b$ to the pot and P_1 only b.

Thus for this game the player who wishes to increase the stakes must risk $2b$ and his opponent only b. Otherwise Game 4 is identical to Game 2, with P_2 having the option of folding if P_1 raises; but if P_1 passes and P_2 raises, P_1 cannot withdraw (but only adds b to the pot, whereas P_2 adds $2b$).

The strategy sets for the two players are the same as those in Games 2 and 3. The 4×4 payoff matrix, as can be easily computed, is:

	(C,R)	(C,P)	(F,R)	(F,P)
(R,R)	$-b/2$	$-b/2$	a	a
(R,P)	0	$b/2$	$-b/2$	0
(P,R)	0	$-b$	$a+b$	a
(P,P)	$b/2$	0	$b/2$	0

This matrix has no saddle points, and there is no domination regardless of the ratio of a to b. The last row of the matrix indicates that the value of the game is at least zero, but furthermore elementary analysis does not seem possible in this case.

However, the simplex method can be used to solve this game, even in this general form with arbitrary positive values for a and b. See Problem 1. The value and optimal strategies for the game are given by

$$v = \frac{ab}{c}$$

$$X_0 = \frac{1}{c}(b, 2a + b, 0, 2a + b)$$

$$Y_0 = \frac{1}{c}(2a, 2a, 0, 3b)$$

where $c = 4a + 3b$.

Thus the value of the game is always positive and therefore the game is favorable to P_1. For example, if $a = 2$ and $b = 1$, $v = \frac{2}{11}$, $X_0 = (\frac{1}{11},\frac{5}{11},0,\frac{5}{11})$, and $Y_0 = (\frac{4}{11},\frac{4}{11},0,\frac{3}{11})$. In this case, then, in order to attain his maximum security level of $\frac{2}{11}$, P_1 should use strategy (R,R) with probability $\frac{1}{11}$, and strategies (R,P) and (P,P) both with probability $\frac{5}{11}$. Hence P_1 should raise with a high card deal $\frac{6}{11}$ of the time and should raise with a low card deal only $\frac{1}{11}$ of the time. Similarly, to keep his security level down to $\frac{2}{11}$, P_2 should respond to a raise by P_1 by calling with probability $\frac{8}{11}$ and folding with probability $\frac{3}{11}$, and should respond to a pass by P_1 by raising with probability $\frac{4}{11}$ and passing with probability $\frac{7}{11}$.

PROBLEM SET 8.7

1. The following refers to Game 4.
(a) Use the simplex method to solve the game when $a = 1$ and $b = 2$.
(b) As in part a, with $a = 2$ and $b = 1$.
(c) Do these two operations suggest a sequence of pivot operations that could be used in the general case (arbitrary a and b)? If they do, try them. (If not, try pivoting for the first step in the first row, fourth column; second step—second column; third step—first column.)

2. Actually, once a pair of strategies is suspected of being optimal, we need not use the methods of Section 8.6 for verification. Prove that the strategies given for Game 4 are optimal by using Theorem 2 of Section 8.5.

3. Provide interpretations for P_1 and P_2 of their optimal strategies in Game 4 if $a = 1$ and $b = 2$.

4. In the standard game of poker, the probability of actually getting a good hand is low. Analyze Game 1 under the assumption that the probability of P_1 being dealt a high card is only $\frac{1}{3}$. [The value of the game is $(a/3)$ $[(b - 2a)/(2a + b)]$.] Intuitively, why does this game become more favorable to P_2 if a is increased and b fixed?

5. In Game 1 suppose $a = b = 1$, and that the probability of P_1 drawing a high card is p. Determine p so that the game is fair.

6. Analyze the following variation of Game 1. Initially one card is dealt to each player. Assume that each of the four possible deals (two high cards dealt; two low cards dealt; high card to P_1 and low to P_2; low to P_1 and high to P_2) are equiprobable. The options for the players and the betting are as in Game 1. However, the pot is won by the player with the higher card and, in the event of a tie, the pot is distributed equally. (Note that now a strategy for P_2 must also include consideration of the card in P_2's hand.)

7. Consider the following card game. Initially both players ante an amount a into the pot. From a three-card deck consisting of a king, queen, and jack, each player is dealt one card. After looking at his card, P_1 can either raise or pass. If he passes, the player with the higher card wins the pot. If P_1 raises, he adds an amount b to the pot, and then P_2, after looking at his card, can either call or fold. If he folds, P_1 wins the pot. If P_2 calls, he adds b to the pot, and then the player with the higher card wins the pot.

(a) Show that there are six possible deals.

(b) A strategy for P_1 must instruct him to either raise or pass in the event that he has either a king, queen, or jack. Show that P_1 has eight pure strategies.

(c) Similarly show that P_2 has eight pure strategies.

(d) Before constructing the game matrix, consider domination. Convince yourself that P_1 has nothing to gain by passing when he has a king, and P_2 has nothing to gain by folding when he has a king. Thus the payoff matrix is essentially reduced to a 4×4 matrix.

(c) Convince yourself that P_2 has nothing to gain by calling when he has a jack, and that then, after these considerations, P_1 has nothing to gain by raising when he has a queen.

(f) Notice that the viable pure strategies for each player have been reduced to 2: P_1 raises on a king, passes on a queen, and either raises or passes on a jack; P_2, in response to a raise by P_1, calls on a king, folds on a jack, and either calls or folds on a queen. Compute the associated 2×2 payoff matrix.

(g) Solve the game.

8. Analyze the following. The only two clothing stores in a shopping center compete for the weekend trade. On a clear day the larger store gets 60% of the business, and on a rainy day, the larger store, being closer to the parking lot, gets 80% of the business. However, either or both retailers may hold a "sidewalk sale" for any given weekend, but the decision to hold such must be made a week in advance and in ignorance of the competitor's plans. If both retailers conduct sidewalk sales, the breakdown in business is just as above. If, however, one holds the sale and the other does not, the one conducting the sale gets 90% of the business on a clear day and 10% on a rainy day. During the present season, it rains 40% of the time. How frequently should the larger and the smaller retailers conduct sales?

9. Analyze the following two-stage games.

(a) Two players play Game 1, with the exception that if P_1 elects to pass, he returns his card to the deck, no payments or additional ante are made, and Game 1 as originally described is played, but with the roles of P_1 and P_2 interchanged. Thus if P_1 passes, the initial ante remains at a for each player, but now P_2 draws the card and makes the first move.

(b) As above, but suppose that now the deck contains only four cards, two high cards and two low cards, and that if P_1 elects to pass, he shows the card he has drawn but does not return it to the deck before P_2 draws his card.

10. Can the following variations of Game 1 be analyzed using the theory of this chapter?

(a) Game 1 played with a deck consisting of $p\%$ high cards, but with p unknown to both players.

(b) As in part a, but with p known by only one player.

(c) P_2 is given a deck of four cards, two high cards and two low cards. He extracts a card of his choice, without revealing his selection to P_1. Then Game 1 is played with this modified deck of the remaining three cards.

9
Other Topics in Game Theory

SECTION 9.1 UTILITY THEORY

One of the basic assumptions that we have made in the study of game theory is that each of the possible outcomes of a game can be assigned a numerical value that represents the value or worth for a particular player of that outcome over the other possible outcomes. However, it is not immediately obvious that this is always possible, as was pointed out in Section 8.2. Even for parlor games in which the payoffs are in terms of money, it may be that these monetary payoffs cannot be used directly to measure a player's preferences. For example, one may derive much more satisfaction in a game of poker from winning $2 by bluffing an opponent as opposed to winning $5 with a hand of four aces. Or, would not the significance of a $5 loss be different for a player already up by $15 on the night as opposed to a player already down by $15? These difficulties are not insurmountable, at least theoretically. The body of knowledge developed to deal with this problem is called *utility theory*. In this section, the only one in this book concerned with the topic, we will provide only a brief introduction into the theory. What we propose to do is to indicate how one might go about assigning appropriate values to three different outcomes, regardless of their nature. The intuitive ideas that we develop form the foundation of utility theory.

Let us denote the three possible outcomes or events by the letters A, B, and C. We want to assign numerical values, or utilities, to each of these events that will in some way represent their relative desirability for an individual. Denote these numbers to be assigned by $u(A)$, $u(B)$, and $u(C)$. Our first step is to order the events linearly, that is, to determine the order of preference between the events. It could be that we are indifferent to two of the events. For example, we may not be able to make any distinction between A and B in that we feel as though we would derive the same amount of satisfaction from either. In that case $u(A)$ should equal $u(B)$, and our problem would reduce to the problem of assigning values to only two distinct events. Thus we assume that the events can be strictly ordered, and that A is preferred over B, and B is preferred over C. (We therefore assume that A is preferred over C because, intuitively, and ordering of events by preference must be transitive.) This ordering demands that $u(A) > u(B) > u(C)$. Our next task is to determine how our preference of A over B compares with our preference of B over C.

The key idea used to make this comparison is a lottery. Consider another event, a lottery, in which there are two possible outcomes, A and C. Suppose the lottery will result in A with probability r and C with probability $1 - r$. For example, suppose the circumference of a wheel is divided into two arcs, one arc of length the fraction r of the entire circumference, and a pointer located at the center of the wheel spun. If the pointer comes to rest in the arc of length r of the whole, A occurs; otherwise, C occurs. This lottery is an event, and we can also assign to it a numerical value that measures its desirability. This value would depend on r. If r is near 0, the outcome of the lottery would more likely be C, and so the value of the lottery would be closer to $u(C)$ then to $u(A)$. And as r increases to 1, this value would approach $u(A)$.

Now the desirability of B lies somewhere between the desirability of C and of A. It seems reasonable to assume that there exists a particular r, $0 < r < 1$, such that we are indifferent to the events B and the lottery with outcomes A and C, A occurring with probability r and C with probability $1 - r$. Let us denote this particular lottery by the symbol $rA + (1 - r)C$.

Since we are indifferent to these two events, the utility of B, $u(B)$, should equal the utility of the lottery, denoted by $u(rA + (1 - r)C)$. We now make another basic assumption concerning the assignment of utilities, in this case, concerning the assignment of utilities to lotteries. Analogous to the definition of expected value in probability theory, as discussed in Section 8.4, it is reasonable to assume that the desirability of the lottery $rA + (1 - r)C$ should be given by $ru(A) + (1 - r)u(C)$, since this can be considered to be the "expected utility value" of the lottery. Thus, to determine $u(B)$, all we need determine is $u(A)$, $u(C)$, and the above r, and then set

$$u(B) = ru(A) + (1 - r)u(C).$$

Consider now the assignment of the values $u(A)$ and $u(C)$. We have seen that the crucial quantity to be measured is how our preference of A over B compares with our preference of B over C. However, these preferences are measured by the differences $u(A) - u(B)$ and $u(B) - u(C)$. This suggests that, unless otherwise restricted, arbitrary values can be assigned to $u(A)$ and $u(C)$, as long as $u(A) > u(C)$. Once these assignments are made, the lottery system described above can be used to determine $u(B)$.

The phrase "unless otherwise restricted" deserves some elaboration. By this we mean that our three events A, B, and C exist by themselves and cannot be compared with any other events or standards, and so we can freely assign the numbers $u(A)$ and $u(C)$. However, in many situations, the values of the outcomes are compared, consicously or subconsciously, to some external standards. For example, if we are involved in a game in which we will either win \$5 (event A) or lose \$5 (event C), the demands of the obvious preferential ordering are satisfied by any assignment of $u(A)$ and $u(C)$ such that $u(A) > u(C)$. In this case, we naturally compare A and C with the standard of not winning or losing anything (event B), intuitively setting $u(B) = 0$. Everything else being equal, we would want $u(A) > 0$ and $u(C) = -u(A)$. Moreover, if the value of the game is to be directly translated into dollars, we would want $u(A) = 5$.

In summary, if we have three events, with no external standards to be imposed, we can assign utilities to these events using the described method. If there are more than three, it should be clear how this method can be extended. Once two events are assigned utilities, the lottery method can be applied to each of the remaining events separately until appropriate utilities have been assigned to all events. (It is not necessary initially to single out the two events at the extremes of the linear preferential ordering. See Problem 1.) In case there are also external standards to be imposed on our system, these events and their associated utilities would simply be added to the set of possible outcomes, and those events with preassigned utilities would provide the starting point.

Although our above discussion has been more suggestive that axiomatic, utility theory can be developed rigorously from a system of axioms. Refer to the books of von Neumann and Morgenstern [33], Luce and Raiffa [23], or Owen [28].

Example

You are involved in a game of chess with Bobby Fischer. There are three possible outcomes, you win(?), event A, you draw, event B, or you lose, event C. Setting $u(A) = 1$ and $u(C) = -1$, what should $u(B)$ be? Because of the abilities of your opponent, much satisfaction would be gained from a draw, and so clearly $u(B)$ should be positive. More precisely, suppose you feel equally disposed to a draw and a lottery in which you have a probability of $\frac{19}{20}$ of being accorded a victory over Fischer and a probability of $\frac{1}{20}$ of being accorded a loss. Then

$$\begin{aligned} u(B) &= u[\tfrac{19}{20}A + \tfrac{1}{20}C] \\ &= \tfrac{19}{20}u(A) + \tfrac{1}{20}u(C) \\ &= \tfrac{19}{20} - \tfrac{1}{20} = \tfrac{9}{10} \end{aligned}$$

PROBLEM SET 9.1

1. Utilities are to be assigned to four events A, B, C, and D. A is preferred over B, B over C, and C over D. $u(B)$ is to be set equal to 1 and $u(C)$ set equal to 0, and it is determined that B is indifferent to the lottery $rA + (1 - r)C$, and C is indifferent to the lottery $sB + (1 - s)D$. Compute $u(A)$ and $u(D)$.

2. Let A be the event that you are given a dollar, B the event that your status quo is maintained, and C the event that you are elected president of the student body. Set $u(A) = 1, u(B) = 0$, and determine your own personal utility for C.

SECTION 9.2 TWO-PERSON, NON-ZERO-SUM GAMES

In the next four sections we will discuss two-person, non-zero-sum games, that is, two-person games for which the sum of the payoffs to the two players for the various possible outcomes of the game is not necessarily always zero. We have seen examples of such games in Section 1.3. Our primary purpose in studying these games is to demonstrate some of the difficulties arising when attempting to develop mathematical models of more complex situations. Indeed, although non-zero-sum games reflect the type of situation encountered much more frequently in real-world applications, the formulation of a well-accepted, all-encompassing mathematical model is, as we will see, not at all straightforward.

For non-zero-sum games, the payoff for one player is not necessarily the negative of the payoff for the other player. Thus, in order to express the payoffs for such games, ordered pairs will be used, where the first component represents the payoff to Player 1 and the second the payoff to Player 2.

Game 1

In the first game described in Section 1.3, both players had two pure strategies and the payoff tableau was given by

	t_1	t_2
s_1	(0,0)	(12, −12)
s_2	(−12,12)	(6,6)

Thus if P_1 uses s_2 and P_2 uses t_1, P_1 would lose -12 units and P_2 would gain 12 units. Note that the (6,6) payoff is the only non-zero-sum payoff.

Zero-sum games are strictly competitive—what one gains the other loses. However, this is not the case for non-zero-sum games. For these games both players may be able to insure for themselves an advantage by cooperating with the other player. Thus the possibility of preplay communication and cooperation adds a new dimension to the study of non-zero-sum games. For example, consider Game 2.

Game 2

	t_1	t_2
s_1	(0,0)	(1,1)
s_2	(1,1)	(0,0)

In Game 2, there is no reason for either player to choose one strategy over the other unless, of course, they can communicate beforehand. Permitted such communication, the players would coordinate their strategies to assure a (1,1) payoff for themselves.

Repeated playing of the same game may achieve the same effect as pregame communication. Certainly, if two people played Game 2 100 times without communication, after several turns a pattern would be established providing a constant (1,1) payoff.

However, cooperation, achieved either through preplay communication or repeated play, does not begin to resolve the difficulties inherent in non-zero-sum games. Consider Game 3.

Game 3

	t_1	t_2
s_1	(10,1)	(2,2)
s_2	(2,2)	(1,10)

In Game 3, P_1 would prefer the (10,1) payoff and P_2 the (1,10) payoff. Following these preferences, P_1 would play s_1 and P_2 would play t_2, resulting in a (2,2) payoff. It is not clear how preplay communication could resolve these conflicting preferences. However, if the game were to be played twice, the players might agree on using strategy pair (s_1,t_1) for the first game and (s_2,t_2) for the second. In this way, both players would gain 11 units. But, after the game is played once with outcome (10,1), what is to prevent P_1 from disregarding the agreement and playing s_1 again, gaining at least a total of 12 units?

Game 1 provides another example of the problems arising from incoercible agreements. In that game, as a result of preplay discussion, both players may agree to the strategy pair (s_2,t_2) with payoff (6,6). But if either player expects the other to abide by this agreement, he can gain 12 units by breaking his part of the bargain and using his first strategy. In fact, notice that strategy s_1 dominates s_2, and t_1 dominates

t_2 (i.e., $0 > -12$ and $12 > 6$). Thus, no matter what the opponent does, each player has more to gain for himself by using his first strategy. The concept of dominance, so reasonable for zero-sum games, leads to the somewhat unreasonable outcome of (0,0) in this case. In general, dominance plays a minor role in non-zero-sum theory. (If you are not yet convinced, consider the game in Problem 1.)

The above examples show that preplay communication can lead to other problems, but in some cases it may not even be desirable. Consider Game 4.

Game 4

	t_1	t_2
s_1	(1,10)	(10,1)
s_2	(0, – 10)	(0, – 9)

In this game P_2 would prefer the (1,10) outcome and P_1 the (10,1) outcome. However, without preplay discussion, P_1 has no reason to use strategy s_2, and so the game would probably result in the (1,10) payoff following the use of the strategy pair (s_1,t_1). However, if the players can communicate, P_1 could demand that P_2 use strategy t_2, threatening the use of s_2 if P_2 does not agree to his demands. Thus, with preplay discussion P_1 can attempt to force P_2 to the (10,1) outcome, and so the notion of a threat becomes a component in the theory of cooperative games.

Another factor for consideration in non-zero-sum games is whether or not utility can be transferred; that is, can one player make a side payment to his opponent after the game is played? Consider Game 5.

Game 5

	t_1	t_2
s_1	(50,0)	(1,5)
s_2	(1,0)	(1,5)

Here, if side payments are not possible, P_2 has no reason to do anything but play t_2, leading to the (1,5) outcome. However, if the players can cooperate and transfer utility, P_1 could offer to share his 50 units in some way with P_2 in return for P_2 using t_1 (and P_1 using s_1). Of course, this assumes that the utility is divisible and transferable and that the utility scales are comparable.

In fact, the comparison of utilities between the players raises various questions that can be critical in the resolution of a game. Does one player derive as much "satisfaction" from a gain of, say, 25 units as the other player does? Is one player concerned not only with his own payoff but also with that of the other player, attempting maybe to make it as small as possible or, on the other hand, to insure that it remain above a certain level? Are negative payoffs always undesirable and positive payoffs always desirable? In any application of the theory, these questions would have to be considered. For the remainder of our discussion of non-zero-sum games, we will assume that the payoffs represent monetary units, say dollars, or $10, and so on, to players in equivalent financial positions.

Given the many variety of difficulties associated with non-zero-sum games, how can we attempt to develop a mathematical model of these games that in some way reflects rational behavior (whatever that is)? To begin, any model needs a precise starting point. Thus, the conditions under which a game is to be played must be made precise. This suggests that for non-zero-sum games, various cases, such as cooperative games versus noncooperative games, be considered separately. Moreover, mathematicians are inclined to study the extreme cases first since, on the one hand, these cases may be the most susceptible to a reasonable theory and, on the other hand, descriptions of the extreme cases may provide insights applicable to the intermediate cases. The study of non-zero-sum games contains thus two major divisions: the study of noncooperative games in which no preplay communication is permitted and the study of cooperative games in which preplay communication and binding agreements are permitted. In the next section noncooperative games will be considered briefly, and in Sections 9.4 and 9.5 an introduction to cooperative games will be presented.

PROBLEM SET 9.2

1. In the following row and column dominance leads to what outcome? Is this outcome reasonable?

	t_1	t_2
s_1	$(-99,-99)$	$(100,-100)$
s_2	$(-100,100)$	$(99,99)$

2. The following game appears at first glance to be symmetric. Show that this is not the case, and that one of the players has a more favorable position.

	t_1	t_2
s_1	$(0,5)$	$(-1,1)$
s_2	$(1,-1)$	$(5,0)$

SECTION 9.3 NONCOOPERATIVE TWO-PERSON GAMES

In this section we assume that no form of communication or cooperation is permitted between the players of the non-zero-sum game. Situations such as a business competition between bitter rivals or a market competition between large companies restricted by antitrust legislation may fall into this category.

In zero-sum games the element of communication and cooperation is not present. In this respect zero-sum and noncooperative non-zero-sum games are similar, and so it would seem that a reasonable starting point for the study of non-zero-sum games would be the two principles set forth in Section 8.2. These two principles, concerning the maximization of security levels and the tendency to equilibrium strategies, provided the foundation for a complete theory for zero-sum games. In this section we will consider the role of security levels and equilibrium strategies in noncooperative non-zero-sum games. However, we will see that in this instance these two principles alone are incapable of leading to a complete theory.

In order to define equilibrium strategies and security levels for non-zero-sum games, it is easiest first to single out the two payoff matrices. For a non-zero-sum game with m pure strategies for P_1 and n pure strategies for P_2, let A be that $m \times n$ matrix with entries equal to the payoffs to P_1, and let B be the corresponding $m \times n$ matrix of payoffs for P_2. As before, by a (mixed) strategy for P_1 (or P_2), we mean an m-component (or n-component) vector with nonnegative coordinates, the sum of which is 1. Again, we will use S and T to denote the sets of strategies for P_1 and P_2, respectively.

A strategy pair is said to be in equilibrium if neither player can gain by deviating from his prescribed strategy as long as his opponent's strategy remains fixed. Formally, for non-zero-sum games, we have the following definition.

Definition The strategy pair (X_0, Y_0) is *in equilibrium* if, for all $X \in S$ and $Y \in T$,

$$XAY_0^t \leq X_0AY_0^t \qquad \text{and} \qquad X_0BY^t \leq X_0BY_0^t$$

To determine the players security levels, we can use the theory of zero-sum games. Consider the zero-sum game with payoff matrix A. The value of this game is the maximum amount P_1 can guarantee himself regardless of the play of P_2 since, as we have seen in the last chapter, the value of this game is equal to P_1's optimal security level v_1, as defined in Section 8.4. And an optimal strategy for the first player of the zero-sum matrix game A would provide P_1 with a *security level strategy*, that is, a strategy that will enable P_1 to realize his security level.

To define P_2's security level, a minor adjustment in the above process must first be made. The matrix B represents the payoffs to P_2, the column player in the non-zero-sum game. Thus, in general, P_2 seeks the larger entries and not the smaller entries, as does the column player in a zero-sum game. P_2's position is identical to the row player of the zero-sum game with matrix B^t, since the transpose operation simply interchanges the rows and columns. Hence P_2's security level is defined in terms of the matrix game B^t.

Definition P_1*'s security level* is the value of the zero-sum matrix game A, and an optimal strategy for the row player of that game provides P_1 with a *security level strategy*.

Similarly, P_2's *security level* is the value of the zero-sum matrix game B^t, and an optimal strategy for the row player of that game provides P_2 with a *security level strategy*.

Example

Consider Game 4 of the last section. The payoff tableau was

	t_1	t_2
s_1	(1,10)	(10,1)
s_2	(0,−10)	(0,−9)

For this game

$$A = \begin{bmatrix} 1 & 10 \\ 0 & 0 \end{bmatrix}, \quad B = \begin{bmatrix} 10 & 1 \\ -10 & -9 \end{bmatrix}, \quad \text{and} \quad B^t = \begin{bmatrix} 10 & -10 \\ 1 & -9 \end{bmatrix}$$

Matrix A has a saddle point at the 1 entry. Therefore P_1's security level is 1, and the associated security level strategy is $X_0 = (1,0)$. Matrix B^t has a saddle point at -9, and so P_2's security level is -9 with the associated security level strategy $Y_0 = (0,1)$, the optimal strategy for the row player of the matrix game B^t. However, this pair (X_0, Y_0) is not in equilibrium; if P_1 uses s_1, P_2 maximizes his gain by using only t_1. That is, if $Y_1 = (1,0)$,

$$10 = X_0 B Y_1{}' \not\leq X_0 B Y_0{}' = 1$$

Moreover, the first row of A dominates the second, and so it follows that any equilibrium strategy pair must have P_1's strategy equal to X_0. But P_2's best response to this is Y_1. Thus (X_0, Y_1) is the only strategy pair in equilibrium. Notice that the payoff (1,10) associated with this pair maintains P_1's security level, provides P_2 with 19 units more than his security level, and seems to be a reasonable resolution of the game if played without communication and cooperation.

In this game, the pair of security level strategies is not in equilibrium, as contrasted with the situation in zero-sum games (see the corollary to Theorem 1 of Section 8.5). So already differences in the results of the application of the two basic principles begin to appear. For the remainder of this seciton, we will consider games in which these differences inhibit the resolution of the games.

Game 6

	t_1	t_2
s_1	(10,1)	(0,0)
s_2	(0,0)	(1,10)

For this game both the strategy pairs ((1,0),(1,0)) and ((0,1),(0,1)) are in equilibrium, as can be easily seen. However, the payoffs associated with these two equilibrium pairs are quite distinct, with P_1 benefiting more from the (10,1) payoff and P_2 from the (1,10) payoff. Thus we come to the major problem of equilibrium pairs: different pairs can provide different payoffs. (Again, this is in contrast to the zero-sum case. See Theorem 2 of Section 8.5.)

This problem cannot be resolved in general. Certainly in Game 6, if we know nothing about the nature of the players themselves, there is no reason to choose one of these equilibrium pairs over the other. P_1's and P_2's security level strategies are $(\frac{1}{11},\frac{10}{11})$ and $(\frac{10}{11},\frac{1}{11})$, respectively, with both players having as security level $\frac{10}{11}$. However, this pair of strategies is not in equilibrium; if P_1 uses s_2 with frequency $\frac{10}{11}$, P_2 would be better off by using t_2 more frequently than $\frac{1}{11}$ of the time, and conversely. On the other hand the strategy pair $((\frac{10}{11},\frac{1}{11}),(\frac{1}{11},\frac{10}{11}))$ is in equilibrium, with the expected payoff of $(\frac{10}{11},\frac{10}{11})$. Would the players tend to this mixed strategy pair?

Notice that if the players in Game 6 could communicate and cooperate, they might agree to coordinate their strategies so that half the time the payoff is (10,1) and the other half it is (1,10), For example, they could agree to flip a coin, and if it turns up heads they both play their first strategy, and if it is tails they play their second stragegy. In this way the (0,0) payoffs would be completely avoided, and the expected outcome for each player then would be $\frac{11}{2}$, a considerable improvement over $\frac{10}{11}$.

Game 7

	t_1	t_2
s_1	(0,0)	(10, −1)
s_2	(−1,10)	(9,9)

This game is of the prisoner's dilemma type, as in Game 1 of the last section. In Game 6, there were three different pairs of equilibrium strategies, and none of these pairs corresponded to the security level strategies. However, in this game there are no such complications. Strategy s_1 dominates s_2 ($0 > -1$ and $10 > 9$) and t_1 dominates t_2. Thus ((1,0),(1,0)) is the only equilibrium pair. The matrices A and B^t associated with this game have saddle points at the 0 entries, so the strategies (1,0) and (1,0) are also the security level strategies, providing the security level of 0 for both players. Hence the two basic principles lead in this game to one unique strategy pair. However, even in this case, is this strategy pair a reasonable resolution of the game, since the (9,9) payoff offers much more to the players? We emphasize this dilemma in the next game.

Game 8

	t_1	t_2	t_3
s_1	(0,0)	(0,−1)	(9,−10)
s_2	(−1,1)	(9,9)	(8,10)

This game is quite similar to Game 7. As can be easily shown, the security level of both players is 0 with security level strategies of $X_0 = (1,0)$ and $Y_0 = (1,0,0)$. Moreover, the strategy pair (X_0, Y_0) is in equilibrium (see Problem 6). However, is not the (9,9) outcome much more likely, especially in this game?

Even if Game 8 is played without cooperation repeatedly between the same two players, it would seem that the (s_1,t_1) strategy pair would occur most infrequently, if ever. Suppose the players start by using the pair (s_2,t_2) several times. If P_2 attempts to increase his payoff by moving to t_3, after several plays of (s_2,t_3), would not P_1 move to penalize P_2 by playing s_1. The pair (s_1,t_3) would result in a -10 payoff to P_2, and so would he not immediately move to t_2 in order to reestablish as quickly as possible the mutually beneficial outcome of (9,9)?

To summarize, these examples show that different equilibrium strategy pairs may provide different payoffs, that a pair of security level strategies may not be in equilibrium, and that in some cases, a resolution of the game in terms of equilibrium pairs may not seem reasonable even when an equilibrium pair is unique and corresponds to the security level strategies. Thus, as adequate and as reasonable as they were for zero-sum games, the two basic principles of Section 8.2 are incapable of providing a complete theory for noncooperative non-zero-sum games.

Even with the above limitations, much of the existing analysis of noncooperative games employs the concept of the equilibrium pair. (See, for example, [23].) It has been shown by J. Nash [26], using the Brouwer Fixed-Point Theorem, that any game has at least one equilibrium pair of strategies (see also [23] or [28]). The role of the players' security levels, however, is not very critical in noncooperative theory. One reason for this is that the payoffs to the two players associated with any equilibrium pair of strategies is at least as great as their security levels. We close this section with a proof of this result.

Theorem *Given a two-person, non-zero-sum game with payoff matrices A and B, let u and v denote the security levels of P_1 and P_2, respectively. Let (X_0, Y_0) be any strategy pair in equilibrium. Then*

$$u \leq X_0 A Y_0^t \qquad \text{and} \qquad v \leq X_0 B Y_0^t$$

Proof Let S and T denote the strategy sets of P_1 and P_2, respectively. Then, for any $X \in S$,

$$\underset{Y \in T}{\text{Min}}\ XAY^t \leq XAY_0^{\,t} \qquad \text{and} \qquad XAY_0^{\,t} \leq X_0AY_0^{\,t}$$

Thus

$$\begin{aligned} u &= \underset{X \in S}{\text{Max}}\ \underset{Y \in T}{\text{Min}}\ XAY^t \\ &\leq \underset{X \in S}{\text{Max}}\ XAY_0^{\,t} \\ &= X_0AY_0^{\,t} \end{aligned}$$

Similarly,

$$\begin{aligned} v &= \underset{Y \in T}{\text{Max}}\ \underset{X \in S}{\text{Min}}\ YB^tX^t \\ &\leq \underset{Y \in S}{\text{Max}}\ YB^tX_0^{\,t} \\ &= \underset{Y \in S}{\text{Max}}\ X_0BY^t \\ &= X_0BY_0^{\,t} \end{aligned}$$

###

PROBLEM SET 9.3

1. Why do we not consider the possibility of cooperation in zero-sum games?

2. Prove that a strategy pair (X_0, Y_0) is in equilibrium if and only if

$$A_{(i)}Y_0^{\,t} \leq X_0AY_0^{\,t} \qquad \text{for } 1 \leq i \leq m$$

and

$$X_0B^{(j)} \leq X_0BY_0^{\,t} \qquad \text{for } 1 \leq j \leq n$$

3. Consider the game with the payoff tableau

	t_1	t_2
s_1	(0,5)	(1,1)
s_2	(−1,−1)	(5,0)

(a) Show that both players have a security level of 0.
(b) Show that $X = (1,0)$ and $Y = (0,1)$ are security level strategies. Is this pair (X,Y) in equilibrium?

(c) Find two obvious pairs of equilibrium strategies that employ pure strategies. Can you find the one other equilibrium pair that uses mixed strategies?
(d) Compare the payoffs associated with the equilibrium pairs to the players' security levels.

4. Consider the game of Problem 2 of Section 9.2. (It is identical to the above game if the 1 and -1 payoffs for P_1 are interchanged.) Show that P_1's security level is greater than P_2's security level, and that the unique equilibrium strategy pair leads to the (5,0) payoff. Thus, for this game these concepts demonstrate P_1's more favorable position.

5. Consider the game with the payoff tableau

	t_1	t_2
s_1	(2,7)	$(-5,-1)$
s_2	$(0,-2)$	(7,2)

(a) Show that P_1's position is the more favorable.
(b) Compute the security levels.
(c) Show that each of the three strategy pairs ((1,0),(1,0)), ((0,1),(0,1)), and $((\frac{1}{3},\frac{2}{3}),(\frac{6}{7},\frac{1}{7}))$ is in equilibrium. Find the associated expected payoffs.
(d) Show that there are no other equilibrium strategy pairs.
(e) Do either of these concepts distinguish P_1's more favorable position?

6. Prove that Game 8 has three strategy pairs in equilibrium. (*Hint.* Use the results of Problem 2.)

7. Do the concepts of security levels and equilibrium strategy pairs assist in the resolution of the following game?
(There are three equilibrium strategy pairs.)

	t_1	t_2
s_1	(1,2)	(0,0)
s_2	$(-2,-1)$	(4,1)

SECTION 9.4 COOPERATIVE TWO-PERSON GAMES

In this and the next section we assume that the two players of the non-zero-sum game can enter into preplay discussion and binding agreements. Conflict situations of this type could be encountered for example, in the negotiations between labor and management over a labor contract or between two countries over a trade agreement.

As pointed out in the discussion of Game 6 in the last section, the players can usually expand the set of possible payoffs of the game by cooperating. In Game 6, if the players coordinate the use of the strategy pairs (s_1,t_1) and (s_2,t_2), using each with probability $\frac{1}{2}$, the expected payoff is $(\frac{11}{2},\frac{11}{2})$. However, if each player would independently use the mixed strategy $X = Y = (\frac{1}{2},\frac{1}{2})$, the expected payoff is

$$(XAY^t,XBY^t) = (\tfrac{11}{4},\tfrac{11}{4})$$

where A and B are defined in the usual manner. Note that the independent use of these strategies would result in a (0,0) payoff with frequency $\frac{1}{2}$.

In fact, since the players are now permitted to discuss the game beforehand and correlate their play, the question of what strategy to use is subservient to the question of what mutually beneficial payoff the players can agree to. Thus we determine first the set of all payoffs possible with the use of cooperation. If the game tableau is $m \times n$, then a payoff, a point in R^2, is produced by the coordinated use of the pure strategy pairs (s_i,t_j), $1 \leq i \leq m$, $1 \leq j \leq n$. To each pair (s_i,t_j) is associated the outcome (a_{ij},b_{ij}), where the a_{ij}'s and b_{ij}'s are the entries of the payoff matrices A and B, respectively. If the players agree to use each pair (s_i,t_j) with probability or frequency r_{ij}, where $0 \leq r_{ij} \leq 1$, then the expected payoff is simply $\sum r_{ij}(a_{ij},b_{ij})$. Thus the set of all possible payoffs, denoted by M, is given by

$$M = \left\{ \sum_{\substack{1 \leq i \leq m \\ 1 \leq j \leq n}} r_{ij}(a_{ij},b_{ij}) \,\middle|\, 0 \leq r_{ij} \leq 1, \sum_{\substack{1 \leq i \leq m \\ 1 \leq j \leq n}} r_{ij} = 1 \right\}$$

This set M, which we will call the *cooperative payoff set*, is the smallest convex region containing all the points (a_{ij},b_{ij}). It can be easily determined from the set of points (a_{ij},b_{ij}), and its boundary consists of line segments with terminal points from this set.

Example 1

For Game 6, the cooperative payoff set M is the hatched region in Figure 9.1.

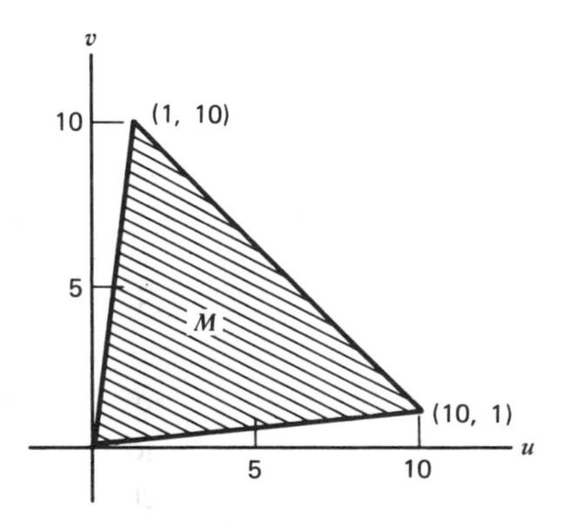

Figure 9.1

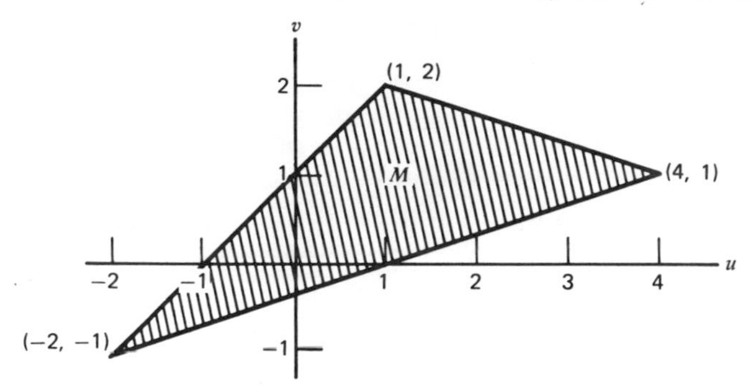

Figure 9.2

Example 2

For the game of Problem 7 of Section 9.3, the set M is sketched in Figure 9.2.

If side payments are permitted, the cooperative payoff set can be easily altered to reflect this fact. For example, suppose the utility units are infinitely divisible and comparable between the players, and that the payoff set without side payments contains points in the first quadrant. If (u',v') is such a point, then the total utility $u' + v'$ can be divided in any way between the two players, and so the set of possible payoffs would contain all the points in the first quadrant of the form (u,v), where $u + v = u' + v'$.

Example 3

For the game of Example 1, if side payments are permitted, the payoff set would be the hatched region of Figure 9.3. For the game of Example 2, if side payments are permitted, the first quadrant of the payoff region would be the set illustrated in Figure 9.4.

In either case, to any cooperative two-person game, there corresponds its cooperative payoff set M. The question we are faced with now is whether or not it is

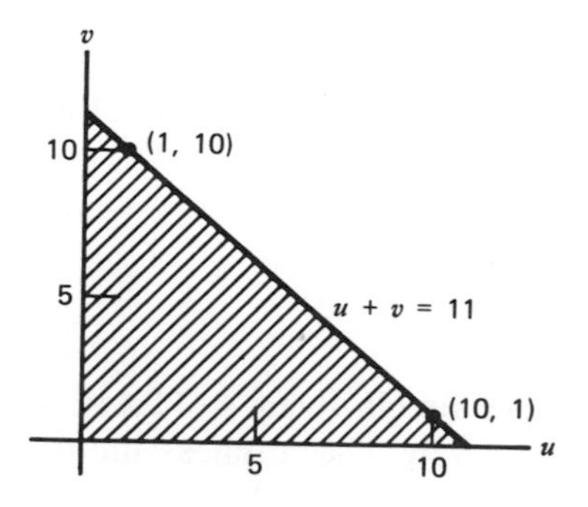

Figure 9.3

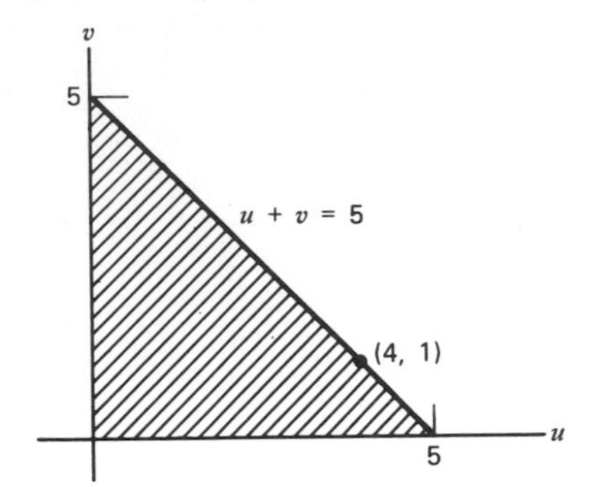

Figure 9.4

possible to develop a theory based on acceptable (to whom?) axioms that will lead to a point, or at least a subset of M, that represents a reasonable (to whom?) resolution of the game. As we have seen in Section 9.2, the concepts of security levels and equilibrium strategies, which worked so well in the zero-sum case, are incapable of producing a conclusive theory for noncooperative non-zero-sum games, and it is reasonable to assume that they again are inadequate by themselves for cooperative games, games that resemble the zero-sum games even less. Thus other principles must be introduced, and associated with such are always the problems of reasonability and acceptability. For the remainder of this section we develop one possible resolution of cooperative games, called the *negotiation set*. It is based on two very plausible principles (one concerning security levels) formulated by von Neumann and Morgenstern, but the resolution leads in general only to a subset of M and not to a unique point.

A point (u,v) of M is said to be *dominated* by the point (u',v') of M if both $u' \geq u$ and $v' \geq v$. Since the players, by acting jointly, can attain as a payoff any point in M, it is reasonable to assume that they would not find acceptable any dominated payoff. Thus we restrict our attention to only the undominated points of M.

Second, although the game is cooperative, each player knows that by using a security level strategy, he can attain at least his security level payoff regardless of the play of the other player. From this it is reasonable to conclude that P_1 would find acceptable only those payoffs (u,v) with the property that $u \geq P_1$'s security level, and similarly for P_2 and his security level.

These two principles together lead to that subset of M consisting of all undominated payoffs (u,v) of M such that $u \geq P_1$'s security level and $v \geq P_2$'s security level. This subset is called the *negotiation set*.

Example 4

Consider the game of Problem 2 of Section 9.2 (see also Problem 4 of Section 9.3.) The security levels of P_1 and P_2 are 1 and 0, respectively. The graph, assuming that

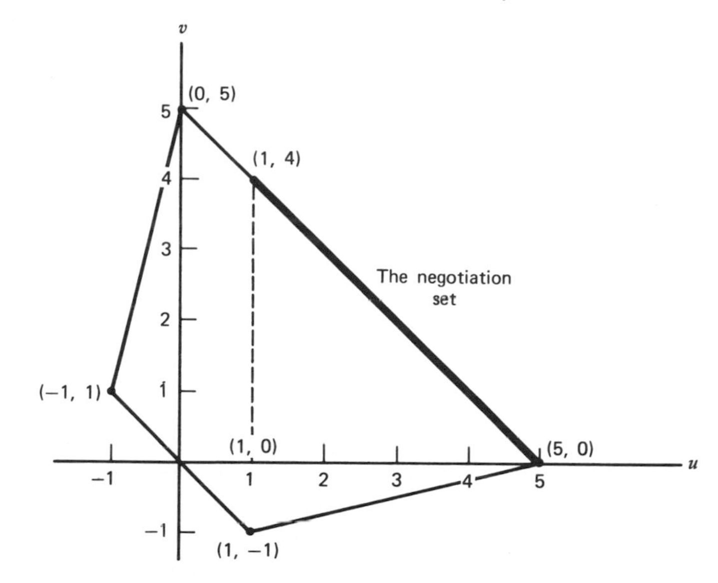

Figure 9.5

side payments are not permitted, is sketched in Figure 9.5. The set of undominated payoffs consists of the entire line segment between the points (0,5) and (5,0). The negotiation set, however, is the subset of that line between the points (1,4) and (5,0).

Now one possible resolution of cooperative two-person games is to say that the solution is simply this negotiation set. It would seem that any "just" or "fair" solution would certainly be contained in this set, but it seems also that we are possibly leaving too much to be decided by the bargaining powers of the two players. In the next section we outline procedures based on a family of axioms by J. Nash that lead to a unique point in the negotiation set for the solution of the cooperative game.

PROBLEM SET 9.4

(Assume in the following problems that side payments are not permitted.)

1. Determine the negotiation set of Game 4 of Section 9.2. (This game was also discussed in the example of Section 9.3.) Does the negotiation set in any way reflect P_1's stronger bargaining position?

2. Determine the negotiation set of Game 8 of Section 9.3.

3. Suppose preplay discussion between the players is permitted, but that agreements reached are not enforceable. Does the negotiation set still provide a reasonable resolution of the game? Consider this question for, say, Game 1 of Section 9.2.

4. The negotiation set may be small. Determine this set for Game 2 of Section 9.2.

5. True or false: If (X,Y) is a strategy pair in equilibrium, then the associated expected payoff (XAY^t,XBY^t) is a point of the negotiation set?

6. Show that if a two-person game is, in fact, zero sum, the negotiation set consists of one point, the solution to the zero-sum game.

SECTION 9.5 THE AXIOMS OF NASH

The axioms of Nash provide a procedure for determining a unique solution to every cooperative two-person game. Mathematically this resolution of such games is quite attractive. The axioms seem reasonable enough, and a simple theorem shows that a unique point, easily determined by considering the maximal value on the set M of an elementary function, satisfies the axioms. A modification of the procedure leads to a scheme that incorporates the familiar notions of security levels and equilibrium strategies and reflects the threat potential of the players. In this section we provide only an outline of the techniques. More details can be found in the papers of Nash [25,27] and the books of Luce and Raiffa [23] and Owen [28].

As we have seen, to every cooperative two-person game we can associate its cooperative payoff set M. We can also associate a point (u^*,v^*) of M, a status quo point, a point consisting of the minimally acceptable payoffs for the two players. For example, the security levels of the players could serve as the components of such a point. Now what we propose to construct is a function, denoted by $F[M,(u^*,v^*)]$, which assigns to the set M and the point (u^*,v^*) a "solution" of the game, a payoff point that is in some way a resolution of the cooperative game. We ask initially—what sort of properties should this function satisfy? The following axioms provide an answer to this question.

Axioms of Nash ([25])

Denote the point $F[M,(u^*,v^*)]$ by (u',v').

1. (u',v') is an undominated point of M such that $u' \geq u^*$ and $v' \geq v^*$.
2. If L is a linear transformation from R^2 to R^2 of the form $L(u,v) = (c_1u + d_1,c_2v + d_2)$, where c_1 and c_2 are positive, then $F[L(M),L(u^*,v^*)] = L(u',v')$.
3. If $N \subset M$, $(u^*,v^*) \in N$, and $(u',v') \in N$, then $F[N,(u^*,v^*)] = F[M,(u^*,v^*)]$.
4. If $(u,v) \in M$ implies that $(v,u) \in M$, and if $u^* = v^*$, then $u' = v'$.

The first axiom states that this bargaining solution F is a feasible payoff, dominated by no other payoff, and at least as acceptable as the status quo point. The second axiom is concerned with changes in the utility functions. Recall, in

Section 9.1, it was pointed out that a utility function is determined once its value for two distinct outcomes is defined, but that the values for these first two outcomes could be arbitrarily set as long as their linear preferential ordering is maintained. If different values are used, the resulting equivalent utility function will differ by a transformation of the form $cu + d$, where $c > 0$. Axiom 2 states that changes such as these should be reflected in the obvious manner in the payoff function F, so that F is essentially invariant under utility transformations.

Axiom 3 is called the "independence of irrelevant alternatives" axiom. It states that if the bargaining solution to a game $[M,(u^*,v^*)]$ is also a point in the subset N of M, then this point is also the solution of the game $[N,(u^*,v^*)]$. Another way of saying this is that if (u',v') is the solution to the game $[N,(u^*,v^*)]$, and if the set N is increased to the set M containing other possible payoffs, the solution to $[M,(u^*,v^*)]$ must be either the former solution (u',v') or a new point in M not in N.

The fourth axiom is straightforward. If M is symmetric about the line $u = v$, and if $u^* = v^*$, then the positions of the players are equivalent and the payoff to P_1 should equal the payoff to P_2.

The remarkable thing about these four axioms is that for any set M and point $(u^*,v^*) \in M$, there exists a unique point of M satisfying the axioms. (Thus they work, admirably!) This point can be easily determined. As long as M contains points (u,v) such that $u > u^*$ and $v > v^*$, $F[M,(u^*,v^*)]$ is the unique point of that subset of M with $u > u^*$ and $v^* > v$ at which the function $(u - u^*)(v - v^*)$ attains its maximal value. [Notice that the level curves of the function $(u - u^*)(v - v^*)$ are hyperbolic, and so uniqueness follows from the convexity of M.] The simple proof that this prescription provides the unique point satisfying the axioms is contained in [25] (see also [23] or [28]).

We now give some examples illustrating this procedure. We again assume that side payments are not permitted and, for the status quo point (u^*,v^*), we use the security levels of the two players.

Example 1

Consider Game 6 of Section 9.3. The security level for both players is 0, and the graph of set M, as described in Example 1 of Section 9.4 is sketched in Figure 9.6. On M, the maximum of the function uv is attained at the point $(\frac{11}{2},\frac{11}{2})$, and so we have $F[M,(0,0)] = (\frac{11}{2},\frac{11}{2})$. This seems to be a reasonable resolution of the game. Note that Axioms 1 and 4 also imply in this case that $F[M,(0,0)]$ is $(\frac{11}{2},\frac{11}{2})$.

Example 2

Consider the game of Problem 2 of Section 9.2. As determined in Problem 4 of Section 9.3, the security levels for P_1 and P_2 are 1 and 0, respectively. The set M, described in Example 4 of Section 9.4, is illustrated in Figure 9.7.

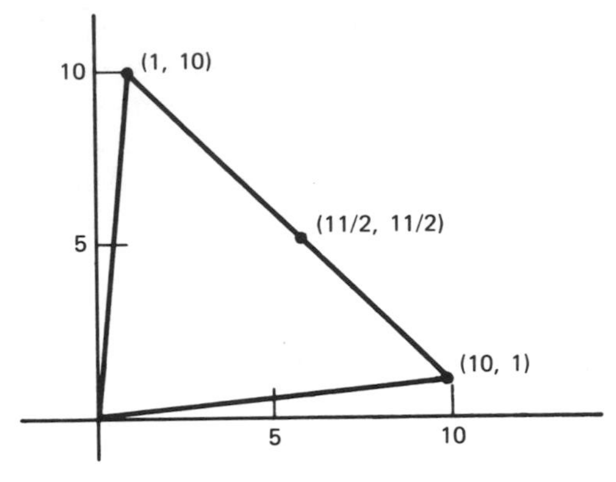

Figure 9.6

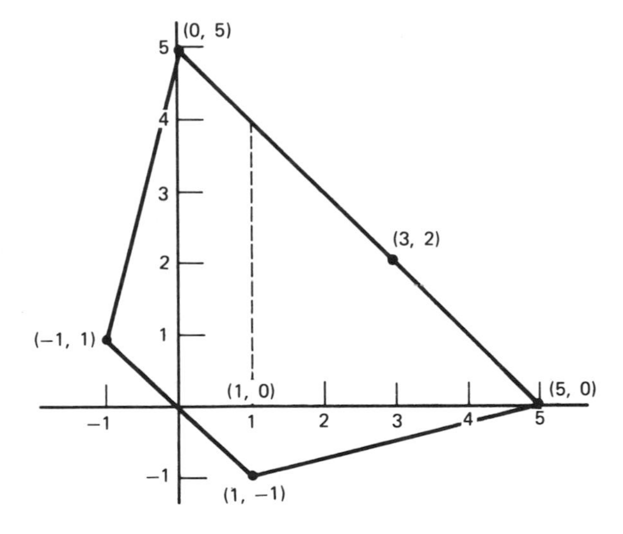

Figure 9.7

The maximal value of the function $(u - 1)v$ on M is attained at the point (3,2). (See Problem 1.) Thus this solution also reflects P_1's stronger bargaining position.

Example 3

Consider Game 4 of Section 9.2. The security levels are 1 and -9 (the example of Section 9.3), and the negotiation set M is depicted in Figure 9.8 (see also Problem 1 of Section 9.4). On M, the maximum of $(u - 1)(v + 9)$ is attained at (10,1) (see Problem 2), and so this resolution of the game, in contrast to the concept of the negotiation set, clearly recognizes the strong bargaining position of P_1.

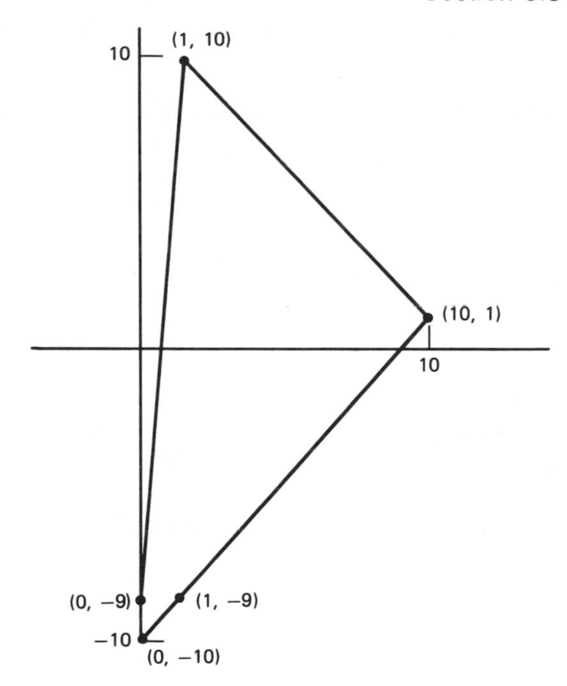

Figure 9.8

Example 4

Consider the game of Problem 5 of Section 9.3. The security level of both players is 1, and the set M is given in Figure 9.9. For this game, $F[M,(1,1)] = (\frac{9}{2},\frac{9}{2})$. Note that the negotiation set is the line segment between the points (2,7) and (7,2). Moreover, as determined in Problem 5, the payoffs associated with the equilibrium strategy pairs do not distinguish the two players. Thus, for this game, all the concepts so far

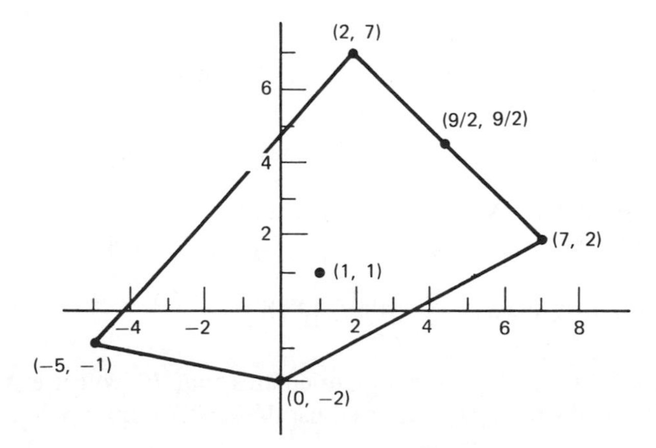

Figure 9.9

developed in no way discriminate P_1 from P_2. Do you still consider P_1's position to be stronger? If so, read on.

John Nash [27] has extended this bargaining procedure to a scheme that considers the full threat potential for the players. Suppose two players are about to play a fixed, non-zero-sum game in which discussion, cooperation, and binding agreements are permitted. The payoff set M is fixed, determined by the payoff tableau. Assume further that the bargaining function F described above accurately represents the payoff mutually agreed on by the two players once a status quo point has been determined. But P_1 questions the use of his security level for determining the status quo point. He wonders—is there a strategy X, a threat strategy, say, such that no matter what strategy Y P_2 uses, if the expected payoff (XAY^t,XBY^t) corresponding to the strategy pair (X,Y) is used as the status quo point in the function F, the payoff to P_1 will be larger? Extending this, P_1 asks—what strategy X will guarantee the largest first component for the function $F[M,(XAY^t,XBY^t)]$ regardless of the strategy Y employed by P_2?

Similarly, P_2 considers his potential threat strategies. Hence each player desires to choose a strategy that he is willing to play and that will provide a status quo point (XAY^t,XBY^t) for the bargaining function $F[M,(XAY^t,XBY^t)]$ that will guarantee for him the most profitable outcome independent of the opponent's strategy.

We can indicate this process as follows. Since M is fixed, $F[M,(XAY^t,XBY^t)]$ can be looked at as a two-step function depending only on $X \in S$ and $Y \in T$. Given a strategy pair (X,Y), first the expected payoff (XAY^t,XBY^t), a point in M, is determined; and second, the value of F is determined using this point as the status quo point. The first component of $F[M,(XAY^t,XBY^t)]$ is the associated payoff to P_1, and the second the payoff to P_2. Denote these two components by $F_1(X,Y)$ and $F_2(X,Y)$. Thus

$$F[M,(XAY^t,XBY^t)] = (F_1(X,Y),F_2(X,Y))$$

Now each player wishes to maximize his security level. (Principle I of Section 8.2.) Paralleling the development for zero-sum games, P_1's and P_2's security levels are given by

$$\operatorname*{Max}_{X \in S} \operatorname*{Min}_{Y \in T} F_1(X,Y) \quad \text{and} \quad \operatorname*{Max}_{Y \in T} \operatorname*{Min}_{X \in S} F_2(X,Y)$$

Suppose $X_0 \in S$ and $Y_0 \in T$ are strategies that realize these security levels; that is,

$$\operatorname*{Max}_{X \in S} \operatorname*{Min}_{Y \in T} F_1(X,Y) = \operatorname*{Min}_{Y \in T} F_1(X_0,Y)$$

and

$$\operatorname*{Max}_{Y \in T} \operatorname*{Min}_{X \in S} F_2(X,Y) = \operatorname*{Min}_{X \in S} F_2(X,Y_0)$$

But now P_1 expects P_2 to use Y_0, and P_2 anticipates that P_1 will use X_0. Thus each asks, is my choice of strategy the best against the strategy anticipated by my opponent? That is, is the pair (X_0, Y_0) stable or in equilibrium, as defined using this scheme? (Principle II of Section 8.2.) In this context, the strategy pair (X_0, Y_0) is in equilibrium if, for any $X \in S$ and $Y \in T$,

$$F_1(X, Y_0) \leq F_1(X_0, Y_0) \quad \text{and} \quad F_2(X_0, Y) \leq F_2(X_0, Y_0)$$

Now it can be shown (see [27]) that there always exist strategies X_0 and Y_0 that realize these security levels (and the use of Max and Min is justified), and that any such strategy pair (X_0, Y_0) is in equilibrium, as defined above. (Actually, the proof is somewhat reversed. It can be shown first that there are equilibrium strategy pairs, and then that they deliver the maximal security levels.) Thus these security levels, the payoff point of $F[M,(X_0 A Y_0{}^t, X_0 B Y_0{}^t)]$, provide another resolution of the cooperative game, a resolution that incorporates the full threat potential of the players. Moreover, in some cases, such as games permittting side payments, the zero-sum theory can be used to compute easily these strategies and the corresponding solution (see, for example, [28]).

Example 4 (continued)

For the game of Example 4, the strategies corresponding to the above can be determined to be $X_0 = (0,1)$ and $Y_0 = (1,0)$. The associated status quo point $(X_0 A Y_0{}^t, X_0 B Y_0{}^t)$ is $(0,-2)$. Using the graph in Figure 9.9 (and Problem 1), we have $F[M,(0,-2)] = (\frac{11}{2}, \frac{7}{2})$. Thus this resolution of the game of Problem 5 of Section 9.3 distinguishes P_1's stronger threat potential.

In fact, these results can be reasoned to by simply considering the game tableau and the corresponding set M. Note that if P_1 threatens to use strategy s_2, no matter what P_2 does, the corresponding expected payoff would lie on the line segment between $(0,-2)$ and $(7,2)$. The bargaining function F maps these points onto the segment between $(\frac{11}{2}, \frac{7}{2})$ and $(7,2)$. Thus P_1's security level is at least $\frac{11}{2}$, and P_2's can be at most $\frac{7}{2}$. If P_2 threatens to use strategy t_1, no matter what P_1 does, the corresponding expected payoff would lie on the line segment between $(0,-2)$ and $(2,7)$. F maps these points onto the segment between $(\frac{11}{2}, \frac{7}{2})$ and $(2,7)$. Hence P_2's security level is at least $\frac{7}{2}$, and P_1's is at most $\frac{11}{2}$. It follows that the payoff $(\frac{11}{2}, \frac{7}{2})$ is the resolution of the game under this scheme. Note also that this strategy pair (s_2, t_1) is in equilibrium under this scheme.

In this section we have outlined briefly two related methods for solving cooperative two-person games. There are other techniques. Some of these are described in the books of Luce and Raiffa [23] and Rapoport [29]. Although these two methods based on the seemingly plausible axioms of Nash lead to unique solutions, the reasonableness and relevance of these axioms must be studied closely. For a critical evaluation of the applicability of these schemes, refer to the excellent book by Luce and Raiffa [23].

PROBLEM SET 9.5

1. (a) Show that if the set M is of the form

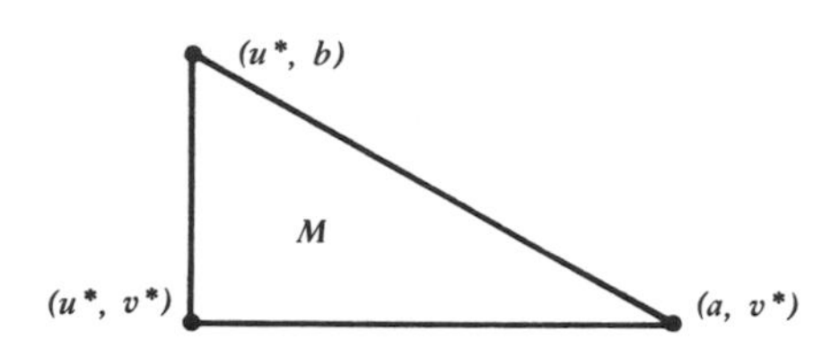

the maximum of the function $(u - u^*)(v - v^*)$ is attained at the point $P = ((u^* + a)/2, (v^* + b)/2)$.

(b) Show that if N is a subset of M containing the point P, then the maximum of that same function on N is also attained at P.

2. Show that in Example 3, the maximum of $(u - 1)(v + 9)$ is attained at the point $(10,1)$.

3. Show that for the game of Example 2 of this section, the pair of threat strategies (s_2, t_1) is in equilibrium according to the second scheme. What is the resolution of this game using this scheme?

4. Determine the solutions of Problem 7 of Section 9.3 (and Example 2 in Section 9.4) under both schemes of this section. For the second, consider the results of P_1 using s_2 as the threat strategy.

SECTION 9.6 AN EXAMPLE

In this section we will develop an example of a two-person game that will illustrate some of the concepts discussed in the last two chapters. We hope that this example will help students to understand how these concepts might be applied and realize the significance of the questions that have been raised. The model should also point out how tenuous the status of a solution to a game may be.

Consider the position of two food wholesalers competing for the public dollar through their respective supermarket chains. Each fall the wholesalers must decide on whether or not they, through their supermarkets, will conduct a promotion campaign the following winter. For example, their stores could offer their customers items such as dinnerware, silverware, encyclopedias, records, and so on, either free of charge or at reduced costs, or they could offer their customers participation in a game or contest with small cash prizes. In general, if only one chain of supermarkets has such a campaign, their business will be increased significantly. However, if both wholesalers decide on promotion campaigns, the effects are nullified. (Nevertheless, such campaigns would still provide an abundance of material for their advertising agencies.) Thus the decision to be made each fall by each wholesaler is to have or not have a promotion campaign that winter.

One wholesaler, the larger of the two, attempts to formulate his decision problem in terms of a two-person game. From past records he knows that in general his chain handles 60% of what he at first considers a fixed segment of the business and his competitor 40%. If he conducts a promotion campaign and his competitor does not, his business increases to 90%. If the situation is reversed, his competitor's volume is increased to 70%. If both offer sales gimmicks, the business breakdown ratio is again 60:40. The wholesaler initially considers this as a zero-sum game, with 30% of the food-purchasing market to be gained or lost on account of promotion campaigns. Thus he sets up the following tableau for the corresponding zero-sum game, with his firm represented by Player 1.

	No Promotion	*Promotion*
No Promotion	0	−30
Promotion	30	0

Upon further reflection, the wholesaler realizes that this model is totally inadequate. For two reasons the game is not zero-sum. On the one hand, he knows that he cannot assume that between these two major wholesalers there is one fixed market. It is true that they control the majority of the market, but there are other small independent food stores from which business can be attracted by an effective promotion campaign. On the other hand, the cost of conducting a promotion campaign—the costs in the items involved, in the associated advertising, and in the labor administrating the campaign—must also be considered.

Thus the wholesaler decides to formulate the problem as a non-zero-sum game. He uses as utility units the volume of business, measured in thousands of dollars, less the cost of a promotion campaign if such is conducted. The result is the following tableau, where again his firm is represented by Player 1.

	No Promotion	*Promotion*
No Promotion	(60,40)	(40,70)
Promotion	(90,20)	(50,30)

The wholesaler believes that this table more accurately reflects the situation. If he conducts a promotion campaign and his competitor does not, he attracts business not only from his competitor, but also from the other independent food stores, and the combined volume of these two major wholesalers is increased by 10 units, with his volume increased 30 units and the competitor's volume decreased 20 units. A similar relationship holds if the competitor is the only one to conduct a campaign. Moreover, the futility of both wholesalers conducting campaigns is now measured. The increases in business, in this case coming only from the independent stores, is not adequate to compensate for the expense of such campaigns, and both wholesalers lose 10 units from their income under routine operations.

The result is a two-person, non-zero-sum game of the prisoner's dilemma type. The "promotion" strategy of each player dominates the other strategy, and so the game has only one equilibrium strategy pair, and the outcome corresponding to this pair is the (50,30) payoff. This payoff also represents the security levels of the two players. Thus if the game is played noncooperatively, the seemingly expected result favors the advertising agencies.

The wholesaler notes, however, that the outcome (60,40) is more beneficial to both parties than the outcome (50,30). Can he bring about this outcome by attempting to cooperate with his competitor? Since they are not yet bitter rivals, this possibility exists; if they could mutually agree not to hold promotion campaigns in the winter, they could realize this advantageous payoff. Note, however, the importance of entering into a binding agreement. If either wholesaler can convince his competitor not to conduct a promotion campaign, that wholesaler has much to gain, at least that winter, by conducting one himself. Acutually, legislation restricting the use of gimmicks by food retailers could have the same effect as the possibility of entering into binding agreements.

Consider now the solution concepts applied to this game for cooperative two-person games. Remember that this theory assumed the possibility of the players entering into binding agreements. Now the negotiation set for this game is the line segment in the plane connecting the points (80,30) and (50,60). The outcome (60,40) is not on this segment. Both solution concepts developed in the last section lead to the outcome (65,45), which is also the midpoint of the negotiation set. To realize this outcome, the wholesalers must enter into an even more involved collusion, each agreeing to conduct promotion campaigns in alternate years and not in competition with each other. In this case such campaigns maintain their novelty potential, and the independent store owners are hurt the most.

This suggests the possibility of the owners of the independent stores uniting and attempting to counteract their weaker position either by promotion campaigns of their own or by adjustments in the selling prices of their products. But this leads to n-person game theory (in this case $n = 3$), and so is beyond the scope of this text. As is frequently the case, models involving human behavior quickly become complicated. We refer the food wholesaler to the references.

Appendix A
Vectors and Matrices

This appendix provides a brief listing of the linear algebra topics used in this book. These topics include the basic concepts of vectors and matrices and the transpose of a matrix.

A point $X = (x_1,x_2,\ldots,x_n)$ of R^n is said to be an *n-dimensional vector* or, simply, a *vector*. Multiplication of vectors by real numbers is defined in the obvious manner; that is, for $r \in R$, rX is defined to be the vector $(rx_1,\ldots,rx_n)$ in R^n. If X and $Y = (y_1,y_2,\ldots,y_n)$ have the same dimension, then we define their *sum* $X + Y = (x_1 + y_1,x_2 + y_2,\ldots,x_n + y_n)$ and *dot product* $X \cdot Y = x_1y_1 + x_2y_2 + \cdots + x_ny_n$. Note that the dot product of two vectors is simply a real number, and that this operation is commutative, that is, $X \cdot Y = Y \cdot X$.

Example

$$\begin{aligned}
\text{For } X &= (3,0,-1,5) \text{ and } Y = (-2,6,7,0),\\
3X &= 3(3,0,-1,5) = (9,0,-3,15)\\
X + Y &= (3 - 2,0 + 6,-1 + 7,5 + 0) = (1,6,6,5)\\
X \cdot Y &= 3(-2) + 0(6) + (-1)(7) + 5(0) = -6 - 7 = -13
\end{aligned}$$

An $m \times n$ *matrix* is simply a rectangular array of real numbers, with the array having m rows and n columns. The matrix

$$A = \begin{bmatrix} a_{11} & a_{12} & \cdots & a_{1n} \\ a_{21} & a_{22} & \cdots & a_{2n} \\ \vdots & & & \\ a_{m1} & a_{m2} & \cdots & a_{mn} \end{bmatrix}$$

can be denoted by (a_{ij}), where a_{ij} is that element of matrix A in the ith row and jth column of the array.

If $A = (a_{ij})$ and $B = (b_{ij})$ are matrices of the same dimensions, we define their sum $A + B = (a_{ij} + b_{ij})$, a matrix with the same dimension. Multiplication of a matrix by a real number is defined as follows. For $r \in R$ and $A = (a_{ij})$, define $rA = (ra_{ij})$.

Example

For

$$A = \begin{bmatrix} 3 & 0 & 1 \\ -1 & 2 & 5 \end{bmatrix} \quad \text{and} \quad B = \begin{bmatrix} 0 & -1 & -2 \\ 4 & 1 & -3 \end{bmatrix},$$

$$A + B = \begin{bmatrix} 3+0 & 0-1 & 1-2 \\ -1+4 & 2+1 & 5-3 \end{bmatrix} = \begin{bmatrix} 3 & -1 & -1 \\ 3 & 3 & 2 \end{bmatrix}$$

$$3A = \begin{bmatrix} 9 & 0 & 3 \\ -3 & 6 & 15 \end{bmatrix}$$

To define multiplication between matrices, suppose $A = (a_{ij})$ is an $m \times n$ matrix and $B = (b_{ij})$ is an $n \times p$ matrix. Then the product $AB = C = (c_{ij})$ is defined to be that $m \times p$ matrix with the ijth element $c_{ij} = \sum_{k=1}^{n} a_{ik} b_{kj}$. Thus the ijth element of the product is the dot product of the ith row of A with the jth column of B.

Example

$$[6 \quad -8]\begin{bmatrix} 1 & -2 \\ 4 & 5 \end{bmatrix} = [6 \cdot 1 - 8 \cdot 4 \quad 6(-2) - 8 \cdot 5] = [-26 \quad -52]$$

$$[6 \quad -8]\begin{bmatrix} 1 & -2 \\ 4 & 5 \end{bmatrix}\begin{bmatrix} -3 \\ 1 \end{bmatrix} = [-26 \quad -52]\begin{bmatrix} -3 \\ 1 \end{bmatrix} = 26$$

$$\begin{bmatrix} 3 & 0 \\ -1 & 2 \\ 6 & -3 \end{bmatrix}\begin{bmatrix} 1 & -2 \\ 4 & 5 \end{bmatrix} = \begin{bmatrix} 3 & -6 \\ 7 & 12 \\ -6 & -27 \end{bmatrix}$$

But

$$\begin{bmatrix} 1 & -2 \\ 4 & 5 \end{bmatrix} \begin{bmatrix} 3 & 0 \\ -1 & 2 \\ 6 & -3 \end{bmatrix}$$

is undefined.

Let I be the $n \times n$ matrix with 1's on the main diagonal and 0's elswhere; that is,

$$I = \begin{bmatrix} 1 & 0 & \cdots & 0 \\ 0 & 1 & \cdots & 0 \\ \vdots & & & \\ 0 & 0 & \cdots & 1 \end{bmatrix}$$

Then for any $n \times n$ matrix A, we have $AI = IA = A$, and I is called the *identity matrix* (of order n). The $n \times n$ matrix A is said to be *invertible* or *nonsingular* if there exists an $n \times n$ matrix B such that $AB = BA = I$. Such a matrix B is called the *inverse* of A and is denoted by A^{-1}. It can be shown that for square matrices A and B of finite order, $AB = I$ implies that $BA = I$ and so $B = A^{-1}$.

Example

If

$$A = \begin{bmatrix} 6 & -2 \\ 4 & -1 \end{bmatrix} \quad \text{and} \quad B = \begin{bmatrix} -\frac{1}{2} & 1 \\ -2 & 3 \end{bmatrix},$$

then

$$AB = BA = \begin{bmatrix} 1 & 0 \\ 0 & 1 \end{bmatrix},$$

and so $A^{-1} = B$, $B^{-1} = A$.

Let $A = (a_{ij})$ be an $m \times n$ matrix. Then the *transpose* of A, denoted by A^t, is that $n \times m$ matrix with ijth element equal to the jith element of A; that is, $A^t = (a_{ji})$. Thus the rows of A^t are simply the columns of A and the columns of A^t are the rows of A. Notice that if this operation is performed twice, the resulting matrix will be the original matrix; that is, $(A^t)^t = A$.

Example

$$\begin{bmatrix} 1 & -2 & 3 \\ -4 & 5 & -6 \end{bmatrix}^t = \begin{bmatrix} 1 & -4 \\ -2 & 5 \\ 3 & -6 \end{bmatrix}$$

$$\begin{bmatrix} 6 \\ 7 \\ 8 \end{bmatrix}^t = [6 \quad 7 \quad 8]$$

Suppose $X = (x_1,x_2,\ldots,x_n)$ and $Y = (y_1,y_2,\ldots,y_n)$ are n-dimensional vectors. Expressed this way, they can also be considered as $1 \times n$ matrices, and their transposes would be $n \times 1$ matrices. In fact, we have $X \cdot Y = Y \cdot X = XY^t = YX^t$, where the last two products are ordinary matrix multiplications.

Example

$$(3,1,-2)\cdot(0,4,8) = (0,4,8)\cdot(3,1,-2) = [3,1,-2]\begin{bmatrix}0\\4\\8\end{bmatrix} = [0 \quad 4 \quad 8]\begin{bmatrix}3\\1\\-2\end{bmatrix} = -12$$

The following result concerning the transpose operator is used in Section 4.3.

Theorem *Suppose A is an $m \times n$ matrix and B an $n \times p$ matrix. Then the two products AB and B^tA^t are defined, and $(AB)^t = B^tA^t$.*

Proof Note first that AB is an $m \times p$ matrix and so its transpose is $p \times m$. Now B^t is a $p \times n$ matrix, A^t an $n \times m$ matrix, and so their product is defined, and is also a $p \times m$ matrix.

We now show that the corresponding elements of these two $p \times m$ matrices are equal. By the definition of the product, the ijth element of AB is the dot product of the ith row of A and the jth column of B; and, by the definition of the transpose, this is the jith element of $(AB)^t$. Now the jith element of B^tA^t is the dot product of the jth row of B^t and the ith column of A^t. But the jth row of B^t is the jth column of B, and the ith column of A^t is the ith row of A. Thus the jith element of B^tA^t is the dot product of the jth column of B and the ith row of A. Since dot product multiplication is commutative, the jith elements of $(AB)^t$ and B^tA^t are equal. # # #

Appendix B
Programming the Simplex Method

Most of the linear programming problems of this text can be solved fairly easily by hand calculations using the simplex method. However, in most applications of the theory, the corresponding mathematical problems contain many variables and equations, and the time and cost involved in solving such problems by hand would be prohibitive. Thus electronic computers are needed to perform the mechanical operations of the simplex method and provide solutions to linear optimization problems. Indeed, many sophisticated programs have been developed to solve such problems, taking into consideration various local factors such as the special nature of a particular type of problem or the storage capacity of a particular piece of computer hardware.

Given this situation, a most useful student exercise is the development and implementation of a computer program to solve linear programming problems. The best way to learn a theory is to teach the theory to someone else, so teaching a computer to use the simplex method also provides an excellent opportunity for developing a thorough understanding of the simplex method and its underlying principles.

This appendix provides assistance by means of flow charts and suggested test problems for students who have access to a computer and the opportunity to develop their own program. It should be mentioned that many computer systems contain programs to solve linear programming problems in the facility's program library. If one has access to such a program, familiarization with its operation might be a helpful first step in the development of one's own program. Such a familiarization, however, should in no way deter a student from the much more rewarding task of the creation of one's own program.

Unless one is an accomplished programmer, the development of a program to determine the solution of linear programming problems is best done in a series of steps, with each step building on the previous steps. This provides the programmer with the opportunity to test and debug her program at each stage, so that programming errors can be more easily traced. The suggested development that follows is divided into three such steps, with the first two involved primarily with linear programming and the third step involved primarily with computer programming. In particular, the first step is concerned with the solving of a linear programming problem already in canonical form and containing no artificial variables. This part of the solution process was discussed in Sections 3.4 and 3.5, and was referred to as the second stage of the simplex method. The second step of our suggested development is concerned with the introduction of artificial variables into the linear programming problem. This was the first stage of the simplex method, and was discussed in Sections 3.6 and 3.7. The final step is concerned with making the program

Figure B.1 Step 1. Start with problem in canonical form [as in (1) of Section 3.4].

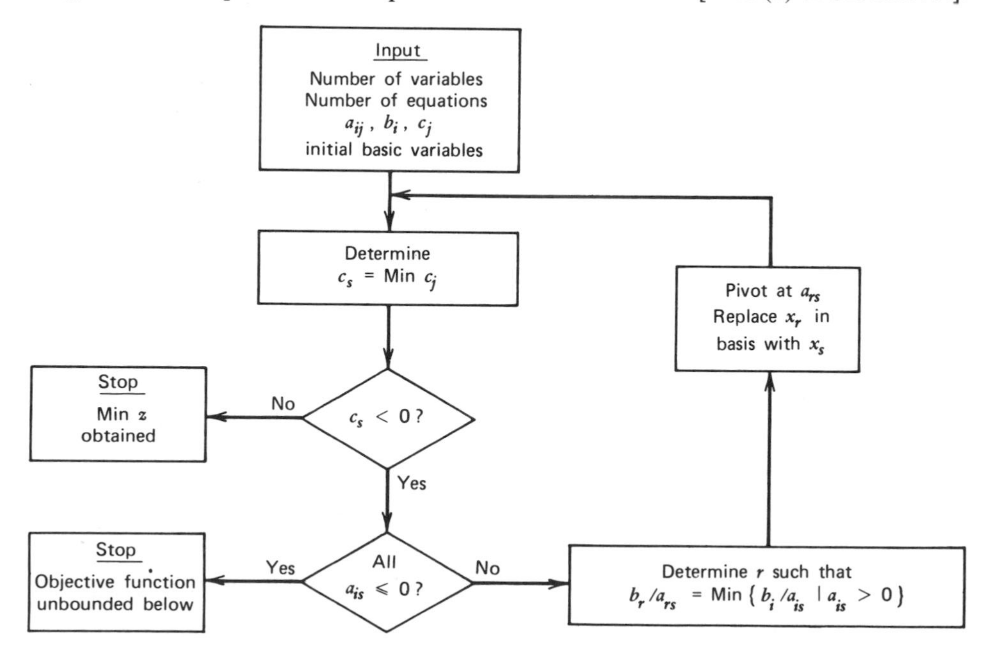

adaptable to general use. Here the understanding of computer programming techniques is probably more essential than the understanding of the simplex method, and the student approaching this programming assignment primarily as an exercise in linear programming should feel free to omit this final step.

In the first step of the development of a program to solve a general linear programming problem, consider only linear programming problems presented in

Figure B.2 Step 2.

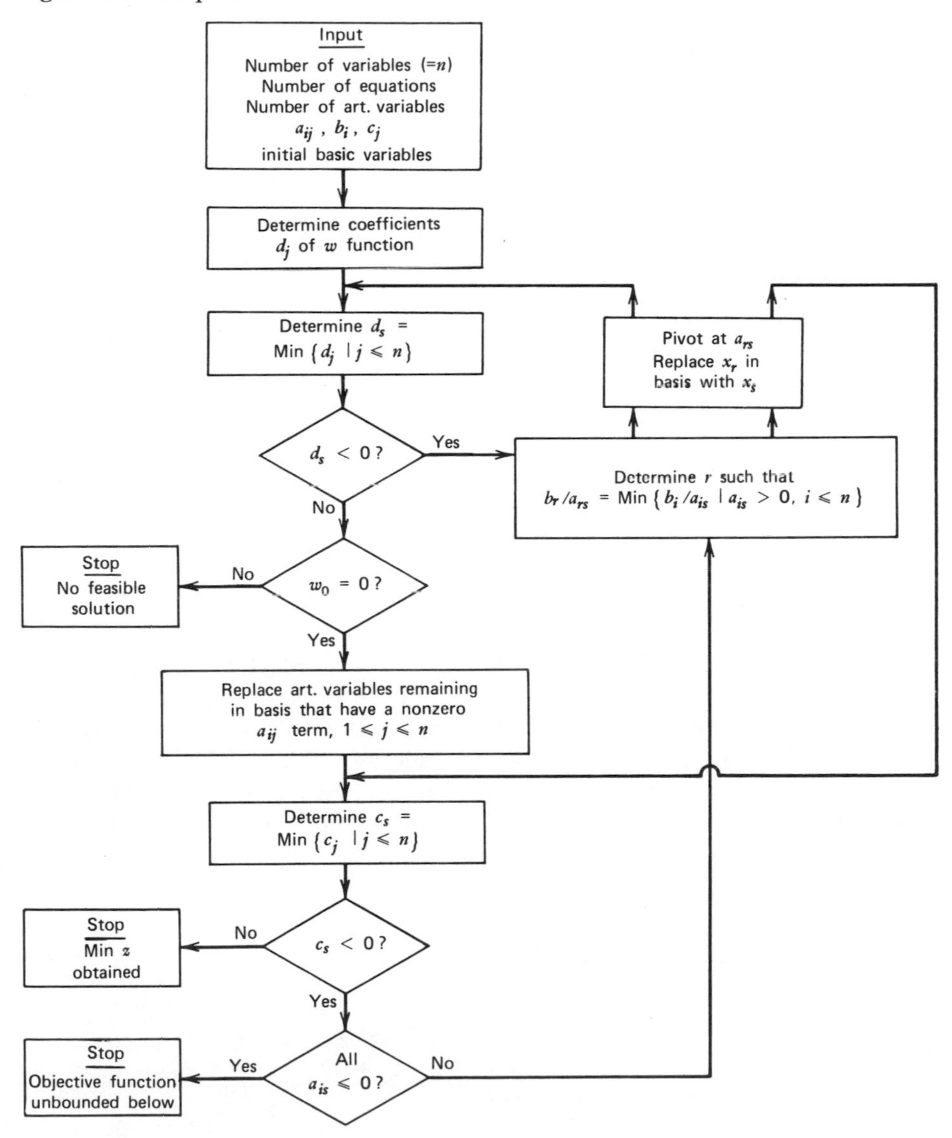

canonical form with a specified set of basic variables and containing no artificail variables. For example, the problem of (1) of Section 3.4 is in this form with, in this case, the first m variables as the basic variables. Develop a program to solve such problems based on the simplex method as outlined at the end of Section 3.4. Figure B.1 provides a flow chart for this step. Test your program on the three examples of Section 3.5. In these problems, there is an example of an unbounded objective function and also an example of degeneracy. Remember first to put these problems into canonical form and, in particular, to express the objective functions in terms of nonbasic variables only.

In the second step consider linear programming problems presented in canonical form through the addition of artificial variables, such as the general problem of (3) of Section 3.6. Develop a program that computes the coefficients d_j for the w function and then eliminates step by step the artificial variables from the set of basic variables using the simplex method applied to the w function, the process described in Section 3.6. Provided a linear programming problem has feasible solutions, this process drives such a problem to a basic feasible solution at which the w function attains the value zero and the second stage of the simplex method, the stage programmed in the first step, can be applied. Figure B.2 provides a flow chart for this combined operation. Test your program on the examples in Sections 3.6 and 3.7. Example 1 of Section 3.6 contains no complications, but in Example 2 of Section 3.6, a full set of artificial variables is not needed, and the original problem has no feasible solutions. In the two examples of Section 3.7, degeneracy is present, with the original system of constraints for the second example containing one redundant equation.

In the final step, the program could be modified so that it is accessible for general use and, in particular, for somebody unfamiliar with the simplex method. For example, the program could accept a system of constraints containing both equalities and inequalities, and an objective function to be either maximized or minimized, with these variables made precise in some way by the user of the program. The program would then first have to put the problem into canonical form by adding the appropriate slack variables, making the constant terms b_i nonnegative, seeking out a set of basic variables and adding artificial variables if needed, and expressing the objective function in terms of the nonbasic variables only. The solution to the problem would also have to be printed out in simple terms.

BIBLIOGRAPHY

1. BEALE, E. M., "Cycling in the Dual Simplex Algorithm," *Naval Res. Logist. Quart., 2* (1955), pp. 269–276.

2. BLAND, R. G., "New Finite Pivoting Rules for the Simplex Method," *Mathematics of Operations Research, 2* (1977), pp. 103–107.

3. DAKIN, R. J., "A Tree Search Algorithm for Mixed Integer Programming Problems," *Computer Journal, 8* (1965), pp. 250–255.

4. DANTZIG, G. B., "Maximization of a Lincar Funciton of Variables Subject to Linear Inequalities," Chap. XXI, in *Activity Analysis of Production and Allocation*, John Wiley and Sons, New York, 1951.

5. DANTZIG, G. B., Inductive Proof of the Simplex Method, The RAND Corporation, Paper P-1851, December 28, 1959. Published in *IBM J. Res. Develop., 4* (1960), pp. 505–506.

6. DANTZIG, G. B., *Linear Programming and Extensions*, Princeton University Press, Princeton, N.J., 1963.

7. DANTZIG, G. B., L. R. FORD, and D. R. FULKERSON, "A Primal-dual Algorithm for Linear Programming," in H. W. Kuhn and A. W. Tucker (eds.), *Linear Inequalities and Related Systems*, Princeton University Press, Princeton, N.J., 1956.

8. DRESHER, M., *Games of Strategy, Theory and Applications*, Prentice-Hall, Inc., Englewood Cliffs, N.J., 1961.

9. FERGUSON, A., and G. DANTZIG, The Allocation of Aircraft to Routes—An Example of Linear Programming under Uncertain Demand, The RAND Corporation, Paper P-727, December 7, 1956. Published in *Management Science, 3* (1956), pp. 45–73.

10. FORD, L. R., and D. R. FULKERSON, "A Simple Algorithm for Finding Maximal Network Flows and an Application to the Hitchcock Problem," *Can. J. Math., 9* (1957), pp. 210–218.

11. FORD, L. R., and D. R. FULKERSON, *Flows in Networks*, Princeton University Press, Princeton, N.J., 1962.

12. GALE, D., H. W. KUHN, and A. W. TUCKER, "Linear Programming and the Theory of Games," in T. C. Koopmans (ed.), *Activity Analysis of Production and Allocation*, John Wiley and Sons, New York, 1951.

13. GARFINKEL, R. S., and G. L. NEMHAUSER, *Integer Programming*, John Wiley and Sons, New York, 1972.

14. GOMORY, R. E., "Outline of an Algorithm For Integer Solutions to Linear Programs," *Bull. Amer. Math. Soc., 64* (1958), pp. 175–278.

15. GOMORY, R. E., "An Algorithm For Integer Solutions to Linear Programs," in R. Graves and P. Wolfe (eds.), *Recent Advances in Mathematical Programming*, McGraw-Hill Book Co., New York, 1963.

16. HADLEY, G., *Linear Programming*, Addison-Wesley, Reading, Mass., 1962.

17. HADLEY, G., *Non-linear and Dynamic Programming*, Addison-Wesley, Reading, Mass., 1964.

18. HITCHCOCK, F. L., "The Distribution of a Product from Several Sources to Numerous Localities," *J. Math. Phys.*, *20* (1941), pp. 224–230.

19. JACOBS, W. W., "The Caterer Problem," *Naval Res. Log. Quart.*, *1* (1954), pp. 154–165.

20. KANTOROVICH, L. V., *Mathematical Methods in the Organization and Planning of Production*, Publication House of the Leningrad State University, 1939. Translated in *Management Science*, *6* (1960), pp. 366–422.

21. LAND, A. H., and A. G. DOIG, "An Automatic Method of Solving Discrete Programming Problems, *Econometrica*, *28* (1960), pp. 497–520.

22. LEMKE, C. E., "The Dual Method of Solving the Linear Programming Problem," *Naval Res. Logist. Quart.*, *1* (1954), pp. 48–54.

23. LUCE, R. D., and HOWARD RAIFFA, *Games and Decisions*, John Wiley and Sons, New York, 1957.

24. McKINSEY, J. C., *Introduction to the Theory of Games*, McGraw-Hill Book Co., New York, 1952.

25. NASH, J. F., "The Bargaining Problem, *Econometrica*, *18* (1950), pp. 155–162.

26. NASH, J. F., "Non-Cooperative Games," *Annals of Mathematics*, *54* (1951), *pp.* 286–295.

27. NASH, J. F., "Two-Person Cooperative Games," *Econometrica*, *21* (195), pp. 128–140.

28. OWEN, G., *Game Theory*, W. B. Saunders Co., Philadelphia, 1968.

29. RAPOPORT, A., *Two-Person Game Theory. The Essential Ideas*, The University of Michigan Press, Ann Arbor, Mich., 1966.

30. STIGLER, G. J., "The Cost of Subsistence," *J. Farm. Econ.*, *27* (1945), pp. 303–314.

31. TAHA, H. A., *Integer Programming: Theory, Applications, and Computations*, Academic Press, New York, 1975.

32. VON NEUMANN, J., "Zur Theorie der Gesellschaftsspiele," *Math. Annal., 100* (1928), pp. 295–320.

33. VON NEUMANN, J., and O. MORGENSTERN, *Theory of Games and Economic Behavior*, John Wiley and Sons, New York, 1944.

34. WOLFE, P., "A Technique for Resolving Degeneracy in Linear Programming," *SIAM J. of Applied Math., 11* (1963), *pp.* 205–211.

Index